Multifunctional Conducting Molecular Materials

Multifunctional Conducting Molecular Materials

Edited by

Gunzi Saito
Kyoto University, Kyoto, Japan

Fred Wudl
University of California, Los Angeles, USA

Robert C. Haddon
University of California, Riverside, USA

Katsumi Tanigaki
Tohoku University, Miyagi, Japan

Toshiaki Enoki
Tokyo Institute of Technology, Tokyo, Japan

Howard E. Katz
Johns Hopkins University, Baltimore, USA

Mitsuhiko Maesato
Kyoto University, Kyoto, Japan

RSCPublishing

The proceedings of the symposium Science and Engineering of the Future with Multifunctional Conducting Molecular Materials held at the 2005 International Chemical Congress of Pacific Basin Societies (Pacifichem 2005) held in Honolulu, Hawaii, USA on 15-17 December 2005.

Cover image reproduced by permission of J. D. Wallis and J.-P. Griffiths from *J. Mater. Chem.*, 2005, **15**, 347.

Special Publication No. 306

ISBN-10: 0-85404-496-5
ISBN-13: 978-0-85404-496-2

A catalogue record for this book is available from the British Library

Published by The Royal Society of Chemistry,
Thomas Graham House, Science Park, Milton Road,
Cambridge CB4 0WF, UK

Registered Charity Number 207890

For further information see our web site at www.rsc.org

Printed by Henry Ling Ltd, Dorchester, Dorset. UK

Preface

The scientific workshop "Science and Engineering of the Future with Multifunctional Conducting Molecular Materials" was held in Honolulu, Hawaii, USA on December 15-17, 2005 as a symposium (#115) in the Area 8 "Materials Chemistry and Nanotechnology" in the 2005 International Chemical Congress of Pacific Basin Societies (PacifiChem 2005). The symposium consisted of 4 half-day sessions, 1 evening session and a poster session with about 150 presentations including 50 oral and about 100 posters.

The symposium was started with the session of "Design and Synthesis of Functional Materials" organized by Professor F. Wudl, followed by the second session entitled "Molecular Conductors and Superconductors" organized by Professors R. C. Haddon and G. Saito, and the third session "Fullerenes, Nanotubes and other Related Nano Materials" organized by Professor K. Tanigaki. The fourth and the evening sessions were devoted to the subjects "Organic/Inorganic Hybrids and Integration & Functionalities" organized by Professors T. Enoki, J-H. Choy, and H. E. Katz. Since Professor Katz moved to Johns Hopkins University during the course of the preparative procedure of this symposium and could not find time to attend this symposium, Professors R. M. Metzger and K. Seki served as co-organizers.

Organization

Organizer: Gunzi Saito (Kyoto University, Japan)

Co-organizers: Fred Wudl (University of California, Los Angels, USA), Robert C. Haddon (University of California, Riverside, USA), Jin-Ho Choy (Soel National University, Korea), Toshiaki Enoki (Tokyo Institute of Technology, Japan), Katsumi Tanigaki (Tohoku university, Japan), Howard E. Katz (Johns Hopkins University, USA), Robert M. Metzger (Alabama University, USA), Kazuhiko Seki (Nagoya University, Japan)

Secretaries: Hideki Yamochi (Kyoto University), Mitsuhiko Maesato (Kyoto University)

Supported by the Chemical Society of Japan, the American Chemical Society, the Canadian Society for Chemistry, the New Zealand Institute of Chemistry, the Royal Australian Chemical Institute, and the Korean Chemical Society. A financial support was made by the 21st Century COE Program Kyoto Alliance for Chemistry.

Scope

"Multifunctional Conducting Molecular Materials" composed of organic, inorganic or molecular compounds or their hybrids in the form of conventional bulk, assemblies or mono-molecule, are a rapidly developing multidisciplinary field of research. These modern materials involve everything from fundamental science to applied science and technology. An important feature of these materials is the almost unlimited possibility to tune their electronic and macroscopic structures and thus functionality in their assemblies. This field is the science of the design, synthesis and fabrication from the combined aspects between the "molecular, crystal, and nano-hybrid structures" in direct-space and the "electronic and band structures and periodicity" in *k*-space. These materials are also important from the viewpoint of nano-fabrication; the science and technology of the future, because nano-fabrication needs very high density, small size integrated and multifunctional properties that can be fabricated under mild condition. We discuss the following topics of molecular conductors in the symposium; 1. Synthesis, design and fabrication, 2. (Super)conductors, 3. Magnetic conductors, 4. Phase transitions, 5. Conducting single molecules, 6. Nano-structure and conducting functions, 8. Integration with other functionality (radiofrequency detection, mechanical response, energy management) including applications. The materials we are concerned with are, for example, organic donor and acceptor molecules, fullerenes, nanotubes, clusters, organic/inorganic hybrids and organic/bio-inorganic hybrids.

Contents

Molecular Conductors and Superconductors

Design and Synthesis of Functional Molecular Materials

Organic/Inorganic Hybrids and photoinduced phenomena

Fullerenes, Nanotubes and Other Related Nano Materials

Integration and Functionalities of Molecular Materials

Molecular Conductors and Superconductors

MESOMERIC FUSED BETAINIC RADICALS AS ORGANIC CONDUCTORS

G. Saito,[1] K. Balodis,[1,2] Y. Yoshida,[1] M. Maesato,[1] H. Yamochi,[1,3] S. Khasanov,[1,4] and T. Murata[1]

[1] Division of Chemistry, Graduate School of Science, Kyoto University, Kyoto 606-8502, Japan
[2] Department of Organic Chemistry, Faculty of Material Science and Applied Chemistry, Riga Technical University, Riga, LV-1048, Latvia
[3] Research Center for Low Temperature and Materials Sciences, Kyoto University, Japan
[4] The Institute of Solid State Physics of Russian Academy of Science, Chernogolovka, 142432, Russia

1   INTRODUCTION

Varieties of one- to two-dimensional (1D, 2D) metallic and superconducting organic solids have been developed based on charge transfer (CT) complexes composed of donor and acceptor molecules, *e.g.* TTF, BEDO-TTF (BO), BEDT-TTF (ET), TCNQ, *p*-chloranil *etc.* (Scheme 1).[1,2] They are multi-component conductors. As for the solids of uni-component, such features have been achieved only by specific physical methodology, namely high pressure,[3-7] except for a few rare cases which include transition metals.[8-11] Their transport properties have essentially been treated by band theory. For uni-molecule, on the other hand, metallic features have not yet been realized, where the transport phenomena are considered theoretically as either tunneling,[12] ballistic[12] or loop current.[13] In the latter two cases the scattering events are of negligible probability, and the elastic mean free path λ of the carriers within a molecule limits such transport regime.[12]

Studies of the electron transport in CT and uni-component solids have given two important issues which may be related with the transport in a uni-molecular conducting wire (*i.e.* molecular wire): 1) the electron correlation U is of crucial importance for the

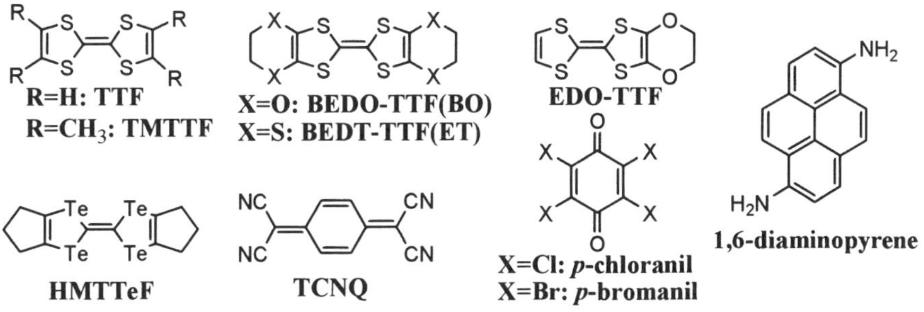

Scheme 1   *Chemicals in text*

organic metal,[1,2] and 2) electron mean free path is not much longer than the lattice constant. As for the latter issue, the estimated intermolecular mean free path λ_{inter} is ~3Å at room temperature (RT) for an organic metal TTF·TCNQ,[2] where dissipation events are mainly originated from defect, electron-phonon interaction, electron-molecular vibration coupling, conformational change of molecule and electron-electron interaction (U). In a molecular wire, the last three events will be the main sources of dissipation. Since the λ is proportional to the square of transfer integral t, the intramolecular mean free path λ_{intra} of TTF and TCNQ type molecules is estimated as 24-39 Å ($\lambda_{intra}/\lambda_{inter} = 0.7$-$0.9$ eV/0.25-0.3 eV), which is not longer than the molecular length of the molecular wire proposed by many authors.[14] Therefore, electron migration will be dissipated within a molecular wire before reacting to anodic electrode and ballistic transport is hardly achieved. That is a remarkable contrast between molecular and atomic wires (*cf.* λ of Cu at RT ~300 Å). Consequently, the transport via the organic molecules should take into account the interaction of the charges within each molecule, which is usually described in terms of U. The development of molecules having small U is indispensable to widen the dissipationless regime of molecular wires.[15]

2. BRIEF REVIEW OF UNI-COMPONENT ORGANIC CONDUCTORS

As is described in the preceding section, the control of U is necessary for not only bulk studies of (super)conductors and magnets but also nanoscience and technology of organic materials. It is noteworthy that U has been well examined quantitatively for bulk molecular conductors.[1] This paper discusses the control of U by employing the conventional methodology of solid state science. Prior to describe our recent results, we briefly review the uni-component solids that exhibit metallic or highly conductive properties, in order to find a good candidate for small U system.

Applying pressure can induce the high conductivity for several molecular solids composed of closed shell molecules. Pentacene is the first example to exhibit metallic phase reported by Drickamer *et al.* in 1964[3] followed by Pt(dimethylglyoxime)$_2$,[4] Pt(benzoquinonedioxime)$_2$,[5] *p*-iodanil[6] and hexaiodobenzene[7] (Scheme 2). The latter two solids exhibit even superconductivity, suggesting that heavy atoms enhance not only the intermolecular interactions but also electron-phonon coupling.

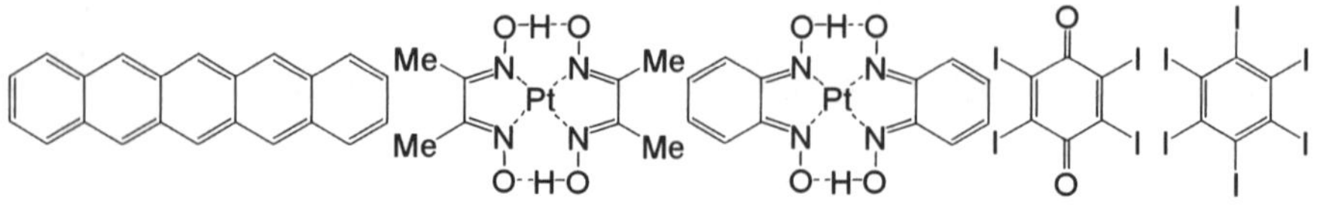

Scheme 2 *Molecules which showed metallic and/or superconducting behaviors under pressure*

Uni- or nearly uni-component conducting solids at ambient condition are classified into four groups: **I**) Organometallic coordination complexes,[8-11] **II**) Neutral closed-shell molecules,[16-18] **III**) Neutral π-radicals,[19-22] and **IV**) Betainic π-radicals.[15,23,24] The advantages of Group **I** stem from the characteristics of the transition metals; 1) mixed valence, as demonstrated by cytochrome-c_3,[10] 2) mixing of π-d orbitals to give rise to a small HOMO-LUMO gap (ΔE), as observed in the metallic dithiolate complexes,[9] and 3)

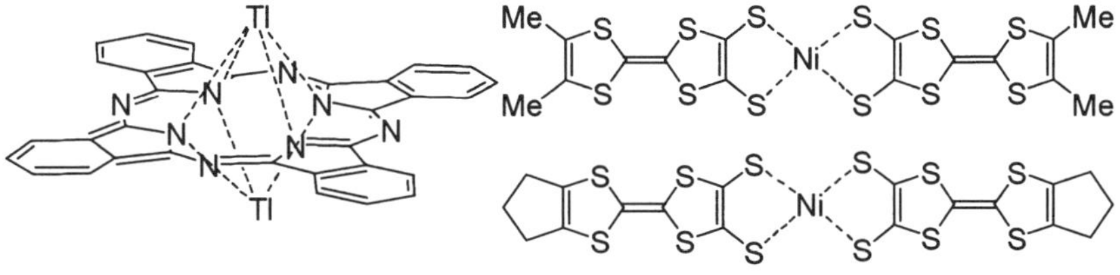

Scheme 3 *Selected molecules in Group I: Organometallic coordination complexes*

3D crystal architecture as found in the phthalocyanine (Pc) solids.[8,11] The metallic behavior down to low temperatures was reported on the Tl_2Pc compound for the first time,[8] having a resistivity at RT (ρ_{RT}) of 10^{-4} Ω cm. A specific 3D packing of the Tl_2Pc molecules is anticipated to form a 3D semimetallic band, typical of the assembly of the molecules with small ΔE. However, the reproducibility to obtain metallic Tl_2Pc is poor.[25]

All purely organic solids of uni-component so far prepared are semiconductors under ambient pressure with the following minimum ρ_{RT}; Group **II** (Scheme 4, 10^3-10^6 Ω cm),[16-18] **III** (Scheme 5, 10^1-10^3 Ω cm),[19-21] and **IV** (Scheme 6, 10^1-10^3 Ω cm).[15,23,24]

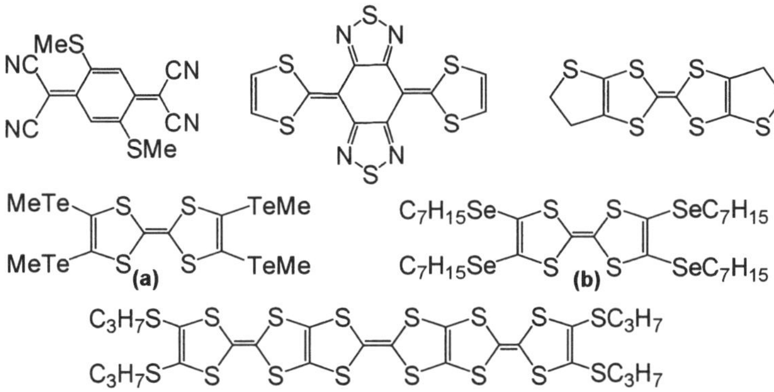

Scheme 4 *Selected molecules in Group II: Closed-shell molecules*

The low-dimensional solids in Group **II** have an energy gap $\Delta\varepsilon$ of *ca.* $\Delta E - 2(t + t')$, here t and t' are the intermolecular transfer energies of HOMOs and LUMOs, respectively. Although a high carrier mobility of 9-28 $cm^2 V^{-1} sec^{-1}$ for solids **a** and **b** in Scheme 4 was achieved by the aid of heteroatomic contacts and/or van der Waals interactions, t and t' were not yet enhanced enough at ambient pressure to afford a semimetal.[16,17]

Within the band theory, Groups **III** and **IV** have a high potential for the creation of metals. In the Group **III** molecules, **a**, **d** and **f** in Scheme 5 are mono-radicals and **b**, **c** and **e** are bi-radicals. Molecules **a**, **b** and **e** may have degenerate SOMOs. For the group **III** solids, the large $U_{eff} (= \Delta\varepsilon + 4t''$, t'': transfer energy of SOMOs), which is expressed as ($U - V$) where V is the nearest-neighbor Coulomb repulsion, dominates over the band width, resulting in the localization of carriers and hence the formation of Mott insulators.[19,20,21,26] Solid of **f** shows a weak ferromagnetism due to spin-canting antiferromagnetic interactions with $Tc = 36$ K.[26]

The betainic (zwitterionic) character of Group **IV**, where the donor (D) and acceptor (A) moieties are bound by π-bond forming D^+-π-A^-, is expected to decrease the U_{eff} according to the LeBlanc's theory[27] as eq. 1

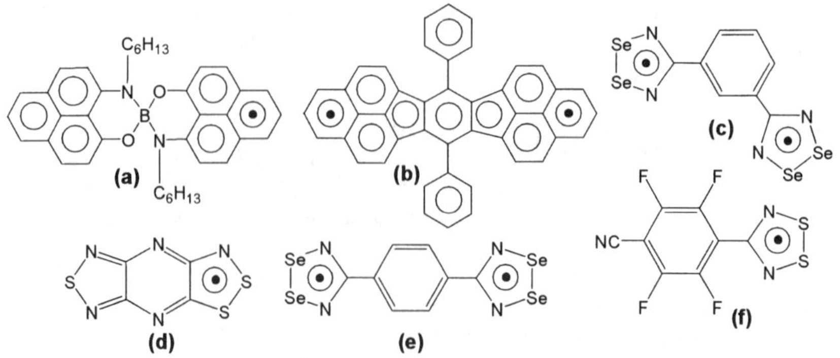

Scheme 5 *Selected molecules in Group **III**: Neutral open-shell molecules*

$$U_{\text{eff}}(\text{Group }\textbf{IV}\text{ solid}) = (1 - \alpha/r^3)(U - V) \tag{1}$$

where α is the molecular polarizability, and r is the average distance between the radical electron and the polarizable media. So far several betainic radicals or precursors have been prepared as shown in Scheme 6 (Molecule **a** in Scheme 5 may also have a betainic character).

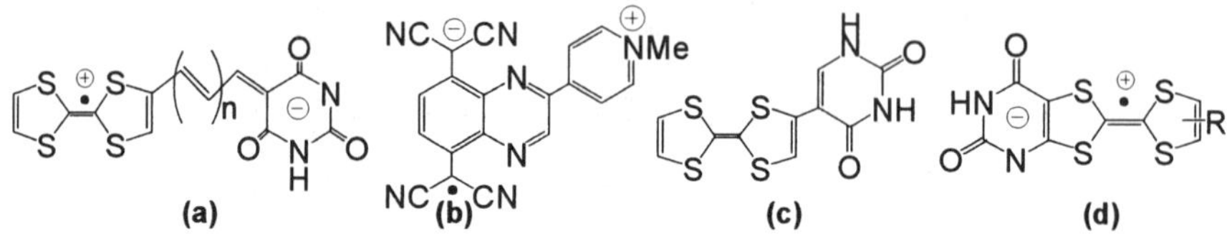

Scheme 6   *Selected betaine radicals (Group **IV**) or precursors in text*

In betainic form, they simultaneously have positive and negative charges located in close proximity. Among them, **a** and **b** have a large r value and hence their U_{eff} values are not much reduced resulting in poor conductivity, which may also be caused by inadequate stacking of the molecules. Even though the betainic radical molecule of **c** have not been prepared yet, a reduction of U_{eff} is expected compared to **a** and **b**. Furthermore, much effective reduction of U_{eff} is expected in the fused mesomeric betaine **d**, keeping the molecular size small, that is critical owing to the short λ value of organic materials.

Our classification of the uni-component organic conductors also evaluates the potential of the molecules for molecular wire. Even if the component molecules in Group **I** and **II** afford a semimetal, the molecule itself has a closed shell electronic structure that may cause a high bias for transport as a molecular wire. The dithiolate complexes in Group **I** (Scheme 3) may be too long for ballistic transport at RT though their exact U_{eff} and λ_{intra} are unknown. In case of Group **III**, the radical electrons bring about the competition among the itinerancy, localization and bond-formation by themselves. In fact, dimerization or polymerization has been observed in some cases, *e.g.* molecules **c-e** in Scheme 5.[19,20] Although the competition is an interesting subject to study switching phenomena,[28] this feature reduces the opportunity to provide ballistic molecular wires. The molecular wire consisting of the molecules in Group **IV** is of special interest since the structural feature, in which each ion radical part is located in the vicinity of the polarizable counter ion, has an analogous aspect to that of the polymer superconductor designed by Little.[29]

In the following, we describe the preparation and electric resistivity, magnetic and optical properties on betaines **2** and **4-6** (Scheme 7) including the structural analysis of the TBA (tetrabutylammonium) salts of **1** and **3**. We aimed 1) the deprotonation at pyrimido N atom and oxidation of TTF moiety to afford fully betainic system, 2) the introduction of ethylenedioxy group in **1** and **2** to provide moderate solubility like BO, 3) the *N*-Me group in **4** to prevent the formation of complementary hydrogen bonds (HBs) that were observed in the TBA salt of **1** (*vide infra*) and may be formed in **2**, **5** and **6**, and 4) the selenium analogue of **5** to improve the transport property of **6**.

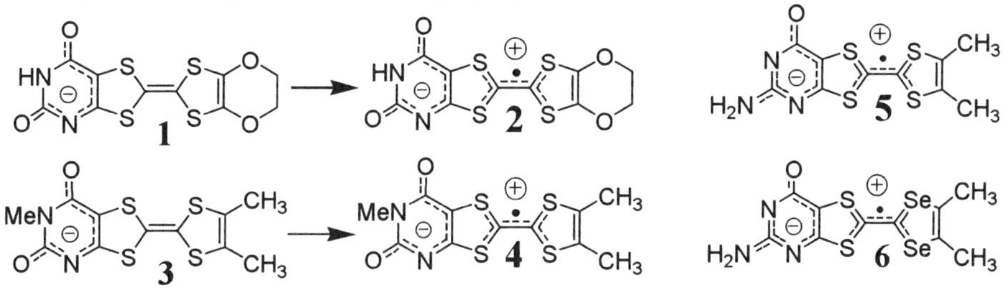

Scheme 7 *Betainic radicals studied in this work*

2   RESULTS AND DISCUSSION

2.1 Synthesis

1-6 were synthesized using the modified methods developed by Neilands' group.[23,24] Figure 1 shows the synthetic procedures of neutral species of **5** and **6**. The neutral species were converted into the TBA salts by treating with TBA·OH in acetonitrile. The betainic radicals were prepared either by electrooxidation of the TBA salts in acetonitrile,

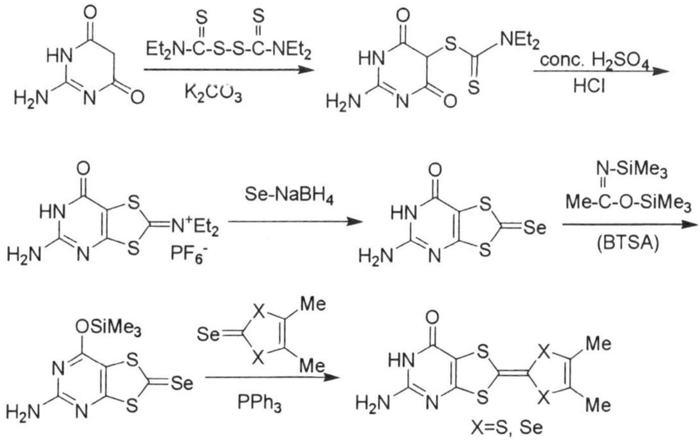

Figure 1 *Synthetic procedures of the neutral species of 5 and 6*

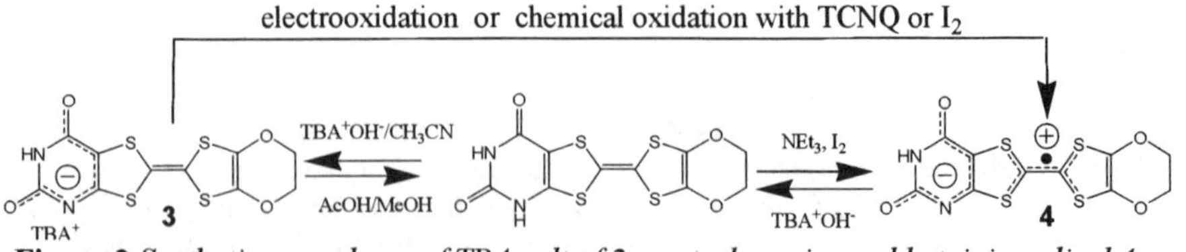

electrooxidation or chemical oxidation with TCNQ or I_2

Figure 2 *Synthetic procedures of TBA salt of 3, neutral species and betainic radical 4*

methanol or DMF, or chemical oxidation by TCNQ or I_2 in acetonitrile. The latter procedure using I_2 sometimes afforded a betainic radical contaminated with cation radical salt with I_3, as has been observed for the first report on the highly conductive betaine **6** by Neilands *et al.*[23,24] Since the betainic radicals, in general are extremely insoluble, the materials were purified by repeating the processes in Figure 2: (TBA salt of anion species) $\rightleftharpoons$ (neutral species) $\rightleftharpoons$ (betainic radical species). Namely the betainic radicals were converted to neutral species by the reduction with hydrazine hydrate and TBA·OH. Also the TBA salts can be returned back to the neutral species by treating with acetic acid in alcohol. The characteristics and preparative methods of the betainic radicals are summarized in Table1.

Table 1   *Charge transfer energy (hv_{CT}), resistivity at room temperature (ρ_{RT}), activation energy for conduction (ε_a) and spin susceptibility (χ) for betainic radicals*

betaine radicals[1)	$h v_{CT}$ /10^3 cm^{-1}	ρ_{RT} /Ω cm	ε_a /eV	χ /%
2 green powder, EO/DMF or TCNQ/MeCN	5.0	2×10^3	0.25	2.8
4 deep green microcrystal, EO/MeCN	7.6	5×10^3	0.20-0.21	2.5
5 green powder, TCNQ/MeCN	4.8	7-10	0.12-0.23	2.5
6 dark green powder, TCNQ/MeCN	4.5	20	0.09-0.12	2.0
EDO-TTF·IBr$_2$	10.0	$>2\times10^9$		

1) EO: electrooxidation. TCNQ: chemical oxidation with TCNQ. DMF: *N,N*-dimethylformamide, MeCN: acetonitrile. MeOH: methanol

2.2 Crystal structures of the tetrabutylammonium salts of 1 and 3

Although the TBA salts of anion species of **5** and **6** or their analogues so far prepared (Scheme 8 shows other pyrimido fused TTF species, the physical properties of their betainic radicals will be reported elsewhere) have not afforded single crystals, single

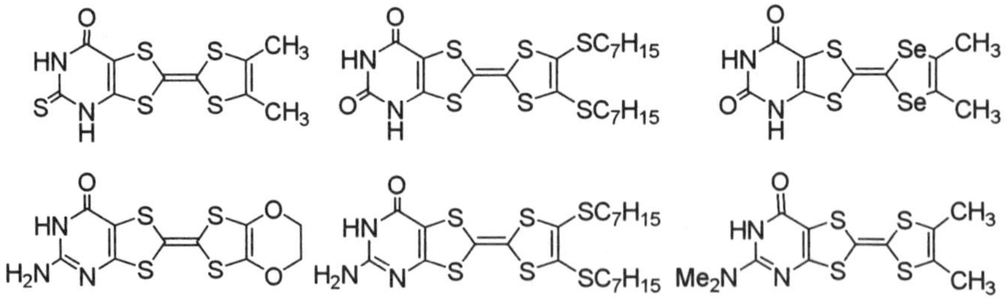

Scheme 8   Other pyrimido fused TTF species for betaine radicals

crystals of **1** and **3** were obtained by recrystallization.

The anion **1** and water molecules in the TBA salt form a layer (Figure 3a) which is sandwiched by the layers of TBA cations along the *b*-axis. Robust HBs, N-H···O (2.881(4), 2.891(4) Å vs. the sum of the van der Waals radii of N and O = 3.07 Å) connect two **1** molecules to form homo-dimer **1=1** constituting the building unit of the crystal (shaded part in Figure 3a). Similar self-complementarity has been observed in uracilbetaine dimer[30] and in the neutral species, dimethyl(2,4-dioxo(1H,3H)pyrimido)TTF[24]; it is expected to be present in **2**, **5** and **6**. Such HB is of importance for molecular recognition of nucleobases[31] and for crystal engineering in supramolecules.[32] However, its role for the electron transport is not clearly understood yet, and it has often been described as a source of high conductivity.[33] It is desirable that HBs are uniform not to create the variation of transfer energies causing dimerized lattice with narrow bands susceptible to undesirable perturbation to give insulating state. Furthermore, it is desirable that hydrogen-bonded structures should provide at least 2D electronic structure to suppress Peierls or Jahn-Teller instabilities typical for low-dimensional conductors.[1,2] In addition to those HBs, short side-by-side S···S atomic contacts (3.426(1) Å vs. the sum of the van der Waals radii of 3.60Å) connect neighboring TTF moieties and contribute to the formation of a stable 2D network (Figure 3a).

Contrary, the anion molecules in the TBA salt of **3**, where two kinds of independent **3** molecules exist, do not exhibit self-complementary HBs as shown in Figures 3b and 3c confirming that *N*-Me group in **3** is effective to protect the pyrimido group from the formation of self-complementary HBs.

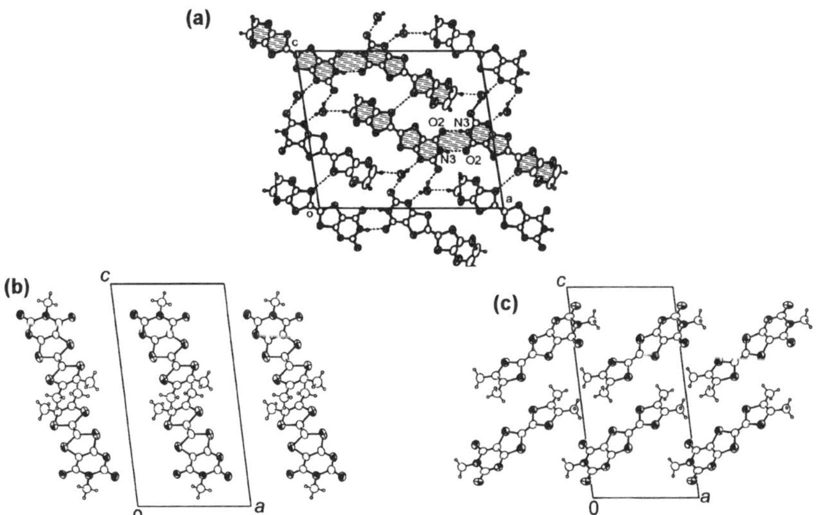

Figure 3 *(a) Anion layer composed of 1 and water molecules in TBA·1·H₂O. Dotted lines indicate hydrogen bonds (N(3)-H···O(2), O-H···O, and C-H···O) and S···S atomic contacts. Two of anion dimers connected by short N(3)-H···O(2) hydrogen bonds (1=1) are shaded. (b,c) One dimensional anion arrangements of 3 viewed along the b-axis at y = 0.5 (b) and 0 (c).*

2.3 Optical spectra of betainic radicals 2 and 4-6.

The UV-Vis-NIR spectra of the betainic radicals dispersed in KBr are compared with that of a CT salt of fully ionized EDO-TTF molecule, EDO-TTF·IBr$_2$ in Figure 4. The betainic radicals exhibit a characteristic low energy band near 4.5-7.6×10^3 cm^{-1} which disappears in diluted solution. This band is known to arise from CT between radical molecules.[34] Its energy is represented by eq. 2 in analogy with that of the dimer with two electrons,[35] where t_d and t_d* are the intra- and inter-dimer transfer integrals, respectively.

$$h\,\nu_{CT} = U_{eff}/2 + (U_{eff}^2/4 + 4t_d^2)^{1/2} - 4t_d* \qquad\qquad (2)$$

The CT energy (Table 1) is related with the electron-correlation U_{eff} and is considerably low compared with those of CT solids of TTF and its alkyl derivatives like TMTTF (10-12×10^3 cm^{-1}),[34] or even of the extended TTF systems like BO or ET (6-10×10^3 cm^{-1}),[36] demonstrating that these betaines belong to low-U_{eff} system. A comparison with the spectrum of EDO-TTF·IBr$_2$ (Figure 4, curve a) or TMTTF·Br ($h\,\nu_{CT}$ = 10×10^3 cm^{-1}) implies that, owing to the betainic nature, the U_{eff} in the solid **4** is reduced by nearly 25%, *i.e.* (1 - α/r^3) $\cong$ 3/4. For the other betainic radicals, which are presumed to have complementary HBs in the solids, the CT band appears near or below 5.0×10^3 cm^{-1} indicating that a dimerization of betaine molecules by complementary HBs further reduces U_{eff} in the solid by nearly 25%.

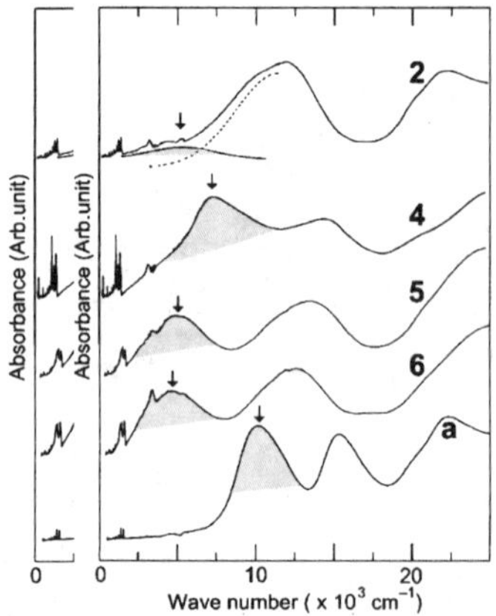

Figure 4. *Absorption spectra of betainic radicals **2**, **4**, **5**, and **6** and **a**) cation radical salt EDO-TTF·IBr$_2$ dispersed in KBr. CT bands are shaded and the peaks are indicated by arrows*

2.4 Resistivity and magnetic properties of betainic radicals 2 and 4-6.

The resistivity at RT (ρ_{RT}) and activation energy for conduction (ε_a) of the pellet samples of these betainic radicals and EDO-TTF·IBr$_2$ are summarized in Table 1. At first the preparation methods to afford pure betainic radicals were studied on **5**, which was contaminated with I$_3$ by 2 mol % and exhibited low resistivity of 5-10 Ω cm in the first report by Neilands *et al.*[23,24] We have found that both methods, electrooxidation and chemical oxidation by TCNQ followed by a through washing with acetonitrile, afforded samples not contaminated with starting reagent(s) and cation radicals. Thus the betainic radicals in Table 1 were prepared by either method. **5** was obtained as green powder either by electrooxidation or chemical oxidation with TCNQ, where the latter method is more preferable to yield large amount of betainic radicals than the former one. The compaction pellet sample of **5** is highly conductive (ρ_{RT} = 7-10 Ω cm, ε_a = 0.12-0.23 eV). Similarly, the selenium analogue **6** was obtained as highly conductive dark green powder with smaller activation energy than that of **5** (ρ_{RT} = 20 Ω cm, ε_a = 0.09-0.12 eV). These are the most conductive uni-component solids so far prepared except transition metal complexes,[8-11] and as far as the activation energy concerns, **6** is the best betainic radical. However, no structure-property relation has been elucidated since no single crystals of **5** and **6** were available.

Although the introduction of an ethylenedioxy group improved the solubility of TBA salt of **1**, only green powder of **2** was obtained with poor transport properties (2×10^3 Ω cm, 0.25 eV). The poor crystallinity of **2**, **5** and **6** was thought to be caused by the presence of robust HBs, and so the *N*-Me derivative (**3** and **4**) was examined in order to block off the HBs. As mentioned in section **2.2**, no HBs were present in the TBA salt of **3**. Also, the crystallinity of **4** was much improved compared with those of **2**, **5** and **6** and single crystal growth of **4** is underway. Within the limited numbers of betainic radicals in Table 1, the 2,4-dioxopyrimido-betaines (**2** and **4**) are less conductive (2-5×10^3 Ω cm) than those of 2-amino-4-oxopyrimido-betaines (**5** and **6**, 7-20 Ω cm) regardless of the HBs. The HBs seem to suppress the activation energy for conduction (0.20-0.21 eV for **4** vs. 0.05-0.11 eV for **2** and **5**). The betainic radicals with high conductivity exhibited the CT band more clearly with stronger intensity than those with poor conductivity as shown in Figure 4.

The last column in Table 1 summarizes the concentration of radical spins in the betainic radicals evaluated from the magnetic susceptibility measurements. Surprisingly only 2.0-2.8 % of total spins show the Curie-Weiss behavior at RT, indicating that *ca.* 97-98 % of spins are strongly dimerized in all the betainic radicals in Table 1. The mechanism of the spin dimerization is not the σ-bond formation as observed in several neutral open-shell molecules in Group **III**,[19,20] since neutral, TBA salt and betainic radicals easily transform to each other (Figure 2) indicating that betainic radicals are not composed of σ-bonded dimer. Furthermore, the spin dimerization is not due to the complementary HBs as observed in Figure 3a since spin dimerization occurs regardless of the presence of complementary HBs. The fact that the majority of spins on the betainic radicals couple antiferromagnetically with very large coupling constant indicates that only the residual spins carry the current.

The transport mechanism of such peculiar organic conductors is nontrivial and not clearly understood yet, though highly conductive CT solids with small carrier concentration have been known, *e.g.* TTF·*p*-chloranil,[37-39] 1,6-diaminopyrene·*p*-haloanil,[40,41] HMTTeF·2,5-diethylTCNQ(THF)$_x$ (x = 0-1.0),[42] and HMTTeF·BTDA-TCNQ(THF)$_x$.[43] All of them are alternating type CT complexes, and the degree of CT is close to the neutral-ionic phase boundary for the former two complexes and is accord with the mixed valence state for the latter two complexes. For 1,6-diaminopyrene·*p*-bromanil (γ-phase), *ca.* 4% of the complexes are ionized (singlet-triplet mechanism) at RT and the resistivity at RT is *ca.* 10 Ω cm (ε_a = 0.10 eV).[41] For HMTTeF·2,5-diethylTCNQ(THF)$_x$ (x~0), *ca.* 3% of the complexes are thermally excited at RT and 0.2% of the complexes being Curie components, and the complex exhibits metallic temperature dependence with σ_{RT} = 10-40 S cm^{-1} down to *ca.* 200 K.[42] The conduction in TTF·*p*-chloranil, in which only Curie components were detected, was explained by the dynamics of charged solitons and/or domain walls (σ_{RT} = 1 S cm^{-1} under 0.9 GPa).[38,39]

2.5 Electron correlation in a dimerized system and future prospects for conductors

For a dimerized system, the gap value $\Delta\varepsilon$ should take into account the on-site Coulomb energy of the dimer unit U_d: it is approximated by the dimerization energy ($\Delta D = 2t_d$) when $U_{eff} > 4t_d$[35] as eq. 3 and the $\Delta\varepsilon$ is expressed by eq. 4.

$$U_d = \Delta D + 1/2[U_{eff} - (U_{eff}^2 + 4\Delta D^2)^{1/2}] \approx \Delta D \tag{3}$$
$$\Delta\varepsilon = U_d - 4t_d^* \approx \Delta D - 4t_d^* \tag{4}$$

In accordance with this, the suppression of ε_a in **5** or **6** as compared to **2** may be related to both the decreased ΔD and increased t_d^* values due to a weakening of dimerization. The latter issue is consistent with the increased oscillator strength f of the CT band in **5** and **6** (Figure 4) compared to that in **2**, since f is approximated as eq. 5.[34]

$$f \propto [2t_d^{*2} / (U_{eff}^2/4 + 4t_d^{*2})^{1/2}] \sim t_d^{*2}/U_{eff}. \tag{5}$$

Based on a condition; $U_{eff}/4 > t_d = U_d/2 > t_d^*$, and all the optical and electric data, U_{eff}, U_d, t_d and t_d^* are numerically estimated as 0.62-0.77, 0.32-0.48, 0.14-0.19 and 0.02-0.08 eV for **2**, **5** and **6**. These U_{eff} and U_d values are nearly 1/2 and 1/6-1/7, respectively, of those estimated for conventional TTF materials (1-2 eV).[1,2]

It is worth noting that the formation of a dimer is very efficient for electron transport since U_d is much smaller than molecular U and even U_{eff}. Such aggregation or self-complementarity is of importance for the design of conductors composed of not only small sized molecules, but also of biological ones[43], and for the aspect of formation of the carrier path of one kind of components (= segregated stack) with small U_{eff}. A semimetallic band picture is possible in the case of $U_d \leq 4t_d^*$ for a dimerized system. Even for the dimerized system, a degenerate system having two donor moieties fused with one acceptor moiety; namely, the betainic radical species represented as symmetric $D^{+\bullet}$-π-A^--π-D or *vice versa*, to share one radical electron between two D moieties may have partial CT state and give rise to high itinerancy of the radical electrons. For other cases, however, the dimerization of the radical molecules always results in both the quenching of carriers and the generation of non-uniform transfer integrals. Consequently, no dimerization and a much reduced U_{eff} are required through the chemical modifications of betainic molecules for the metallic nature on uni-component solids as well as for the dissipationless conduction on a molecular wire near RT.

3   CONCLUSION

Molecular wires with less dissipation are necessary for highly sophisticated molecular devices operative at RT. We demonstrated that the mesomeric zwitterionic (betainic) radicals, having fused positive and negative moieties, are promising for creating conductors in the form of a uni-component material originating from the diminished effective on-site Coulomb repulsive energy U_{eff}. For one such class of compounds, pyrimido-fused TTF betainic radicals, we describe the preparation, purification and characteristic optical, electric and magnetic properties. Self-complementary hydrogen bonds were observed in the 2,4-dioxo-pyrimido substituted betainic anion. Such hydrogen bonds were prevented by blocking the N atom at 3-position of pyrimido ring by methyl group. The radical electrons on betainic radicals were strongly coupled regardless of the hydrogen bonds, and only 2-3% of total spins were uncoupled and gave rise to high conductivity. The U_{eff} and on-site Coulomb energy for a dimer U_d were evaluated together with the intra- and interdimer transfer energies. Based on these data, further molecular design of betainic molecules for the metallic nature on uni-component solids as well as for the dissipationless conduction on a molecular wire were discussed.

Acknowledgement

We thank Ministry of Education, Science, Sports, Culture and Technology, Japan, CREST of JST, JSPS, and NEDO for financial support.

References

1   T. Ishiguro, K. Yamaji and G. Saito, *Organic Superconductors*, 2nd Edn, Springer-Verlag, Berlin 1998.

2   D.O. Cowan, *New aspects of organic chemistry-I. in Proceedings of the 4th International Kyoto Conference on New Aspects of Organic Chemistry*, eds. Z. Yoshida, T. Shiba, and Y. Ohshiro, Kodansha Ltd, Tokyo, 1989.

3   R.B. Aust, W.H. Bentley and H.G. Drickamer, *J. Phys. Chem.*, 1964, **41**, 1856.

4   I. Shirotani, A. Kawamura, K. Suzuki, W. Utsumi, T. Yagi, *Bull. Chem. Soc. Jpn.*, 1991, **64**, 1607.

5   I. Shirotani, K. Suzuki, T. Suzuki, T. Yagi, M. Tanaka, *Bull. Chem. Soc. Jpn.*, 1992, **65**, 1078.

6   T. Yokota, N. Takeshita, K. Shimizu, K. Amaya, A. Onodera, I. Shirotani and S. Endo, *Czech. J. Phys.*, 1996, **46**, 817.

7   E. Iwasaki, K. Shimizu, K. Amaya, A. Nakayama, K. Aoki and R.P. Carlon, *Synth. Met.*, 2001, **120**, 1003.

8   J. Janczak, R. Kubiak, A. Zaleski and J. Olejniczak, *Chem. Phys. Lett.*, 1994, **225**, 72.

9   H. Tanaka, Y. Okano, H. Kobayashi, W. Suzuki and A. Kobayashi, *Science*, 2001, **291**, 285.

10 K. Kimura, Y. Nakahara, T. Yagi and H. Inokuchi, *J. Chem. Phys.*, 1979, **70**, 3317.

11 K. Morimoto and T. Inabe, *J. Mater. Chem.*, 1995, **5**, 1749.

12 C. Joachim, J.K. Gimzewski and A. Aviram, *Nature*, 2000, **408**, 541.

13 S. Nakanishi and M. Tsukada, *Surf. Sci.*, 1999, **438**, 305.

14 J.M. Tour, *Acc. Chem. Res.*, 2000, **33**, 791.

15 G. Saito, A. Ota, Y. Yoshida, M. Maesato, H. Yamochi, K. Balodis, O. Neilands, S. Khasanov and M. Tanatar, *Mol. Cryst. Liq. Cryst.* 2004, **423**, 99.

16 H. Inokuchi, K. Imaeda, T. Enoki, T. Mori, Y. Maruyama, G. Saito, N. Okada, H. Yamochi, K. Seki, Y. Higuchi and N. Yasuoka, *Nature*, 1987, **329**, 39.

17 G. Saito, *Pure Appl. Chem.*, 1987, **59**, 999.

18 Y. Yamashita, S. Tanaka, K. Imaeda, H. Inokuchi and M. Sano, *J. Org. Chem.*, 1992, **57**, 5517.

19 A.W. Cordes, R.C. Haddon, R.T. Oakley, L.F. Schneemeyer, J.V. Waszczak, K.M. Young and N.M. Zimmerman, *J. Am. Chem. Soc.*, 1991, **113**, 582.

20 T.M. Barclay, A.W. Cordes, R.C. Haddon, M.E. Itkis, R.T. Oakley, R.W. Reed and H. Zhang, *J. Am. Chem. Soc.*, 1999, **121**, 969.

21 X. Chi, M.E. Itkis, B.O. Patrick, T.M. Barclay, R.W. Reed, R.T. Oakley, A.W. Cordes and R.C. Haddon, *J. Am. Chem. Soc.*, 1999, **121**, 10395.

22 T. Kubo, A. Shimizu, M. Sakamoto, M. Uruichi, K. Yakushi, M. Nakano, D. Shiomi, K. Sato, T. Takui, Y. Morita and K. Nakasuji, *Angew. Chem. Int. Ed.*, 2005, **44**, 6564.

23 O. Neiland, V. Tilika, I. Sudmale, I. Grigorjeva, A. Edzina, E. Fonavs and I. Muzikante, *Adv. Mater. Opt. Electron.*, 1997, **7**, 39.

24 O. Neilands, *Mol. Cryst. Liq. Cryst.*, 2001, **355**, 331. The betaine **5** (ρ_{RT} = 5-10 Ω cm) prepared by the chemical oxidation with I_2 might be contaminated with I_3 by 2 mol % based on the elemental analysis.

25 T. Suga, S. Isoda and T. Kobayashi, *J. Porphyrins Phthalocyanines*, 1999, **3**, 397.

26 F. Palacio, G. Antorrena, M. Castro, R. Burriel, J. Rawson, J.N.B. Smith, N. Bricklebank, J. Novoa and C. Ritter, *Phys. Rev. Lett.*, 1997, **79**, 2336.

27 O.H. LeBlanc, Jr., *J. Chem. Phys.*, 1965, **42**, 4307.

28 M.E. Itkis, X. Chi, A. W. Cordes and R.C. Haddon, *Science*, 2002, **296**, 1443.

29 W.A. Little, *Phys. Rev., A*, 1964, **134**, 1416.

30 A. Schmidt, M.K. Kindermann, P. Vainiotalo and M. Nieger, *J. Org. Chem.*, 1999, **64**, 9499.

31 S. Metzger and S.B. Lippert, *Angew. Chem. Int. Ed. Engl.*, 1996, **35**, 1228.

32 W. Jones, *Organic Molecular Solids, Properties and Applications*, ed. W. Jones, CRC Press, New York, 1997, ch. 6.

33 M. Fourmigue and P. Batail, *Chem. Rev.*, 2004, **104**, 5379.

34 C.S. Jacobsen, *Semiconductors and Semimetals*, Vol. 27, ed. E. Conwell, Academic Press, New York, 1988, ch. 5.

35 B. Harris and R.V. Lange, *Phys. Rev.*, 1967, **157**, 295.

36 G. Saito, H. Izukashi, M. Shibata, K. Yoshida, L.A. Kushch, T. Kondo, H. Yamochi, O.O. Drozdova, K. Matsumoto, M. Kusunoki, K. Sakaguchi, N. Kojima and E.B. Yagubskii, *J. Mater. Chem.*, 2000, **10**, 893.

37 T. Mitani, G. Saito, Y. Tokura and T. Koda, *Phys. Rev. Lett.*, 1984, **53**, 842.

38 N. Nagaosa and J. Takimoto, *J. Phys. Soc. Jpn.*, 1986, **55**, 2735; *J. Phys. Soc. Jpn.*, 1986, **55**, 3488.

39 Y. Tokura, H. Okamoto, T. Koda, T. Mitani and G. Saito, *Phys. Rev.*, 1988, **B38**, 2215.

40 Y. Matsunaga, *Nature*, 1966, **211**, 183.

41 T. Fujinawa, H. Goto, T. Naito, T. Inabe, T. Akutagawa and T. Nakamura, *Bull. Chem. Soc. Jpn.*, 1999, **72**, 21.

42 G. Saito, S.S. Pac and O.O. Drozdova, *Synth. Metals*, 2001, **120**, 667.

43 G. Saito and Y. Yoshida, *Bull. Chem. Soc. Jpn.*, submitted.

44 B. Giese, *Acc. Chem. Res.*, 2000, **33**, 631.

TTF BASED MATERIALS: FROM EXOTIC ORGANIC SUPERCONDUCTORS TO TRANSISTORS

T. Mori,[1] T. Kawamoto,[1] Y. Bando,[1] B. Noda,[1] H. Wada,[1] T. Matsuzawa,[1] T. Taguchi,[1] M. Katsuhara,[1] I. Aoyagi,[1] T. Kambayashi,[1] K. Ishikawa,[1] H. Takezoe,[1] S. Uji,[2] K. Takimiya,[3] and T. Otsubo[3]

[1]Department of Organic Materials, Tokyo Institute of Technology, O-okayama, Tokyo 152-8552, Japan
[2]National Institute for Mateirals Science, Tsukuba, Ibaraki 305-0003, Japan
[3]Department of Applied Chemistry, Hiroshima University, Higashi Hiroshima 739-8527, Japan

1 INTRODUCTION

Since the first synthesis in 1970,[1] tetrathiafulvalene (TTF) has been the most widely exploited component of organic conductors, and particularly TTF has been almost the only skeleton that realizes organic superconductivity, except for a few metal-dithilate-based superconductors. Recently TTF has found another potential application as a constituent of organic field-effect transistors (OFET).[2] It is another recent surprise that a charge-transfer salt works as an "organic thyristor" on the basis of the giant nonlinear conduction coming from the competing charge ordered states.[3-5] In the present proceedings, we report our recent works on TTF-based OFET and incommensurate organic superconductors.

2 RESULTS

2.1 TTF-based OFET

In comparison with oligothiophenes which are widely used as OFETs, TTF has strong aggregation power, and to fabricate good thin-film transistors, such molecular design is necessary as to control the morphology by introducing alkyl and/or phenyl groups. Along this line, we have explored mono-alkyl substituted BEDT-TTF (bis(ethylenedithio) tetrathiafulvalene) molecules,[6] and found that an alkyl and cyano substituted TTF (C_6EDT-DC, Scheme 1) shows good transistor characteristics (mobility 0.02 cm^2/Vs) together with the liquid crystalline (SmA) properties.[7] Here the cyano group strongly stabilizes liquid crystalline phases; this tendency has been proved in newly prepared α,ω-cyanoalkyloligothiophenes (Scheme 1).[8] These oligothiophenes form very stable liquid crystalline phases.

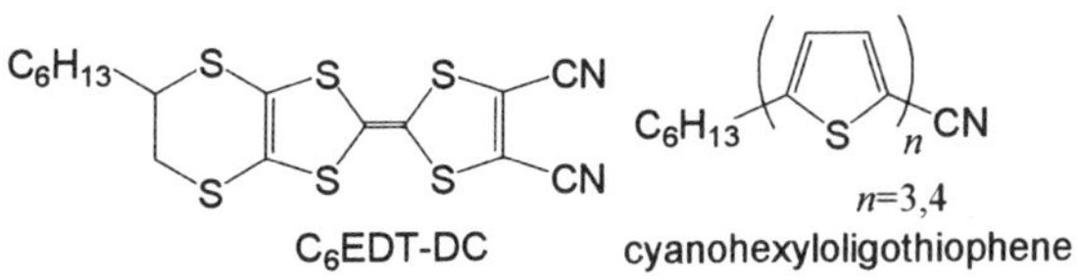

Scheme 1

Table 1 Redox potentials[a] and melting points.

Compound	$E_1^{1/2}$(V)	$E_2^{1/2}$(V)	m.p. (°C).
TTF	0.38	0.88	120-123
C$_6$EDT-DC[7]	0.76	1.08	145
C$_6$TDC-TTF	0.91	1.18	84
TCN-TTF	1.39	1.62	265-266

[a]vs. Ag/AgCl in n-Bu$_4$NPF$_6$ / C$_6$H$_5$CN at a Pt working electrode.

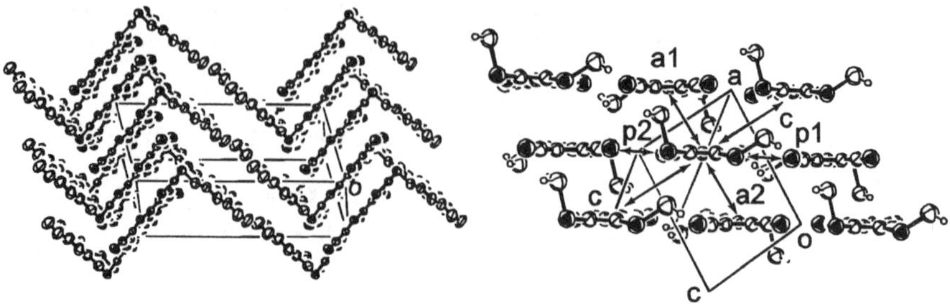

Figure 1 Crystal structure of C$_6$TDC-TTF. The overlap integrals are a1 = 1.2, a2 = 3.6, c = 0.57, p1 = -0.02, and p2 = 0.05 x10^{-3}.[9]

In this connection, we have prepared TTF with alkylthio and cyano groups (C$_n$TDC-TTF, n=6 and 10, Scheme 2). These compounds are weak donors like C$_6$EDT-DC (Table 1). C$_n$TDC-TTF makes a β"-type structure similar to TTC$_n$-TTF (tetraalkylthio-TTF), but the column is dimerized.[9] C$_6$TDC-TTF makes good thin films on SiO$_2$ substrates by vacuum evaporation (Figure 2). The d-spacing 19.9 Å corresponds to the crystallographic b axis (Figure 2), and the molecules are arranged perpendicular to the substrates, but FET characteristics are not observed.

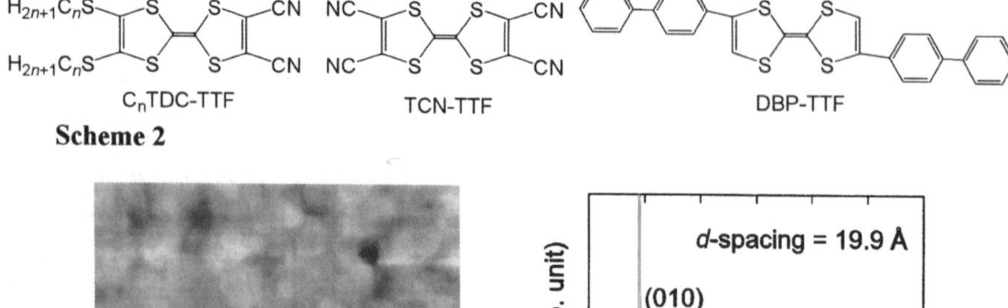

Scheme 2

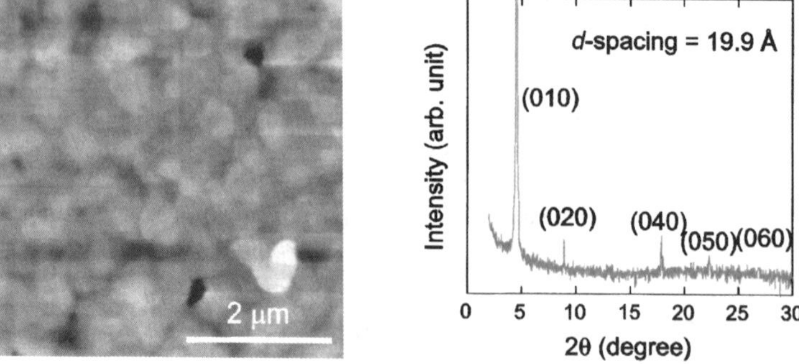

Figure 2 (a) AFM image and (b) XRD profile of C$_6$TDC-TTF.

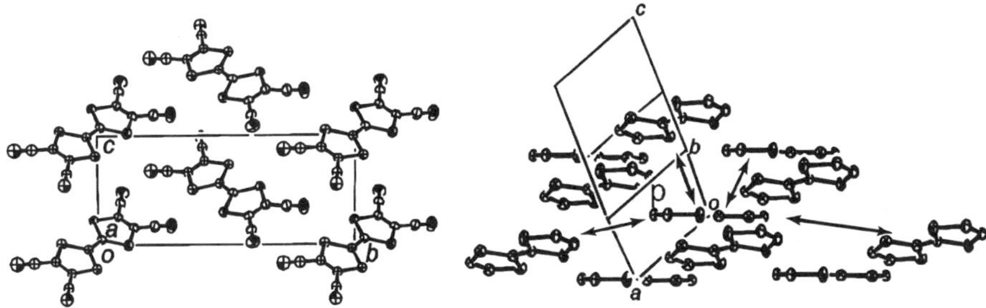

Figure 3 Crystal structure of TCN-TTF.[11] The overlap integrals are a = -2.3, p = -2.6, q = 0.7, and r = 0.3 x10^{-3}.

We have also investigated tetracyano-TTF (Scheme 2).[10] This compound forms a uniformly stacked two-dimensional structure (Figure 3),[11] but the vacuum evaporated film is composed of small grains.

Because biphenyl substituted TTF (DBP-TTF, Scheme 2) has exhibited improved transistor properties (mobility 0.11 cm^2/Vs),[12] we have prepared a variety of alkylphenyl-substituted TTFs, and investigated the FET properties as well as the bulk crystal structures.[13] When the bulk crystal structure consists of strongly dimerized units, the material does not show transistor properties, but when the crystal has a regular stacking structure, transistor properties are observed. On the other hand, since acceptor-based organic superconductors have been realized only in metal dithilate complexes, we have investigated phenyl substituted nickel dithilates, and observed n-type transistor characteristics.[14]

2.2 Incommensurate Superconductors

Although more than one hundred organic superconductors have been reported so far, what is most characteristic of organic superconductors is its definite composition and the uncontrollability of the carrier number.[15] Most organic superconductors have donor : anion = 2:1 composition, and only very few are known to have other composition like 3:1. In this context, Takimiya's finding of MDT-TSF (methylenedithiotetraselenafulvalene, Scheme 3) series superconductors has opened an entirely new era,[16] because it has such a "nonstoichimetric" composition as (MDT-TSF)(AuI$_2$)$_{0.436}$,[17] which is far from a "stoichiometric" composition. Furthermore, a series of compounds are superconducting; this has prompted new aspects of investigations.

There are several combinations of selenium substitutions as shown in Scheme 3. The

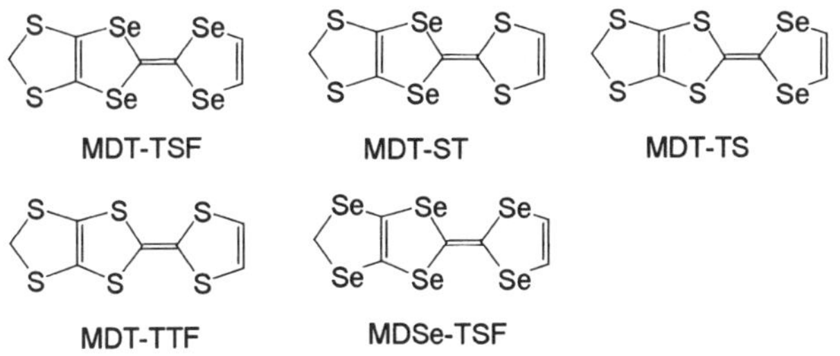

MDT-TSF　　　　　MDT-ST　　　　　MDT-TS

MDT-TTF　　　　　MDSe-TSF

Scheme 3

Table 2 MDT-TSF series superconductors

Compounds	x	T_c	$\xi_a/\,\xi_b/\,\xi_c$ (Å)[a]	A_F/A_{BZ} (%)	Ref.
(MDT-TSF)(AuI$_2$)$_{0.436}$	0.436	4.5 K	319/121/61	43.7/10.4/2.5/1.2	16,17,26,28
(MDT-TSF)(I$_3$)$_{0.422}$	0.422	4.9 K	261/166/31	42.0/9.30	21,22
(MDT-TSF)I$_{1.19}$Br$_{0.08}$	(0.423)[b]	5 K			21
(MDT-ST)(I$_3$)$_{0.417}$	0.417	4.3 K	278/205/25	42.3/9.44/8.14	23,24,25,27,29
(MDT-ST)I$_{1.07}$Br$_{0.20}$	(0.423)[b]	3.2 K			23
(MDT-TS)(AuI$_2$)$_{0.441}$	0.441	4.0 K (11.4 kbar)			30,31

[a] 1.6 K [b] Assuming trihalides.

(a)   (b)

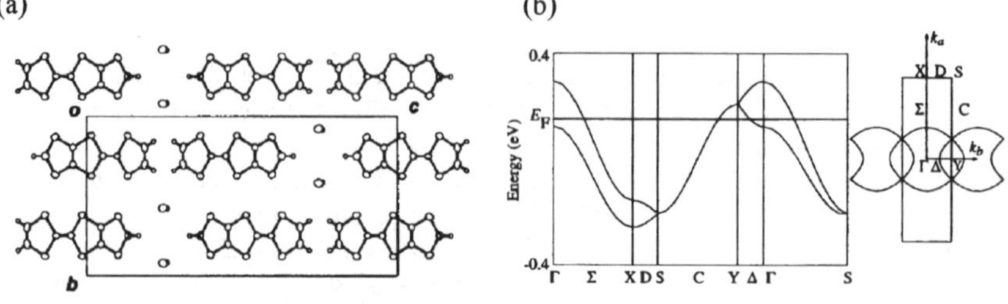

Figure 4 (a) Crystal structure of (MDT-TSF)(AuI$_2$)$_{0.436}$ and (b) Energy band structure and the Fermi surface of (MDT-ST)(I$_3$)$_{0.417}$. The transfer integrals are $t_a = 104$ $t_{p1} = -8.8$, and $t_{p2} = -34.1$ meV.[25]

all sulphur analogue forms a 2:1 superconductor κ-(MDT-TTF)$_2$AuI$_2$ (T_c = 3.5 K) with a κ-structure.[18] This compound is well known because the Hebel-Slichter peak has been observed in the ^{1}H-NMR measurement,[19] and this is a rare example of organic superconductors in which s-wave pairing has been reported. The all-selenium analogue also forms a stoichiometric superconductor κ-(MDSe-TSF)$_2$Br (T_c = 4 K).[20] Other upper three donors shown in Scheme 3 contain both sulphur and selenium atoms, and make the incommensurate phase (Table 2). In the naming of these donors, the MDT part is located on the left, and the TTF part is designated as ST or TS according to the sequence of the selenium and sulphur atoms from the left to the right.[23]

First, superconductivity of (MDT-TSF)(AuI$_2$)$_{0.436}$ has been observed at T_c = 4.5 K at ambient pressure.[16] The donors construct uniform columns along the crystallographic a axis, and the anions form linear chains parallel to the donor columns (Figure 4(a)). The exact composition has not been revealed until the complicated pattern of the x-ray oscillation photograph is analyzed.[17] The photograph shows a d-spacing of a' =9.221 Å originating from the anions, in addition to the donor spacing of a = 4.016 Å. From the ratio of these lattice spacings, we can evaluate the exact composition as $x = a/a'$ = 0.436. Although this is a non integer number, this number does not change for a given combination of the donor and the anion.

Other linear anions also form superconductors like (MDT-TSF)I$_{1.27}$ (T_c = 4.9 K) and (MDT-TSF)I$_{1.19}$Br$_{0.08}$ (T_c = 5 K).[21,22] The determination of the precise charge-transfer degree in an iodine salt is carried out for the MDT-ST salt prior to the MDT-TSF salt,[23-25] because it gives a better crystal, (MDT-ST)I$_{1.251}$. This is an ambient-pressure superconductor at T_c = 4.3 K. The AuI$_2^-$ anion has a definite unit, whereas it is not obvious that these halide anions have the trihalide units such as I$_3^-$, because there are other possibilities like I^- and I_5^-. Actually, the oscillation photograph of the AuI_2 salt exhibits a clear $h' = 1$ layer corresponding to $a' = 9.221$ Å, while other halide anions do not show the $h' = 1$ layer, and only the average atomic spacing corresponding to $h' = 3$ (3.13-3.19 Å) is clear. Therefore the charge-transfer degree has not been determined until the Raman investigation demonstrates the real anion species to be I_3^-.[25] Accordingly we can write as $(MDT-ST)(I_3)_{0.417}$ instead of $(MDT-ST)I_{1.251}$. In general, the x number of the I_3 salt is smaller than that of the AuI_2 salt, in agreement with the longer anion length of I_3 than that of AuI_2.

2.3 Superconductivity, Fermi Surface, and Shubnikov-de Haas Oscillation

Since we have determined the exact charge-transfer degree, we can calculate the energy band structure and the Fermi surface (Figure 4(b)). Although the structure consists of a one-dimensional stack, the interstack interactions are significant ($t_\perp/t_{//} \sim 1/3$) so that the Fermi surface is closed. The reflectance spectra of $(MDT-TSF)(AuI_2)_{0.436}$ gives the ratio of the plasma frequency to be $\omega_{p\perp}/\omega_{p//} \sim 0.62$,[26] indicating the considerable interstack interactions. The Ginzburg-Landau coherence lengths ξ evaluated from the measurements of the critical fields show small anisotropy (< 2) (Table 2), for $(MDT-TSF)(AuI_2)_{0.436}$,[17] $(MDT-TSF)(I_3)_{0.422}$,[22] and $(MDT-ST)(I_3)_{0.417}$.[27] Therefore the present compounds are regarded as a two-dimensional system with small (2~3) in-plane anisotropy. Not only within the conducting sheet but also in the perpendicular direction, the angular dependence of the critical fields follows the anisotropic three-dimensional model rather than the two-dimensional model,[22,27] so that these compounds are regarded as anisotropic three-dimensional superconductors.

The existence of the closed Fermi surface has been verified by the observation of the Shubnikov-de Haas (SdH) oscillations.[28] Four fundamental frequencies are observed for $(MDT-TSF)(AuI_2)_{0.436}$ (Table 2). The largest frequency $A_F/A_{BZ} = 43.7\%$, called the δ-orbit, corresponds to the large elliptical Fermi surface. Therefore, this is equal to the charge-transfer degree x, and is the ultimate evidence of the nonstoichiometric composition. The γ orbit (10.4%) comes from the overlapped area of this δ-orbit. The other two small pockets (β: 2.5% and α: 1.2%) are generated from the overlapped areas when the original Fermi surface is transposed by three times of the incommensurate anion potential ($3q$) (Figure 5).[28] The q transpose does not generate any new overlapped areas, but the overlapped Fermi surface by $2q$ is not observed. This is because the $3q$ potential, corresponding the average atomic spacing (3.040 Å), is actually the strongest periodic potential which is observed in the x-ray measurement.[17,28]

The SdH oscillations are also investigated for $(MDT-ST)(I_3)_{0.417}$.[29] Since the x number is slightly small, the δ-orbit as well as the derived γ orbit is small ($A_F/A_{BZ} = 42.3\%$ and 9.44%) (Table 2). Consequently, the reconstructed α orbit is too small to be observed, and only the β orbit is observed. Since the situation of the Fermi surface reconstruction changes sensitively according to the x number, this is a very delicate probe of the nonstoichiometry and the $3q$ incommensurate potential.

Angular dependent magnetoresistance oscillations (AMRO) of $(MDT-TSF)(AuI_2)_{0.436}$ have detected an elliptical Fermi surface corresponding to the γ orbit.[26] This is surprising because the YS zone boundary is degenerated, and the large δ-orbit should be observed similarly to the degenerated κ-phase compounds. This apparent breakdown of degeneracy has been attributed to the influence of the incommensurate anion potential.

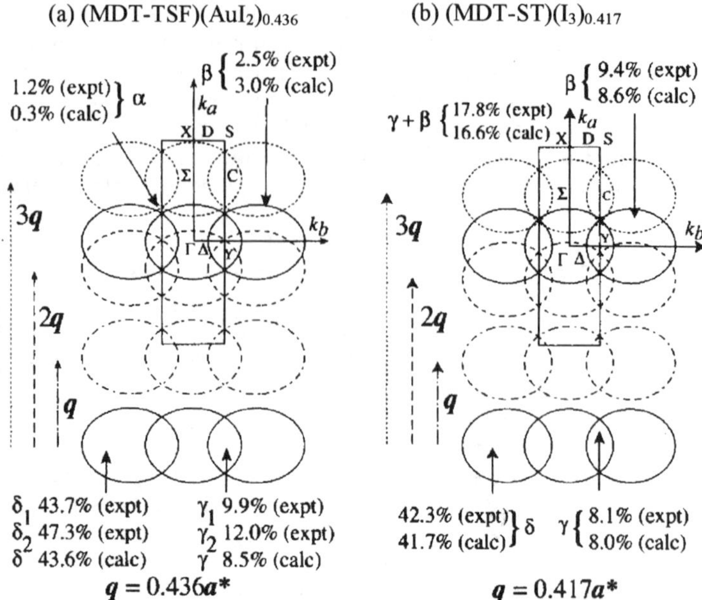

Figure 5 Fermi surface reconstruction in (a) (MDT-TSF)(AuI$_2$)$_{0.436}$ and (b) (MDT-ST)(I$_3$)$_{0.417}$.[28,29]

2.4 Incommensurate Antiferromagnetic State in MDT-TS

All the above donors construct ambient-pressure superconductors, but only the MDT-TS salt undergoes a metal-insulator transition under ambient pressure.[30] In the MDT-TS crystals, there are no interstack Se-Se contacts on account of the molecular geometry, and the bandwidth of the MDT-TS salts are 26% smaller than that of the MDT-ST salts. As a result, the MDT-TS salts are the most correlated system. The firstly investigated compound is (MDT-TS)(AuI$_2$)$_{0.441}$, whose metal-insulator transition is gradually suppressed under pressure, and superconductivity appears at $T_c = 4.0$ K under the pressure of 11.4 kbar.[31] As the measurement of magnetic susceptibility shows, the insulating state is antiferromagnetic.[31] Since the composition is nonstoichiometric, we have to imagine an incommensurate antifferromagnetic insulating state, but this is not a simple state like a spin density wave state, because the band structure is two dimensional. The phase diagram of this compound (Figure 6) demonstrates that the superconducting phase appears bordering on the incommensurate antifferromagnetic insulating state.

3 CONCLUSION

We have surveyed the various new aspects of incommensurate MDT-TSF superconductors. The physics and chemistry of TTF will further increase the importance in the near future as a source of organic superconductors, particularly a new type of exotic superconductors, and of organic conductors incorporated in such devices as organic transistors and organic thyristors.

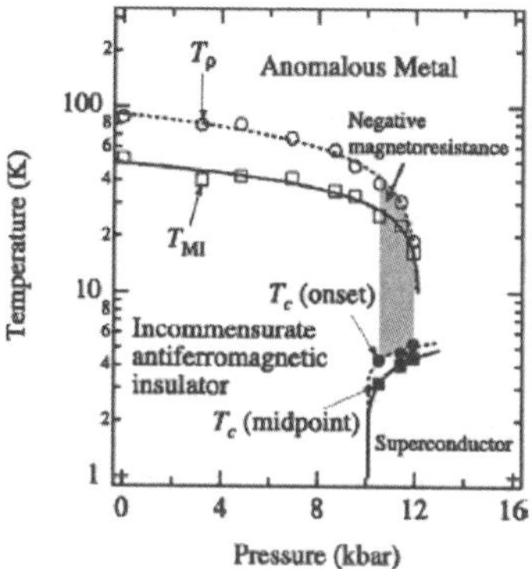

Figure 6 Phase diagram of $(MDT-TS)(AuI_2)_{0.441}$.[31]

References

1   F. Wudl, G. M. Smith, and E. J. Hufnagel, *J. Chem. Soc., Chem. Commun.* **1970**, 1453; D. L. Coffen, J. Q. Chambers, D. R. Willians, P. E. Garrett, N. D. Canfield, *J. Am. Chem. Soc.* 1971, **93** 2258; S. Hunig, G. Kisslich, D. Sceutzow, R. Zharandik, P. Carsky, *Int. J. Sulfur Chem.*, Part C, **1971**, 109.

2   M. Mas-Torrent, M. Durkut, P. Hadley, X. Ribas, and C. Rovira, *J. Am. Chem. Soc.* 2004, **126**, 8546; M. Mas-Torrent, P. Hadley, S. T. Bromley, N. Crivillers, J. Veciana, and C. Rovira, *Appl Phys. Lett.* 2005, **86**, 012110.

3   F. Sawano, I. Terasaki, H. Mori, T. Mori, M. Watanabe, N. Ikeda, Y. Nogami, Y. Noda, *Nature,* 2005, **437**, 522.

4   K. Inagaki, I. Terasaki, H. Mori and T. Mori, *J. Phys. Soc. Jpn.* 2004, **73**, 3364.

5   T. Mori, *J. Phys. Soc. Jpn.* 2003, **72**, 1469.

6   I. Aoyagi, M. Katsuhara, and T. Mori, *Sci. Tech. Adv. Mater.* 2004, **5**, 443.

7   M. Katsuhara, I. Aoyagi, H. Nakajima, T. Mori, T. Kambayashi, M. Ohfuji, Y. Takanishi, K. Ishikawa, H. Takezoe, and H. Hosono, *Synth. Met.* 2005, **149**, 219.

8   H. Wada, T. Taguchi, T. Kambayashi, T. Mori, K. Ishikawa, and H. Takezoe, to be published.

9   Crystallographic data for $C_6TDC-TTF$: $P1$-, $a = 8.552(5)$, $b = 20.447(9)$, $c = 7.547(3)$ Å, $\alpha = 97.11(3)$, $\beta = 109.23(4)$, $\gamma = 78.15(4)$ °, $V = 1217(1)$ Å^3, $Z = 2$, and $R = 0.040$. $C_{10}TDC-TTF$: $P1$-, $a = 8.655(4)$, $b = 26.275(7)$, $c = 7.423(2)$ Å, $\alpha = 90.14(2)$, $\beta = 109.02(3)$, $\gamma = 86.88(3)$ °, $V = 1593.3(9)$ Å^3, $Z = 2$, and $R = 0.044$.

10 G. Cooke, A. K. Powell, S. L. Heath, *Synthesis* **1995**, *11*, 1411.

11 Crystallographic data for TCN-TTF: $P2_1/c$, $a = 5.222(2)$, $b = 16.823(7)$, $c = 7.209(5)$ Å, $\beta = 108.0$ $(4)°$, $V = 600.4(4)$ Å^3, $Z = 2$, and $R = 0.027$.

12 B. Noda, M. Katsuhara, I. Aoyagi, T. Mori, T. Taguchi, T. Kambayashi, K. Ishikawa, and H. Takezoe, *Chem. Lett.* 2005, **34**, 392.

13 B. Noda, M. Katsuhara, I. Aoyagi, T. Mori, T. Taguchi, T. Kambayashi, K. Ishikawa, and H. Takezoe, in these proceedings.

14 H. Wada, T. Taguchi, T. Kambayashi, T. Mori, K. Ishikawa, and H. Takezoe, in these proceedings.

15 T. Mori, *Chem. Rev.* 2004, **104**, 4947.

16 K. Takimiya, Y. Kataoka, Y. Aso, T. Otsubo, H. Fukuoka, and S. Yamanaka, *Angew. Chem. Int. Ed.* 2001, **40**, 1122.

17 T. Kawamoto, T. Mori, K. Takimiya, Y. Kataoka, Y. Aso, and T. Otsubo, *Phys. Rev. B* 2002, **65**, 140508(R).

18 G. C. Papavassiliou, G. A. Mousdis, J. S. Zambounis, A. Terzis, T. A. Hountas, B. Hilti, C. W. Mayer, J. Pfeiffer, *Synth. Met.* 1988, **27**, B379.

19 T. Takahashi, Y. Kobayashi, T. Nakamura, K. Kanoda, B. Hilti, and J. S. Zambounis, *Physica C*, 1994, **235**, 2461.

20 M. Kodani, A. Takamori, K. Takimiya, Y. Aso, T. Otsubo, *J. Solid State Chem.* 2002, **168**, 582.

21 K. Takimiya, M. Kodani, Y. Kataoka, Y. Aso, T. Otsubo, T. Kawamoto, and T. Mori, *Chem. Mater.* 2003, **15**, 3250.

22 T. Kawamoto, T. Mori, T. Konoike, K. Enomoto, T. Terashima, S. Uji, H. Kitagawa, K. Takimiya, and T. Otsubo, to be published.

23 K. Takimiya, A. Takamori, Y. Aso, T. Otsubo, T. Kawamoto, and T. Mori, *Chem. Mater.* 2003, **15**, 1225.

24 T. Kawamoto, T. Mori, T. Terashima, S. Uji, H. Kitagawa, K. Takimiya, A. Takamori, and T. Otsubo, *J. Phys. IV France* 2004, **114**, 517.

25 T. Kawamoto, T. Mori, S. Uji, J. Yamaura, H. Kitagawa, A. Takamori, K. Takimiya, and T. Otsubo, *Phys. Rev. B*, 2005, **71**, 172503.

26 T. Kawamoto, T. Mori, C. Terakura, T. Terashima, S. Uji, H. Tajima, K. Takimiya, Y. Aso, and T. Otsubo, *Eur. Phys. J. B*, 2003, **36**, 161.

27 T. Kawamoto, T. Mori, T. Terashima, S. Uji, A. Takamori, K. Takimiya, and T. Otsubo, *J. Phys. Soc. Jpn.*, 2005, **74**, 1529.

28 T. Kawamoto, T. Mori, C. Terakura, T. Terashima, S. Uji, K. Takimiya, Y. Aso, and T. Otsubo, *Phys. Rev. B*, 2003, **67**, 020508(R).

29 T. Kawamoto, T. Mori, K. Enomoto, T. Konoike, T. Terashima, S. Uji, A. Takamori, K. Takimiya, and T. Otsubo, *Phys. Rev. B*, in press.

30 K. Takimiya, M. Kodani, N. Niihara, Y. Aso, T. Otsubo, Y. Bando, T. Kawamoto, and T. Mori, *Chem. Mater.* 2004, **16**, 5120.

31 T. Kawamoto, Y. Bando, T. Mori, K. Takimiya, and T. Otsubo, *Phys. Rev. B*, 2005, **71**, 052501.

TWO ISOMERIC METHYLENEDISELENO-DISELENADITHIAFULVALENES (MDSE-TS AND MDSE-ST): SYNTHESIS AND PROPERTIES OF NEW SELENIUM-CONTAINING FULVALENE-TYPE ELECTRON DONORS

K. Takimiya, Y. Nakamura, Y. Aso and T. Otsubo

Department of Applied Chemistry, Graduate School of Engineering, Hiroshima University, Kagamiyama, Higashi-Hiroshima 739-8527 Japan

1 INTRODUCTION

It has been long known that MDT-TTF (methylenedithiotetrathiafulvalene)[1] gives a superconducting AuI_2 salt (T_c = 4.1 K) with a conventional κ-type molecular arrangement. Its selenium homologues, MDT-TSF (methylenedithiotetraselenafulvalene)[2] and MDT-ST (methylenedithiodiselenadithiafulvalene)[3] also form superconductors with linear anions such as AuI_2^- and I_3^-, but their crystal structures are quite different from the AuI_2 salt of all-sulfur donor, MDT-TTF: they consist of uniform donor stacks without dimerization, and the anion lattice is incommensurate with the donor subcell, leading to nonstoichiometric donor/anion ratio. This unconventional structural type is not only unique among the organic superconductors, but also intriguing as a novel prototype for superconductivity in radical cation salts.[4] Additional characteristic feature in these superconductors is that they have an anisotropic quasi-three dimensional character elucidated by detailed studies of their electronic state.[5]

Considering these experimental results, it is apparent that introduction of selenium atom into the donor framework plays an essential role to produce the radical cation salts with the unconventional crystal structure. Taking advantage of the flexibility in designing asymmetrical donor molecules, we have focused on two novel selenium-containing donors (MDSe-TS and MDSe-ST).[6] We here report the selective synthesis of these two donors, and the structures and properties of their radical cation salts.

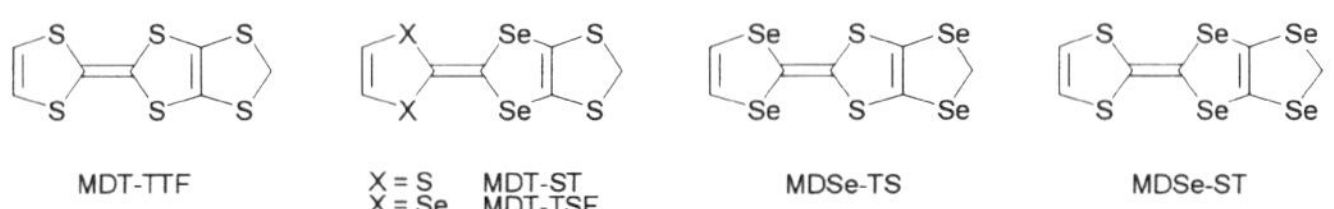

Figure 1 *Structure of MDT-TTF and its selenium analogues*

2 METHOD AND RESULTS

2.1 Synthesis and Characterization of MDSe-TS and MDSe-ST

The synthesis of MDSe-TS was readily achieved as shwon in Scheme 1. A cross-coupling reaction between 1,3-diselenol-2-one (**1**)[7] prepared from the corresponding selone[8] with mercury (II) acetate and 4,5-bis[(2-methoxycarbonyl)ethylseleno]-1,3-dithiole-2-thione (**2**)[9] gave 2,3-bis[(2-methoxycarobnyl)ethylseleno]diselenadithiafulvalene (**3**) (68% yield). Then, deprotection of the methyl propionate protecting group in **3** by caesium hydroxide generated diselenadithiafulvalene diselenolate dianion (**4**),[10] which was realkylated by diiodomethane to give MDSe-TS in 75 % isolated yield.

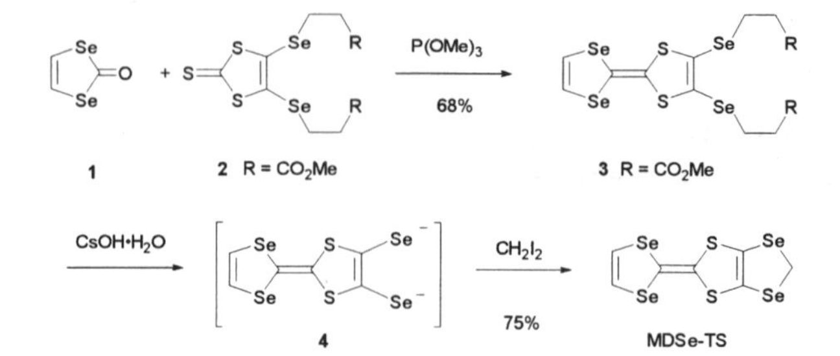

P(OMe)$_3$ 68%

1 **2** R = CO$_2$Me **3** R = CO$_2$Me

CsOH·H$_2$O CH$_2$I$_2$ 75%

4 MDSe-TS

Scheme 1 *Synthesis of MDSe-TS*

The synthesis of MDSe-ST was carried out with a modified route as outlined in Scheme 2 employing 4,5-bis(methoxycarbonyl)-1,3-dithiole-2-thione (**5**)[11] instead of the parent 1,3-dithiole-2-thione.[12] A cross-coupling reaction of **5** and **6** gave the desired unsymmetrical intermediate **7** in 69% yield. Formation of the outer selenium-containing ring was carried out in the same manner as that for MDSe-TS. In this case, however, the reaction product contains diester (**8**) and monoester (**9**), the latter of which was a partially deesterificated product probably effected by concomitant caesium iodide in the realkylation reaction. The precursors **8** and **9** were converted into MDSe-ST using a standard deestrerification conditions.[13]

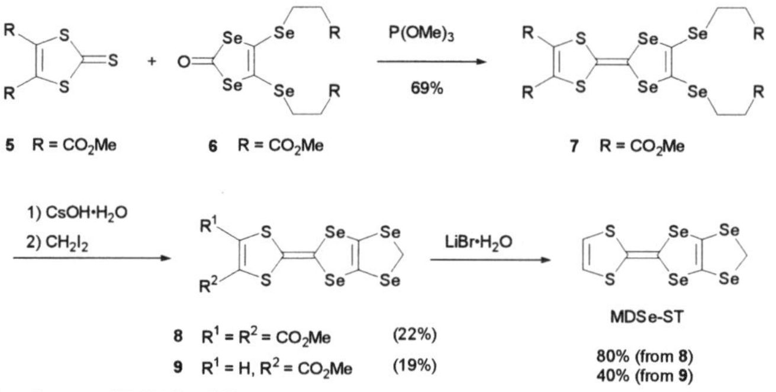

P(OMe)$_3$ 69%

5 R = CO$_2$Me **6** R = CO$_2$Me **7** R = CO$_2$Me

1) CsOH·H$_2$O 2) CH$_2$I$_2$ LiBr·H$_2$O

8 R^1 = R^2 = CO$_2$Me (22%)
9 R^1 = H, R^2 = CO$_2$Me (19%)

MDSe-ST
80% (from **8**)
40% (from **9**)

Scheme 2 *Synthesis of MDSe-ST*

Electrochemical properties of the new donors were evaluated by cyclic voltammetry. Both compounds showed two reversible redox couples; MDSe-TS: $E^{ox}_{1/2}(1)$ = +0.42 V, $E^{ox}_{1/2}(2)$ = +0.70 V, MDSe-ST: $E^{ox}_{1/2}(1)$ = +0.44 V, $E^{ox}_{1/2}(2)$ = +0.71 V (vs. Ag/AgCl in PhCN). The first oxidation potentials of the present donors are fairly lower than that of MDT-TSF ($E^{ox}_{1/2}(1)$ = +0.57 V under the identical condition) indicating that these donors have an enhanced electron donating ability.

Molecular structures of the neutral donors elucidated by an X-ray crystal structural analysis are shown in Figure 2. MDSe-TS has a relatively planar diselenadithiafulvalene core with a bent outer selenium-containing ring, whereas MDSe-ST has a folded structure at the central five-membered ring in the diselenadithiafulvalene core. Although the planarity of these donors are quite different, the intramolecular Se–Se (3.18–3.19 Å) and S–S distances (2.95–2.97 Å) in five-membered rings are almost constant. These fixed chalcogen–chalcogen distances result in the different molecule shapes of the two isomeric donors, which will affect the donor arrangement as well as intermolecular interaction in the transverse direction through the chalcogen–chalcogen contacts in crystal of radical cation salt (*vide infra*).

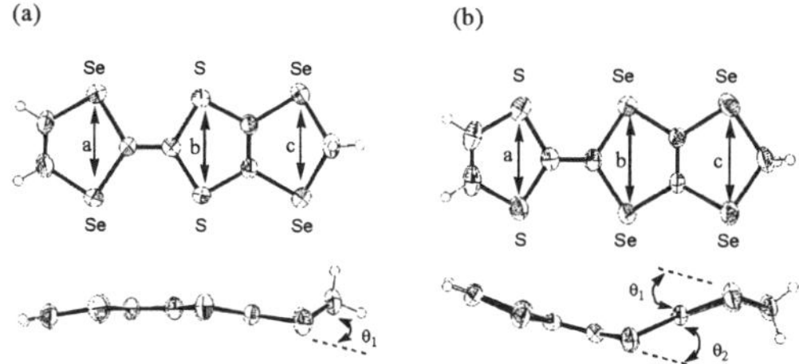

Figure 2 *(a) Molecular structure of neutral MDSe-TS: θ_1 = 40.9°. Intramolecular Se-Se and S-S distances are; a = 3.19, b = 2.97, c = 3.18 Å. (b) Molecular structure of neutral MDSe-ST: θ_1 = 34.62 (36.67) and θ_2 = 28.82(34.48)°. Intramolecular Se-Se and S-S distances are; a = 2.95 (2.95), b = 3.19 (3.18), c = 3.19 (3.19) Å. The values in the parentheses are those of crystallographically independent molecule.*

2.2 Radical Cation Salts of MDSe-TS and MDSe-ST

Electrocrystallization of MDSe-TS and MDSe-ST in the presence of tetrabutylammonium salts ($^nBu_4NX$, X = Br, AuI_2,[14] I_3) in chlorobenzene gave radical cation salts as shown in Table 1. Crystallization of MDSe-TS with AuI_2^- anion was not successful. On the contrary, preparation of a Br salt of MDSe-ST was failed. The structural phase and the donor/anion ratio of the radical salts were determined by X-ray studies. The electrical resistivity of these radical salts measured down to 1.5 K is shown in Figure 3.

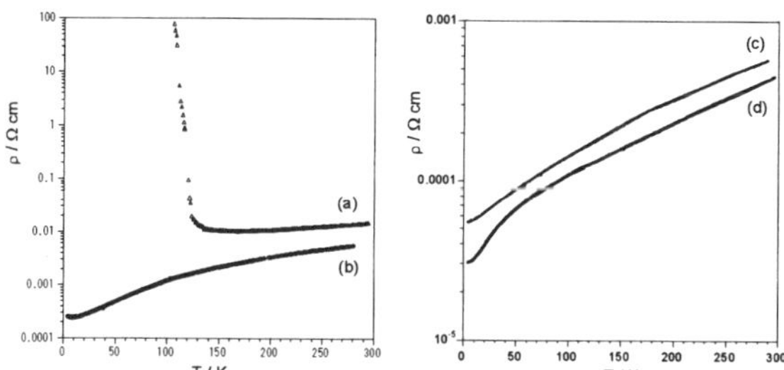

Figure 3   *Temperature dependence of resistivity; (a) κ-(MDSe-TS)$_2$Br, (b) θ-(MDSe-TS)$I_{1.26}$, (c) (MDSe-ST)(AuI$_2$)$_{0.44}$, (d) (MDSe-ST)$I_{1.26}$.*

κ-(MDSe-TS)$_2$Br   The Br salt of MDSe-TS adopts a typical κ-phase with a 2:1 donor/anion ratio. Stacked donor dimmers in head-to-tail overlap manner form the conducting sheets (Figure 4a), which are separated by insulating anion layer (Figure 5a). This structure is isostructural with κ-(MDSe-TSF)$_2$Br, where MDSe-TSF is

Table 1.   *Radical cation salts of MDSe-TS and MDSe-ST*

Radical salt	appearance	σ_{rt}/S cm^{-1} [a]	Remarks
κ–(MDSe-TS)$_2$Br	black plates	100–200	T_{MI} = 120 K
θ–(MDSe-TS)I$_{1.26}$ [b]	bronze thin plates	150–400	metallic to 1.5 K
(MDSe-ST)I$_{1.26}$ [b]	bronze thin plates	1500–2000	metallic to 1.5 K
(MDSe-ST)(AuI$_2$)$_{0.44}$ [b]	black thin pates	2000–4000	metallic to 1.5 K

[a] *Measured on a single crystal with a four-probe method.* [b] *The composition was determined on the basis of the X-ray oscillation photograph.*

methylenediselenotetraselenafulvalene.[15] Comparing these two salts, the intermolecular overlap integrals in the MDSe-TS salt are fairly smaller than those in the MDSe-TSF salt (Figure 5b).[16,17] This is explained by an effect of substitution of selenium atoms by sulfur atoms at the central five membered-ring in MDSe-TSF; in the methylenedichalcogeno-tetrachalcogenafulvalene donors, the coefficients of HOMO on the chalcogen atoms in the tetrachalcogenafulvalene core are generally greater than those in the outer methylenedichalcogeno-ring. Thus, in the MDSe-TS salt, intermolecular interaction via chalcogen–chalcogen contacts is less effective compared with those in κ–(MDSe-TSF)$_2$Br.

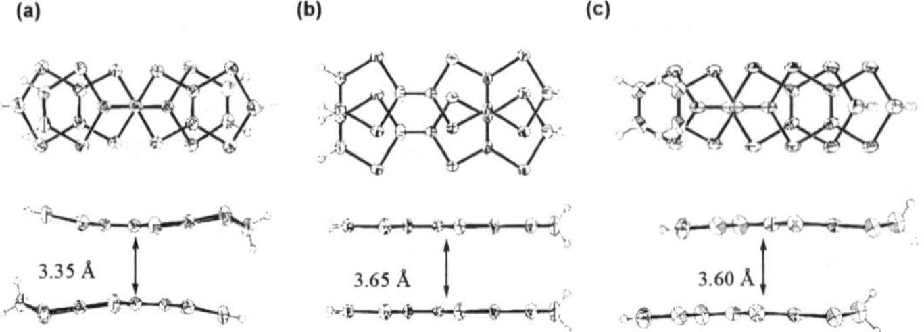

Figure 4   *Molecular structure and overlap mode of donor molecules in radical cations salts: (a) κ-(MDSe-TS)$_2$Br. (b) θ-(MDSe-TS)I$_{1.26}$. (c) (MDSe-ST)I$_{1.26}$.*

The conducting behaviour of κ–(MDSe-TS)$_2$Br reflects the above structural aspects; room temperature conductivity of the present salt (~200 S cm^{-1}) is lower than that of κ–(MDSe-TSF)$_2$Br (~1000 S cm^{-1}), and furthermore, a metal-insulator transition occurs at around 100 K, in sharp contrast to the latter salt being a stable metal down to 4.2 K.

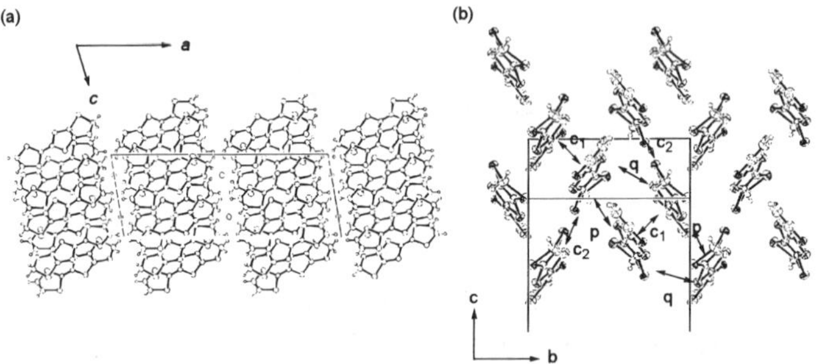

Figure 5   *Crystal structure of κ-(MDSe-TS)$_2$Br: (a) b-axis projection. (b) donor arrangement in the conducting layer. Calculated overlap integrals are c_1 = 25.4 (34.5), c_2 = 3.27 (5.74), p = 20.74 (23.4), and q = –6.10 (–6.57)(× 10^{-3}). The values in the parentheses are those of κ-(MDSe-TSF)$_2$Br*

θ–(MDSe-TS)I$_{1.26}$ As shown in Figure 6, the I$_3$ salt of MDSe-TS is classified into θ–phase in contrast to the I$_3$ and AuI$_2$ salts of the isomer MDSe-ST, which are isostractural with the superconducting MDT-TSF and MDT-ST salts (*vide infra*). As seen in ordinary θ–phase salts, the present salt consists of uniform donor stacks along *a*-axis with a "ring-over-atom" overlap mode (Figure 4b). The dihedral angle between the neighbouring stacks is ca. 126°, which falls in the insulating regime of the universal diagram of θ–phase salts of BEDT-TTF.[18] However, this salt showed stable metallic behaviour down to 1.5 K without any transition (Figure 3) as similar to other θ–phase salts derived form MDSe-TSF[15] and MDT-TSF.[19] In these metallic θ–phase salts derived from the selenium-containing MDT-TTF donors, the intermolecular overlap integrals between the neighbouring columns are fairly large owing to selenium atoms introduced in the donor framework, despite the dihedral angles are larger than the criteria for metallic salts for θ–BEDT-TTF salts.

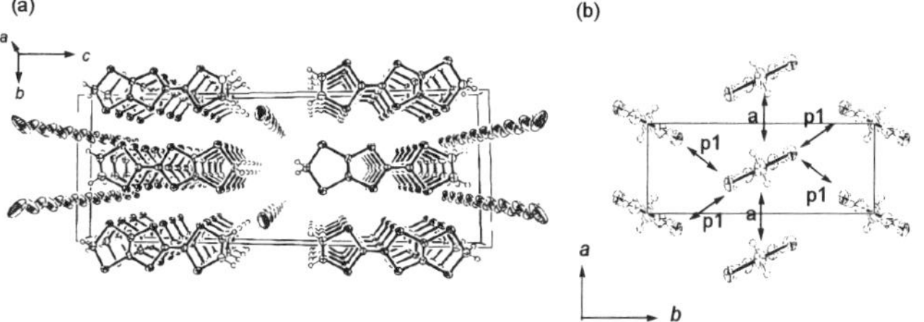

Figure 6   *Crystal structure of θ-(MDSe-TS)I$_{1.26}$. a-axis projection (a) and donor arrangement in the conducting layer (b). Calculated overlap integrals are a = 4.7, p1 = -16.3 (× 10^{-3}).*

In the present structural analysis, the iodine atom has a fairly large thermal displacement parameter along the *a*-axis, resulting in a cigar-shaped thermal ellipsoid in the ORTEP plots. In addition, the optimized occupancy of the iodine atom deviates from a usual 2:1 donor/anion ratio. These are indicative of the presence of an anion lattice incommensurate with the donor lattice. An X-ray oscillation photograph along the crystallographic *a*-axis clearly shows the existence of such incommensurate anion subcell (Figure 7a). In the photograph, there can be seen incommensurate lines with the donor lines at *a/d* = 1.263 and 2.519 together with the Bragg spots originating from the donor lattice. The distance for the first incommensurate line of the anion lattice corresponds to 3.173 Å, which is in good agreement with the mean distance of I–I bond in the infinite polymeric iodine chain.[20] Thus the composition of the present salt is exclusively determined as (MDSe-TS)I$_{1.26}$. Because of the lack of clear lines for charged iodine lattice, i.e. I$^-$ (ca. 4 Å), I$_3^-$ (ca. 9 Å), or I$_5^-$ (ca. 15 Å),[20] the actual charged unit in the polyiodide chain is not determined. However, the I$_3^-$ lattice is most likely, because the related MDT-TSF[2b] and MDT-ST[3] have I$_3^-$ unit as a charged iodine species.[21]

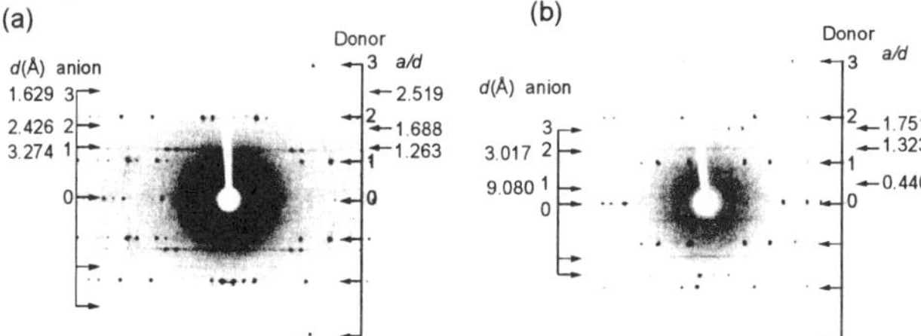

Figure 7   *X-ray oscillation photograph of (a) θ-(MDSe-TS)I$_{1.26}$ and (MDT-ST)(AuI$_2$)$_{0.44}$ (b). The vertical direction corresponds to the a-axis.*

AuI₂ and I₃ salts of MDSe-ST   The structures of these two salts are of isostructural with the superconducting MDT-TSF and MDT-ST salts; i.e. uniform donor stacks with a ring-over-bond overlap (Figure 4c) and incommensurate anion lattice with the donor's. Figure 8 shows the crystal structure of the I_3 salt with the calculated overlap integrals. The overlap integrals of these salts are similar to those of the corresponding MDT-TSF salt. The donor/anion ratio determined on the basis of the X-ray photograph is almost equal to those of the corresponding MDT-TSF salts as summarized in Table 1. However, the profiles of X-ray photographs are somewhat different: in the photograph of the AuI_2 salt of MDSe-ST, the first line corresponding to AuI_2 anion lattice ($a/d = 0.440$) is diffused, indicating that the anion lattice is not well-ordered (Figure 7b). This is in contrast to that of (MDT-TSF)(AuI_2)$_{0.436}$ showing clear spots at $a/d = 0.436$.[5a] Such diffused anion lines are also seen in the photograph of I_3 salt of MDSe-ST.

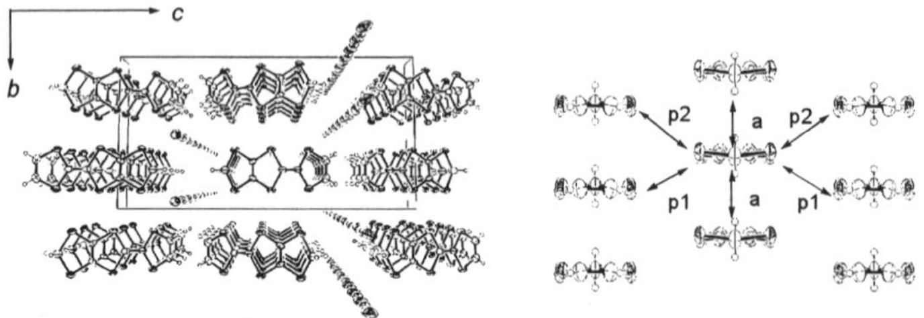

Figure 8   *Crystal structure of (MDSe-ST)I$_{1.26}$. a-axis projection (a) and donor arrangement in the conducting layer (b). Calculated overlap integrals are a = -26.73, p1 = 11.97, p2 = 2.73 ($\times$ 10^{-3}).*

These two salts are highly conductive at room temperature ($2\sim4 \times 10^3$ S cm^{-1}), being almost comparable with or slightly exceeding those of MDT-TSF salts. In spite of these similarities in crystal structure, donor/anion ratio, and room temperature conductivity, the present salts did not show superconducting transition down to 1.5 K (traces c and d in Figure 3). The reason for the lack of superconductivity of the MDSe-ST salts are not clear. However, we speculate that subtle changes in intermolecular interaction caused by the different allocation of selenium and sulfur atoms in the donor skeleton are most likely. As a result these two salts are the first examples of non-superconducting radical cation salts having the crystal structure of the (MDT-TSF)(AuI_2)-type.

3 CONCLUSION

Employing the appropriate selenium-containing building blocks, two isomeric donors, MDSe-TS and MDSe-ST are selectively synthesized. These two donors have fairly good electron donating ability among the selenium-modified MDT-TTF series. Difference of the donor structure brings different crystal phase of the salts obtained by electrocrystallization with I_3^- anion. The I_3^- salt of MDSe-TS has a θ–phase structure in contrast to that of MDSe-ST salt isostructural with the superconducting MDT-TSF and MDT-ST salts. θ–(MDSe-TS)I$_{1.26}$, despite of its relatively large dihedral angles between neighbouring donor stacks, showed stable metallic behaviour down to 1.5 K without any transition. On the other hand, the I_3 and AuI_2 salts of MDSe-ST are of isostructure with the corresponding superconducting MDT-TSF salts. However, these salts did not show a superconducting transition and are simply metallic down to 1.5 K.

References

1. (a) G.C. Papavassiliou, G.A. Mousdis, J.S. Zambounis, A. Terzis, A. Hountas, B. Hilti, C.W. Mayer and J. Pfeiffer, *Synth. Met.*, 1988, **27**, B379. (b) A.M. Kini, M.A. Beno, D. Son, H.H. Wang, K.D. Carlson, L.C. Porter, U. Welp, B.A. Vogt, J.M. Williams, D. Jung, M. Evain, M.-H. Whangbo, D.L. Overmyer and J.E. Schirber, *Solid State Commun.*, 1989, **69**, 503.
2. K. Takimiya, Y. Kataoka, Y. Aso, T. Otsubo, H. Fukuoka and S. Yamanaka, *Angew. Chem., Int. Ed.*, 2001, **40**, 1122. (b) K. Takimiya, M. Kodani, Y. Kataoka, Y. Aso, T. Otsubo, T. Kawamoto and T. Mori, *Chem. Mater.*, 2003, **15**, 3250.
3. K. Takimiya, A. Takamori, Y. Aso, T. Otsubo, T. Kawamoto and T. Mori, *Chem. Mater.*, 2003, **15**, 1225.
4. Related electron donor, MDT-TS [5*H*-2-(1,3-diselenol-2-ylidene)-1,3,4,6-tetrathiapentalene] gave a superconducting AuI_2 salt (T_c = 4.8 K at 11.4 Kbar). K. Takimiya, M. Kodani, N. Niihara, Y. Aso, T. Otsubo, Y. Bando, T. Kawamoto and T. Mori, *Chem. Mater.*, 2004, **16**, 5120.
5. (a) T. Kawamoto, T. Mori, K. Takimiya, Y. Kataoka, Y. Aso, and T. Otsubo, *Phys. Rev. B*, 2002, **65**, 140508. (b) T. Kawamoto, T. Mori, C. Terakura, T. Terashima, S. Uji, K. Takimiya, Y. Aso, and T. Otsubo, *Phys. Rev. B*, 2003, **67**, 020508.
6. The IUPAC names for MDSe-TS and MDSe-ST are 5*H*-2-(1,3-diselenol-2-ylidene)-1,3-dithia-4,6-diselenapentalene and 5*H*-2-(1,3-dithiol-2-ylidene)-1,3,4,6-tetraselenapentalene, respectively.
7. H. Poleschner and E. Fanghänel, *J. Pract. Chem.*, 1982, **324**, 691.
8. K. Takimiya, A. Morikami and T.Otsubo, *Synlett*, 1997, 319.
9. Compound **2** was prepared from tetrabutylammonium 4,5-bis(2-thioxo-1,3-dithiole-4,5-diselenolate)zincate with methyl 3-bromopropionate: (a) R.-M. Olk, A. Röhr, B. Olk and E. Hoyer, *Z. Chem.*, 1988, **28**, 304. (b) G.C. Papavassiliou, V.C. Kakoussis, J.S. Zambounis and G.A. Mousdis, *Chem. Scr.*, 1989, **29**, 123. (c) F. Ogura and K. Takimiya, in *Organoselenium Chemistry; A practical Approach*, ed., T.G. Back, Oxford University Press, New York, 1999, pp. 269–270.
10. (a) N. Svenstrup, K.M. Rasmussen, T.K. Hansen and J. Becher, *Synthesis*, 1994, 809. (b) K.B. Simonsen, N. Svenstrup, J. Lau, O. Simonsen, P. Mørk, G.J. Kristensen and J. Becher, *Synthesis*, 1996, 407. (c) J. Becher, J. Lau, P. Leriche, P. Mørk and N. Svenstrup, *J. Chem. Soc., Chem. Commun.*, 1994, 2715. (d) L. Binet, J.M. Fabre and J. Becher, *Synthesis*, 1997, 26. (e) K. Takimiya, A. Oharuda, A. Morikami, Y. Aso and T. Otsubo, *Eur. J. Org. Chem.*, 2000, 3013. (f) M. Kodani, K. Takimiya, Y. Aso, T. Otsubo, T. Nakayashiki and Y. Misaki, *Synthesis*, 2001, 1614.
11. B.R. O'Connor and F.N. Jones, *J. Org. Chem.*, 1970, **35**, 2002.
12. A cross coupling reaction of 1,3-dithiole-2-thione and **6** gave a complex mixture, and a tedious chromatographic separation gave the desired cross-coupling product in 7% yield.
13. S. Yoneda, T. Kawase, Y. Yasuda and Z. Yoshida, *J. Org. Chem.*, 1979, **44**, 1728.
14. P. Braunstein, J.H. Clark, W. Ramsay, R. Foster and C. Ingold, *J. Chem. Soc., Dalton Trans.*, 1973, 1845.
15. M. Kodani, A. Takamori, K. Takimiya, Y. Aso and T.Otsubo, *J. Solid State Chem.*, 2002, **168**, 582.
16. T. Mori, A. Kobayashi, Y. Sasaki, H. Kobayashi, G. Saito and H. Inokuchi, *Bull. Chem. Soc. Jpn.*, 1984, **57**, 627.
17. T. Mori and M. Katsuhara, *J. Phys. Soc. Jpn.*, 2002, **71**, 826.
18. H. Mori, S. Tanaka and T. Mori, *Phys. Rev. B*, 1998, **57**, 12023.

 Multifunctional Conducting Molecular Materials

19. K. Takimiya, Y. Kataoka, M. Kodani, Y. Aso and T. Otsubo, *Synth. Met.*, 2003, **133–134,** 185.
20. T.J. Marks and D.W. Kalina, in *Extended Linear Chain Compounds*, ed., J.S. Miller, Plenum, New York, 1983, Vol. I, Chap. 6.
21. M. Mas-Torrent, E. Ribera, V. Tkacheva, I. Mata, E. Molins, J. Vidal-Gancedo, S. Khasanov, L. Zorina, R. Shibaeva, R. Wojciechowski, J. Ulanski, K. Wurst, J. Veciana, V. Laukhin, E. Canadell, E. Laukhina and C. Rovira, *Chem. Mater.*, 2002, **14**, 3295.

CORRELATION AND FRUSTRATION EFFECTS IN MOLECULAR CONDUCTORS
Et$_x$Me$_{4-x}$Z[Pd(dmit)$_2$]$_2$ (dmit=C$_3$S$_5^{2-}$, Z = P, As, Sb, x = 0, 1, 2)

Reizo Kato, Akiko Tajima, Akiko Nakao, Naoya Tajima, and Masafumi Tamura

RIKEN, JST-CREST, 2-1, Hirosawa, Wako-shi, Saitama, 351-0198, Japan

1 INTRODUCTION

Competition between the kinetic and repulsion energies of electrons provides a variety of exotic physical properties. In systems with a half-filled energy band, where each unit in the crystal has one conduction electron, strong on-site Coulomb energy (U) and narrow band width (W) lead to the Mott-insulating state. If the conduction electrons are situated on a geometrically frustrated lattice structure, the ground state can be nontrivial, because the spin entropy is not easily released by simple symmetry breaking such as the antiferromagnetic (AF) order, and a possibility of some exotic quantum liquid state is theoretically proposed.[1] Many of molecular conductors belong to the (effectively) half-filled band system with the narrow band width and the strong on-site Coulomb repulsion. An isostructural series of anion radical salts β'-(Cation)[Pd(dmit)$_2$]$_2$ (dmit = 1,3-dithiol-2-thione-4,5-dithiolate, Cation=Me$_4$Z$^+$, Et$_2$Me$_2$Z$^+$ where Z=P, As, Sb and EtMe$_3$Z$^+$ where Z=As, Sb) are Mott insulators at ambient pressure.[2] An important character of the β'-Pd(dmit)$_2$ salts is a two-dimensional quasi triangular lattice formed by [Pd(dmit)$_2$]$_2$ dimers. The Mott insulator can turn metallic, when the correlation parameter U/W is reduced or the half-filled condition is removed by the doping of holes or electrons. In addition, theoretical studies suggest that the frustration can diminish the stability of the insulating state by destroying the AF correlation and induce the paramagnetic metallic state.[1,3] The application of pressure (including hydrostatic pressure and uni-axial strain) can effectively tune the correlation parameter, the degree of the frustration, and possibly the band filling. The pressure effects on the β'-type Pd(dmit)$_2$ salts indicate how the correlation and frustration work in the strongly correlated electron system with the quasi triangular lattice. Another important aspect of the Pd(dmit)$_2$ salts is a HOMO-LUMO interplay which comes from strong dimerization and a small HOMO-LUMO energy splitting.[4] This two-MO

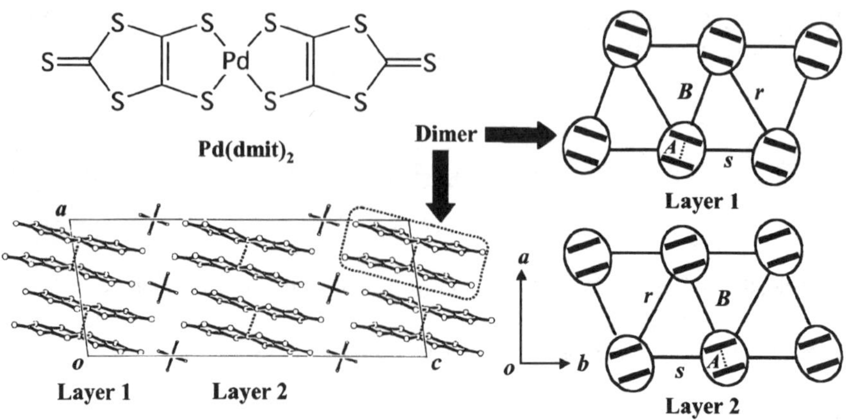

Figure 1   *Crystal structure of β'-(Cation)[Pd(dmit)$_2$]$_2$ (Cation=Me$_4$As) and quasi triangular lattice of Pd(dmit)$_2$ dimers. A, B, r, s are intra- and inter-dimer transfer integrals (see text).*

nature can drive a spontaneous charge separation (2 Dimer⁻ → Dimer⁰ + Dimer²⁻).[5] This is a new type of charge ordering transition which is not expected for the conventional single-MO systems. We demonstrate here a variety of electronic properties of the unique strongly correlated electron system, β'-(Cation)[Pd(dmit)$_2$]$_2$.

2 CRYSTAL AND ELECTRONIC STRUCTURES

Single crystals of β'-(Cation)[Pd(dmit)$_2$]$_2$ were prepared by the air oxidation of (Cation)$_2$[Pd(dmit)$_2$] in an acetone solution containing acetic acid at 5-10 °C. X-ray diffraction measurements were performed using graphite monochromatized Mo Kα (λ = 0.71070 Å) radiation. A Weisenberg-type imaging plate system (DIP 320S, Mac Science Co., Inc) equipped with a closed-cycle helium refrigerator (V202C5LZR, DAIKIN Co.) was used for data collection. The structures were solved by the direct method, and refined with the full-matrix least-squared method (SHELXL-97).[6] Anisotropic temperature factors were applied for the non-hydrogen atoms. All calculations were performed using teXsan crystallographic software package of the Molecular Structure Co.

Figure 1 shows crystal structure and molecular arrangement within the conduction layer. The monoclinic unit cell with the space group *C*2/*c* consists of two crystallographically equivalent Pd(dmit)$_2$ layers (1 and 2) parallel to the crystallographic *ab* plane. The non-conducting cations are located between the conducting Pd(dmit)$_2$ layers. The EtMe$_3$Sb cation shows disorder taking two possible orientations with each 50 % occupancy. Pd(dmit)$_2$ molecules are strongly dimerized in an eclipsed mode and stack along the *a+b* direction in Layer 1 and along the *a−b* direction in Layer 2. All the dimers [Pd(dmit)$_2$]$_2$⁻ are crystallographically equivalent. Within the isolated dimer, each of HOMO and LUMO in the monomer generates bonding and antibonding combinations. Since the dimerization is strong and the HOMO-LUMO energy splitting is small, the antibonding combination of the HOMOs is higher than the bonding combination of the LUMOs and forms a conduction band. The electronic structure near the Fermi level can be described by the tight-binding band calculation using a HOMO-HOMO transfer integral in the dimer (A), and transfer integrals among the antibonding combinations of the HOMOs (B, s, r).[2]

3 MAGNETIC PROPERTIES

For the β'-type Pd(dmit)$_2$ salts, the calculated conduction band is narrow and half-filled. At ambient pressure, all these salts are non-metallic due to the strong on-site Coulomb repulsion. In this Mott insulating state, one electron is located on each dimer [Pd(dmit)$_2$]$_2^-$. The localized electrons exhibit various magnetic behaviors depending on the counter cation.

Magnetic susceptibilities of the randomly oriented polycrystalline samples were measured by use of a Quantum Design MPMS XL7 magnetometer. Above 10 K, the magnetization varied almost linearly with field up to 5 T. The data, collected over 5-290 K at 5 T, were corrected for the diamagnetic contribution as described elsewhere.[7]

Figure 2 shows three types of temperature dependences of the magnetic susceptibility. The EtMe$_3$Sb salt, where the deviation from the regular triangular lattice is small, shows no sign of magnetic ordering down to the lowest temperature. The temperature dependence can be explained by the model of the spin-1/2 Heisenberg triangular antiferromagnet,[7,8] which indicates that the frustration operates in this system. The competition between the spin frustration and the antiferromagnetic ordering in the Mott insulators is an important subject. The EtMe$_3$As salt with larger deviation from the regular triangular lattice

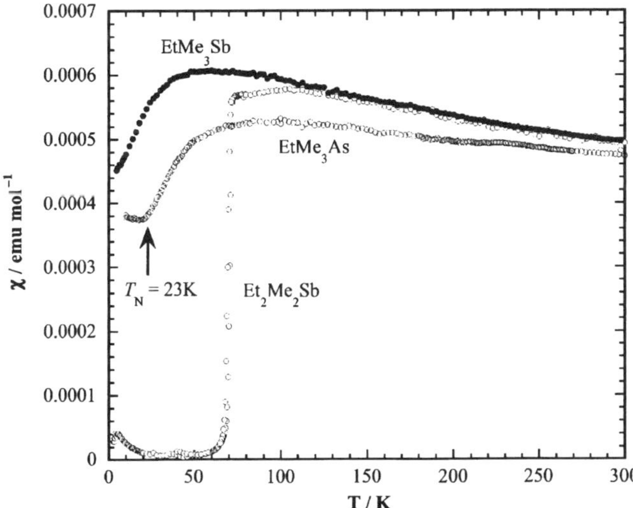

Figure 2 *Temperature dependence of magnetic susceptibility for β'-(Cation)[Pd(dmit)$_2$]$_2$ (Cation=EtMe$_3$Sb, EtMe$_3$As, Et$_2$Me$_2$Sb).*

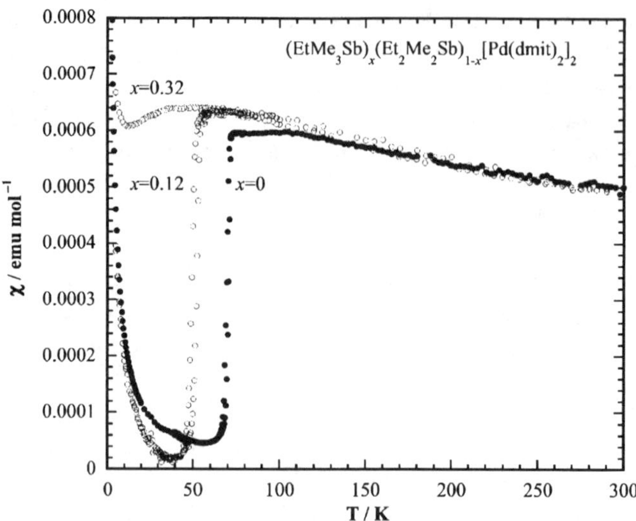

Figure 3   *Temperature dependence of magnetic susceptibility for β'-(EtMe$_3$Sb)$_x$ (Et$_2$Me$_2$Sb))$_{1-x}$[Pd(dmit)$_2$]$_2$. The x values were determined by X-ray crystal structure analysis. For each salt, a Curie component is not subtracted.*

undergoes an antiferromagnetic transition at 23 K.

On the other hand, the Et$_2$Me$_2$Sb salt, which has an almost triangular lattice of the dimers and thus cannot release the spin frustration by growing the AF ordering, exhibits a first-order transition to a non-magnetic state at about 70 K.[9] Low-temperature X-ray crystal structure analysis has revealed a structural transition which is accompanied by a break in the C-centered monoclinic symmetry and a doubling of the b-axis.[10] In the non-magnetic phase, two crystallographically independent Pd(dmit)$_2$ dimers are arranged alternately along the b axis. These two dimer units show an obvious difference in the intra-dimer distance; the one has a shorter distance, and the other a longer distance, compared with the dimer at room temperature. The temperature dependence of polarized reflectivity spectra exhibits a remarkable splitting for the bonding-antibonding excitation of the dimer in the non-magnetic phase.[11] All these results indicate that this transition is the charge ordering transition, 2 Dimer$^-$ → Dimer^0 + Dimer^{2-}, where the neutral dimer with the stronger dimerization is stabilized by a HOMO-LUMO double bond.[5]

An interesting point is that this charge ordering transition in the Et$_2$Me$_2$Sb salt can be suppressed by impurities at the cation site. We have prepared mixed salts, β'-(EtMe$_3$Sb)$_x$ (Et$_2$Me$_2$Sb))$_{1-x}$[Pd(dmit)$_2$]$_2$. Figure 3 shows that an increase of the EtMe$_3$Sb content shifts the transition temperature downward and then removes the charge ordering transition. It should be noted that the transition at about 50 K (x=0.12) is as sharp as that in the pure sample. This impurity effect is associated with the steric effect between the alkyl groups of the cations and the terminal S atoms of the Pd(dmit)$_2$ units.[10] The terminal S atoms push the alkyl groups to cause the geometric changes of the cations, as a result of the

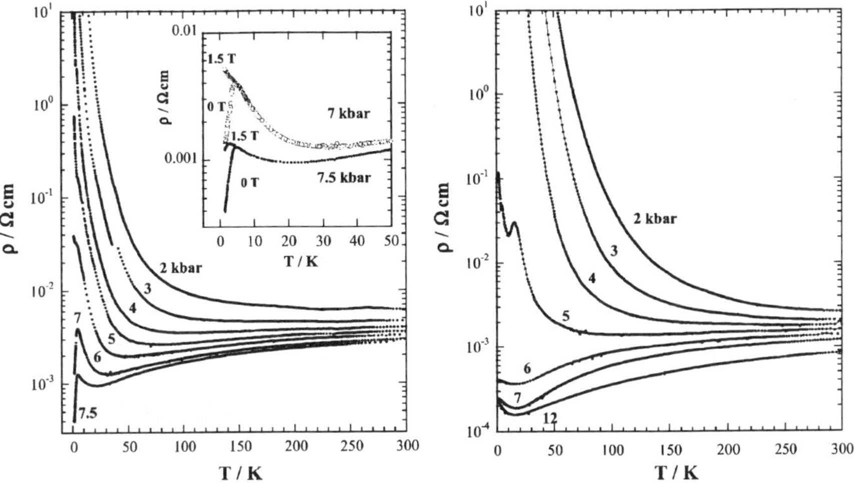

Figure 4 *Hydrostatic pressure effects on resistivity for β'-(Cation)[Pd(dmit)$_2$]$_2$ (left; Cation = EtMe$_3$Sb, right; Cation = EtMe$_3$As). The inset shows magnetic field dependences of low-temperature resistivities.*

displacement of the Pd(dmit)$_2$ units. The steric anomaly around the impurity prevents the cooperative and spontaneous structural change required for the phase transition. The absence of the charge ordering transition in the EtMe$_3$Sb salt would be explained by the cation disorder.

4 PRESSURE EFFECT

4.1 Hydrostatic Pressure

The Mott insulating state of the β'-type Pd(dmit)$_2$ salts can be suppressed by the application of hydrostatic pressure. This hydrostatic pressure effect depends on the choice of the counter cation. When cation = Me$_4$Sb,[2] Et$_2$Me$_2$Z (Z=P, As),[2] and EtMe$_3$As, the system turns metallic and shows superconductivity under hydrostatic pressure (Fig. 4 left). The EtMe$_3$Sb salt also shows a metallic behavior, but no superconductivity is observed up to the highest pressure (Fig. 4 right). On the other hand, in the case of the Me$_4$P and Me$_4$As salts, the application of hydrostatic pressure cannot suppress the non-metallic behavior.[2, 12] The charge ordering transition in the Et$_2$Me$_2$Sb salt is accompanied by a sharp increase of the resistivity at ambient pressure. The application of hydrostatic pressure turns this resistivity anomaly to a metal-insulator transition.

4.2 Uni-axial Strain

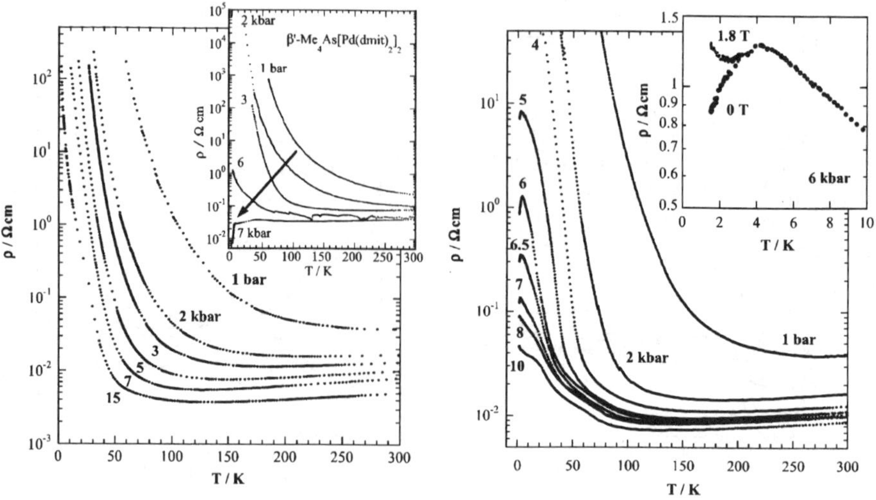

Figure 5   *Uni-axial strain effects on resistivity for β'-Me₄P[Pd(dmit)₂]₂ (left; P//b, right; P//~ a+3b). The insets show the b-axis strain effect for β'-Me₄As[Pd(dmit)₂]₂ (for comparison) and magnetic field dependences of low-temperature resistivity at 6 kbar, respectively.*

The uni-axial strain can change only the lattice parameter parallel to the direction of pressure application and regulate the intermolecular interaction selectively. Uniaxial strain was applied using the epoxy-crystal method.[13] The samples embedded in the epoxy (Stycast #1266 and catalyst B) were inserted in a Cu-Be clamp cell. A remarkable uni-axial strain effect for the β'-type Pd(dmit)$_2$ salts was first observed in the Me$_4$As salt (the inset in Fig. 5 left).[14] While the non-metallic behavior of this salt cannot be suppressed under the hydrostatic pressure up to 17 kbar, the uni-axial strain along the b-axis suppresses the non-metallic behavior and induces the superconductivity. On the other hand, the Me$_4$P salt, which shows a hydrostatic pressure effect similar to that for the Me$_4$As salt, remains non-metallic even under the b-axis strain up to 15 kbar (Fig. 5 left). The uni-axial strain (approximately) along the $a+3b$ direction, however, readily reduces low-temperature resistivity and induces an abrupt decrease of the resistivity at 3 K around 6 kbar (Fig. 5 right). A recovery of the resistivity under the magnetic field suggests that this is an appearance of the superconductivity.

5   INTERPLAY OF CORRELATION AND FRUSTRATION

Above mentioned pressure and cation effects on the electronic state of the β'-Pd(dmit)$_2$ salts have led us to propose a schematic phase diagram which shows roles of the correlation and frustration in this system (Fig. 6).[2] Since the effective Coulomb repulsion energy on the dimer has a local linear dependence on the intra-dimer transfer integral A,[15] the ratio A/W can be used to measure the strength of the correlation. In the tight-binding

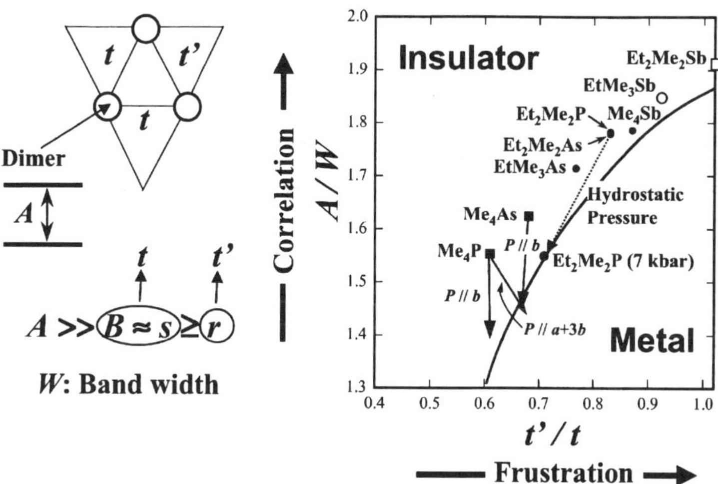

Figure 6 *Schematic phase diagram for β′-(Cation)[Pd(dmit)₂]₂ based on the room temperature structures (see also Fig. 1). Solid arrows indicate simulated uni-axial effects with a 0.1 Å reduction in the inter-dimer distance along the corresponding direction. For the Et₂Me₂P salt under the hydrostatic pressure of 7 kbar, the reduction in the b-axis is about 0.2 Å*

approximation, the band width W is mainly determined by the larger inter-dimer transfer integrals B and s. The deviation from the regular triangular character in the inter-dimer interaction is indicated by the ratio $r/s(\approx B) = t'/t$. Based on the room temperature crystal structures, the β'-Pd(dmit)₂ salts are mapped in a t'/t– A/W diagram (Fig. 6). An important point is that the frustration plays a crucial role for the stabilization of the metallic state in this strongly correlated system. The crystal structure analysis under hydrostatic pressure for the Et₂Me₂P salt has revealed that the application of the hydrostatic pressure decreases both correlation and frustration effects as indicated by the dotted arrow in Fig. 6.[16] This trade-off between correlation and frustration explains that the application of hydrostatic pressure cannot suppress the non-metallic behavior in the systems with smaller frustration effect (Me₄Z; Z=P, As). The b-axis strain, however, induces the metallic state for the Me₄As salt. The b-axis strain gives an equivalent effect on both conduction layers (1 and 2) and its direction corresponds to that of the inter-dimer interaction s. A simulation of the band parameters under the b-axis strain based on a simple distortion model for the Me₄As salt suggests that the b-axis strain effectively enhances (mainly) the inderdimer transfer integral s, which results in an enhancement of W.[16] Therefore, we can ascribe an essential role of the b-axis strain to more effective reduction of the correlation than the hydrostatic pressure. The Me₄P salt exhibits the largest deviation from the regular triangular lattice. The non-metallic hebavior of this salt cannot be suppressed even by the b-axis strain. We have demonstrated that the uni-axial strain (approximately) along the $a+3b$ direction tends to suppress the non-metallic behavior. A simulation suggests that the strain along this direction effectively enhances the inter-dimer interaction r and thus the frustration effect within Layer 2 (Fig. 6). In addition, it should be noted that the $a+3b$ strain gives a different effect on Layer 1 and thus makes two conduction layers non-equivalent. This allows two energy bands associated with Layers 1 and 2 to have difference in the band filling, which removes the half-filled state and induce the metallic state. This is a kind of self-doping. Therefore, the $a+3b$ strain effect would contain these two aspects, the enhancement of the frustration and the self-doping.

In conclusion, the electronic state of the β'-Pd(dmit)$_2$ salts is governed by both correlation and frustration which can be tuned by the pressure and cation effects.

Acknowledgements

This work was partially supported by a Grant-in-Aid for Scientific Research (No. 16GS0219) from the Ministry of Education, Culture, Sports, Science and Technology of Japan.

References

1   H. Morita, S. Watanabe, and M. Imada, *J. Phys. Soc. Jpn.*, 2002, **71**, 2109.

2   R. Kato, *Chem. Rev.*, 2004, **104**, 5319.

3   H. Kondo and T. Moriya, *J. Phys. Soc. Jpn.*, 1999, **68**, 3170.

4   E. Canadell, I. E. –I. Rachidi, S. Ravy, J-P. Pouget, L. Brossard, and J-P. Legros, *J. Phys. France* 1989, **50**, 2967.

5   M. Tamura and R. Kato, *Chem. Phys. Lett.*, 2004, **387**, 448.

6   G. M. Sheldrick, SHELXL97, Program for the Refinement of Crystal Structures. University of Gottingen, Germany (1997).

7   M. Tamura and R. Kato, *J. Phys.: Condens. Matter.*, 2002, **14**, L729.

8   N. Elstner, R. R. P. Singh, and A. P. Young, *Phys. Rev. Lett.*, 1993, **71**, 1629.

9   M. Tamura and R. Kato, *Polyhedron* 2005, **24**, 2817.

10 A. Nakao and R. Kato, *J. Phys. Soc. Jpn.*, 2005, **74**, 2754.

11 M. Tamura, K. Takenaka, H. Takagi, S. Sugai, A. Tajima, and R. Kato, *Chem. Phys. Lett.*, 2005, **411**, 133.

12 Y. Ishii, M. Tamura, R. Kato, M. Hedo, Y. Uwatoko, and N. Môri, *Synth. Met.*, 2005, **152**, 389.

13 M. Maesato, Y. Kaga, R. Kondo, and S. Kagoshima, *Rev. Sci. Instrum.* 2000, **71**, 176.

14 R. Kato, N. Tajima, M. Tamura, and J. Yamaura, *Phys. Rev. B*, 2002, **66**, 020508(R).

15 M. Tamura and R. Kato, *J. Phys. Soc. Jpn.*, 2004, **73**, 3108.

16 R. Kato, A. Tajima, A. Nakao, N. Tajima, M. Tamura, and J. Yamaura, *Synth. Met.*, 2005, **153**, 397.

ELECTRIC FIELD EFFECT ON QUANTUM OSCILLATIONS IN FEW-LAYER
GRAPHENE IN HIGH MAGNETIC FIELDS

E. Jobiliong, J. G. Park, R. Vasic and J. S. Brooks

NHMFL and Physics Department, Florida State University, Tallahassee, 32310, USA

1 INTRODUCTION

Graphene is a single layer of carbon atoms densely packed into a benzene-ring structure with a covalent bond between nearest neighbors.[1] This term is widely used for carbon-based materials, including graphite, carbon nanotubes, etc. (e.g. carbon nanotubes are graphene sheets rolled up into nanometer sized cylinders). The electronic properties of few-layer graphene (FLG) have been intensively investigated both theoretically[2,3] and experimentally[4,5,6,7]. In a recent investigation by Zhang et al.[4], it has been shown that both the carrier concentration and the type of carriers (holes or electrons) in FLG can be modified by applying a gate voltage. Moreover, the integer quantum hall has been observed in this material.[5,6] However, the electronic properties of FLG in very high magnetic fields (B > 20 T) have not, to our knowledge, been examined before. Hence the motivation for this work has been to explore the nature of FLG at low temperature and in magnetic fields higher than in previous studies[5,6] to see how the quantum oscillation properties change as a function of gate voltage.

2 EXPERIMENT AND RESULTS

We have examined the behavior of the in-plane resistance of FLG of highly oriented pyrolitic graphite (HOPG) at high magnetic field (up to 33T). Bulk HOPG samples were cleaved in a conventional manner with adhesive tape and transferred onto a SiO_2/n-doped Si substrate by pressing mechanically by hand. In this manner, samples with a lateral size of ~ 20 μm and thickness between 5 to 50 nm can be obtained. Once the samples are on the substrate, Cr/Au electrodes with the thickness of 5/25 nm are fabricated onto the sample by photolithography and thin film evaporation. Figure 1 shows the image of the sample taken by both optical and atomic force (AFM) microscopes. From the AFM analysis, we determined the thickness of this sample to be ~ 5 nm (~ 15 layers of graphene). In addition, the doped silicon substrate is used as a gate electrode

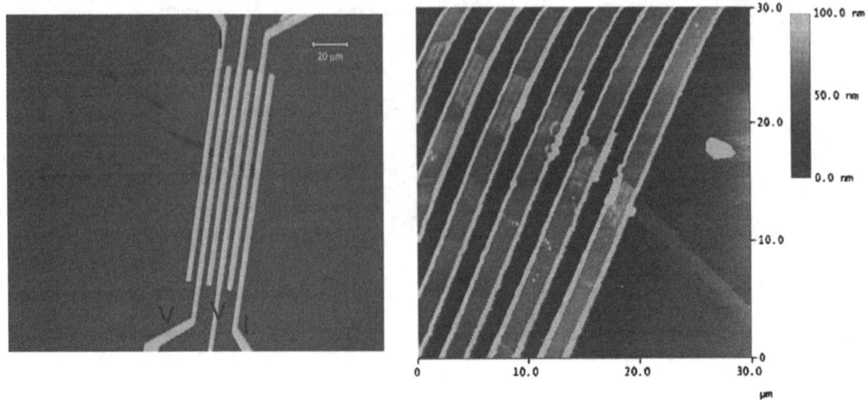

Figure 1   *Left: The optical microscope image of the FLG. The gap between two leads is 2 μm. I and V are the current and voltage leads, respectively. Right: The AFM image of the FLG. The thickness of this sample is ~ 5 nm.*

with 100 nm of SiO_2 acting as the gate dielectric. A positive or negative gate voltage is applied to the doped Si substrate, and the ground for the gate potential is referenced to the ground potential of the 4-terminal current source and ac lock-in sample voltage amplifier. The 4-terminal lead arrangement indicated by I and V, as shown in Fig. 1 is used to obtain both longitudinal R_{xx} and transverse Hall R_{xy} resistance. The total voltage signal at constant current (typically 100 nA) is measured by applying forward and reverse magnetic fields perpendicular to the graphite layers at constant temperature vs. magnetic field for different gate voltage. From the total signal, R_{xx} and R_{xy} signals were derived from the sum and difference of the forward-reverse field sweeps. It is interesting to note that even though the probe geometry should in principle favor a predominant longitudinal R_{xx} signal, we found that often the Hall R_{xy} signal could be comparable, or even larger than R_{xx}. Hence the precise nature of the current paths in the FLG sample appear to be highly dependent on the in-plane shape and orientation of the sample, the electrodes used for the measurements, and perhaps also the quality of the electrode-to-sample contact, or to variations in the number of graphene sheets (thickness) of the FLG sample.

The normalized resistivity $(\rho(T)/\rho(295K))$ of the sample, shown in Fig. 2, increases as the temperature decreases. In contrast, in bulk samples show metallic properties $(d\rho/dT>0)$.[8] This behavior can be understood by considering two competing factors, namely the temperature dependence of the mobility and carrier density. In the two band model, which has been successfully explain the electronic properties of graphite, the carrier density, n for the fully compensated material can be written as[1]

$$n = AT \ln\left(1 + e^{\frac{E_0}{T}}\right) \tag{1}$$

where A is a constant independent of T and E_0 is half the band overlap energy between the electron and hole bands. We use the mobility obtained experimentally by Kawamura et al.,[9]

$$\frac{1}{\mu} = \frac{1}{\mu_0} + BT^{1.6} \tag{2}$$

where μ_0 is related to the contribution from static scattering and B is a constant that depends on electron-phonon scattering. Hence, the resistivity of the sample can be expressed as

$$\rho \propto \left(\frac{1}{B\mu_0} + T^{1.6} \right) \bigg/ \left\{ T \ln\left(1 + e^{E_0/T} \right) \right\}. \tag{3}$$

We used this expression to fit the temperature dependence of the normalized resistivity, which described the data reasonably well above 30K. As the result of the fit, we obtain E_0 of ~106 K (9.1 meV), which is smaller than the value found for bulk sample ($E_0 \sim 20$ meV).[10] However, this value is consistent with the recent investigation of E_0 for samples of different thickness.[11] This clearly indicates that E_0 is reduced with a decreasing number of layers, and is consistent with a trend where the thinner samples approach the zero-gap semiconductor regime for a single graphene layer[12]. We note that by using a $T^{1.6}$ temperature dependence for the inverse mobility in our data, the fitting result is much better than for the linear T dependence used in Ref. 11.

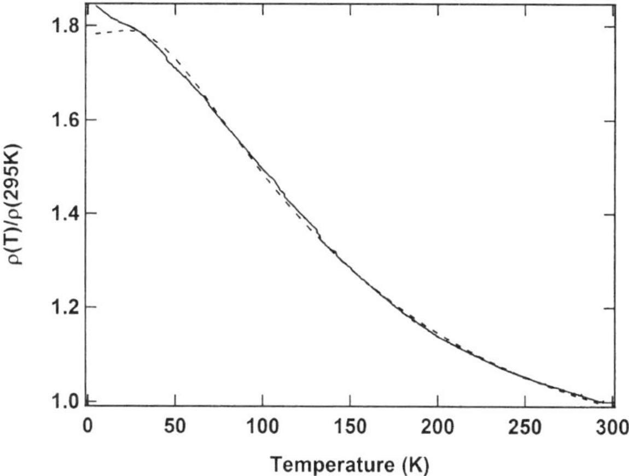

Figure 2   *The temperature dependence of the normalized resistivity (ρ(T)/ρ(295K)). The solid and dashed lines indicate the experimental and fitting data, respectively (see text).*

In Figure 3 we present typical results for the magnetoresistance of the FLG system in high magnetic fields at low temperatures. In the left panel the in-plane resistance of FLG up to 33T at 0.6 K is presented for different gate voltages (curves are offset). In the right panel the effects of the gate V_G on the resistance are shown as the difference between $R(V_G)$ and $R(V_G=0)$. Our results indicate that the sample resistance increases (decreases) with positive (negative) gate voltage. Also evident are Shubnikov-de Haas (SdH) oscillations with a frequency of about ~15T, which appear to shift in phase and/or frequency with gate voltage, although a detailed analysis has not yet been carried out. In Figure 4 we show R_{xx} and R_{xy} at 1.75 K for both +12 V and − 12 V gate voltages, where we find a significant shift in the Hall signal above about 22 T.

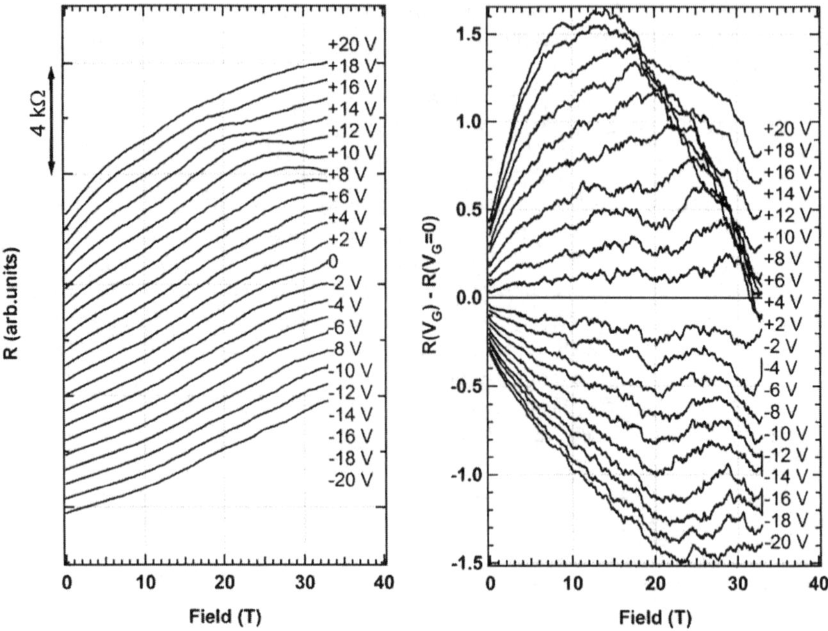

Figure 3		*Left: The magnetic field dependence of the resistance of the FLG sample in Fig. 1 for different gate voltage. The curves are offset for clarity. Right: The difference of the magnetic field dependent resistance of the FLG sample for $V_G=0$ and $V_G \neq 0$.*

3 DISCUSSION

The preliminary results we present here have characteristics that in some respects are consistent with previous studies of these systems, but in other respects seem quite different. In all cases to date for either FLG or single graphene systems, that is, systems with 50 nm or less thickness, the SdH frequency seems to be in the range of 5 to 130 T, and the appearance of the SdH is significantly different from that seen in the bulk materials[9,13,14,15] where only lower frequencies (< 20 T) appear. At high fields above 20 T in bulk samples a field-induced CDW-type transition (but no SdH oscillations) is observed. However, for the FLG, SdH oscillations persist to the highest field, with no evidence for any of the CDW behavior observed in the bulk materials. We do, however, see gate dependent anomalies in the MR and Hall effect above about 22 T, which is in the vicinity of the field induced CDW phase, but the details of its behavior do not follow the temperature dependence of the CDW phase boundary as previously reported[13,15]. One significant difference in our results for HOPG with other FLG and single layer graphene studies is the trend of the MR with gate voltage. As mentioned above, we find that the MR increases monotonically with V_G in all samples and configurations we have measured to date, but in all other reports the MR decreases with V_G. Since our gate arrangement seems nearly identical to that of Zhang et al.[4], we do not at present understand the source of this discrepancy.

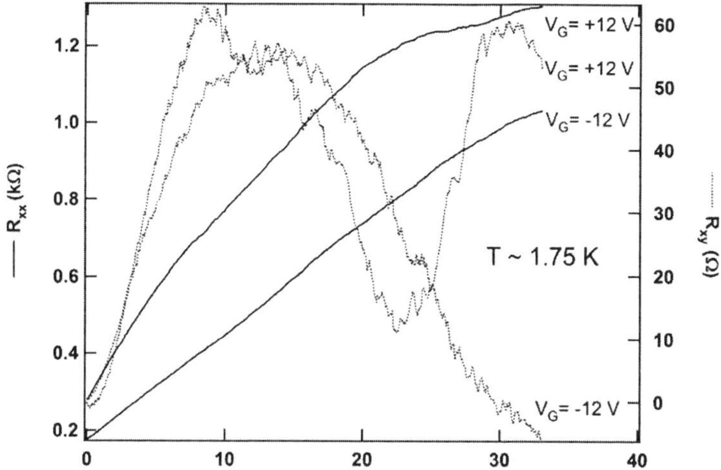

Figure 4 *The field dependence of the R_{xx} and R_{xy} signal for $V_G=+12$ V and -12 V at 1.75 K obtained from forward and reverse magnetic field sweeps.*

Further work is presently underway to optimize the sample preparation, lead and gate contact configurations for further studies of few-layer materials including graphite and other systems that show high field-induced phase transitions in the bulk single crystalline state. Here we note that there are at least two possible, major effects that may alter these ground states. The first is that the nano-scale size may compete with the special order parameters of the ground states, such as density wavelengths, or coherence lengths, and the second is that the electric field can alter the carrier density.

ACKNOWLEDGEMENTS

We would like to acknowledge support from the NSF-DMR 0203532 and the National Nuclear Security Administration under the Stewardship Science Academic Alliances program through DOE Research Grant #DE-FG03-03NA00066. A portion of this work was performed at the National High Magnetic Field Laboratory, which is supported by NSF Cooperative Agreement No. DMR-0084173, by the State of Florida, and by the DOE. We are grateful to the facilities at MARTECH-FSU, and also at the FSU College of Engineering, where some aspects of processing and characterization of the samples were carried out.

References

1   B. T. Kelly, *Physics of Graphite*, Applied Science, Englewood, NJ 1981.
2   V. P. Gusyin and S. G. Sharapov, *Phys. Rev. Lett.*, 2005, **95**, 146801.
3   I. A. Luk'yanchuk and Y. Kopelevich, *Phys. Rev. Lett.*, 2004, **93**, 166402.
4   Y. Zhang, J. P. Small, M. E. S. Amori and P. Kim, *Phys. Rev. Lett.*, 2005, **94**, 176803.

5   K. S. Novoselov, A. K. Geim, S. V. Morozov, D. Jiang, M. I. Katsnelson, I. V. Grigorieva, S. V. Dubonos, A. A. Firsov, *Nature*, 2005, **438**, 04233.
6   Y. Zhang, Y. W. Tan, H. L. Stormer, and P. Kim, *Nature*, 2005, **438**, 201.
7   J. S. Bunch, Y. Yaish, M. Brink, K. Bolotin, and Paul L. McEuen, *Nanoletters*, 2005, **5**, 287.
8   T. Tokumoto, E. Jobiliong, E. S. Choi, Y. Oshima and J. S. Brooks, *Solid State Commun.*, 2004, **129**, 599.
9   K. Kawamura, T. Saito, and T. Tsuzuku, *J. Phys. Soc. Jpn.*, 1977, **42**, 574.
10 N. B. Brandt, S. M. Chudinov and Y. G. Ponomarev, *Semimetals I: Graphite and its compounds*, North-Holland, Amsterdam, 1988.
11 Y. Zhang, J. P. Small, W. V. Pontius and P. Kim, Appl. Phys. Lett., 2005, **86**, 073104.
12 M. S. Dresselhaus and G. Dresselhaus, Adv. Phys., 2002, **51**, 1.
13 S. Uji, J. S. Brooks and Y. Iye, *Physica B*, 1998, **246-247**, 299.
14 H. Yaguchi and J. Singleton, *Phys. Rev. Lett.*, 1998, **81**, 5193.
15 Y. Iye, P. M Tedrow, G. Timp, M. Shayegan, M. S. Dresselhaus, G. Dresselhaus, A. Furukawa and S. Tanuma, *Phys. Rev. B*, 1982, **25**, 5478.

NEW SINGLE-COMPONENT MOLECULAR CONDUCTORS WITH DISELENADITHIAFULVALENE FRAMEWORKS

E. Fujiwara,[1] A. Kobayashi,[1] Y. Okano,[2] H. Kobayashi,[2] Y. Fujishiro,[3] E. Nishibori,[3] M. Takata[4] and M. Sakata[3]

[1] The University of Tokyo, Hongo, Bunkyo-ku, Tokyo 113-0033, Japan
[2] Institute for Molecular Science, Myodaiji, Okazaki 444-8585, Japan
[3] Nagoya University, Nagoya 464-8603, Japan
[4] Japan Synchrotron Radiation Research Institute, Hyogo 679-5198, Japan

1 INTRODUCTION

A novel single-component metal complex with a stable metallic state down to very low temperature, [Ni(tmdt)$_2$] (tmdt = trimethylenetetrathiafulvalenedithiolate), was reported in 2001.[1] Recently, the existence of three-dimensional (3D) Fermi surfaces was experimentally confirmed by the observation of the de Haas–van Alphen (dHvA) oscillation.[2] Unlike the case of neutral bis(dithiolato)nickel complexes, neutral bis(dithiolato)gold complexes have one unpaired electron per complex, which makes the electromagnetic properties of such neutral gold complexes very attractive.[3, 4] In addition, these unpaired electrons in the bis(dithiolato)gold complexes have a possibility to form a metallic band. From such a viewpoint, the gold analogue [Au(tmdt)$_2$] was developed in 2003, which maintained its high conductivity even when an antiferromagnetic transition takes place around 100 K.[4] The *ab* initio plane-wave pseudo-potential calculations indicated that the nesting of the Fermi surfaces is not perfect and Fermi-surface pockets remain below the magnetic phase transition temperature.[5] On the other hand, the introduction of selenium atoms to the metal complexes with TTF skeletons are expected to increase intermolecular interactions in solid states and to widen conduction bands. In the present report, we will discuss the preparation, structures and physical properties of newly synthesized [M(tmstfdt)$_2$] (M = Ni, Au; tmstfdt = trimethylenediselenadithiafulvalenedithiolate) 1 and 2, which are the selenium analogues of [M(tmdt)$_2$] (M = Ni, Au).[1, 4]

2 RESULTS AND DISCUSSION

2.1 Syntheses

Tetramethylammonium complexes 6 and 7 were synthesized as shown in Scheme 1. Thus, 4, 5-bis(2'-cyanoethylthio)-4', 5'-trimethylene-STF (5) was obtained in 12% yield by a trimethyl phosphite-mediated cross-coupling reaction of 3[6] and 4[7] at 110°C, followed by the purification by column chromatography (silica gel, dichloromethane), and recrystallization from dichloromethane/*n*-hexane. A deprotection of 5 was performed with

25 wt% methanol solution of tetramethylammonium hydroxide in dry THF at room temperature, and the air-unstable $(Me_4N)_2$tmstfdt ligand was generated as powder in reaction solution.   Then, nickel and gold complexes **6** and **7** were obtained by the successive treatment with a methanol solution of $NiCl_2 \cdot 6H_2O$ or $HAuCl_4 \cdot 4H_2O$ at -78°C to room temperature, respectively.   The resulting powders were collected by filtration, washed three times with a mixed solvent of dry THF and dry MeOH (v/v = 10/1) at room temperature under argon atmosphere to afford pale reddish brown powder for **6** and brown powder for **7**, respectively.   Electrocrystallization of **6** and **7** in the presence of tetra-*n*-butylammonium perchlorate in THF or MeCN produced single-component metal complexes **1** and **2** as black microcrystals within 3-4 weeks, respectively.

Scheme 1   *Reagents and conditions: i, P(OMe)₃, 110 °C, 1.5 h; ii, 25 wt% Me₄NOH in MeOH (4 equiv.), dry THF, room temperature, 30 m, then NiCl₂·6H₂O (0.5 equiv.) or HAuCl₄·4H₂O (0.5 equiv.), -78°C to room temperature, overnight; iii, electrocrystallization, ⁿBu₄N·ClO₄, THF or MeCN, 0.3 or 0.5 μA, room temperature*

2.2 Crystal Structure Analyses

Synchrotron radiation X-ray powder experiments were performed on **1** and **2** to obtain their structural information.   Their X-ray powder patterns indicated that the crystal structures of **1** and **2** are isostructural to those of $[M(tmdt)_2]$ (M = Ni, Au).[1, 4]   The structures were determined by a MEM/Rietveld method.[8]   X-Ray crystallographic data of $[Ni(tmstfdt)_2]$: $C_{18}H_{12}S_8Se_4Ni$, triclinic, space group $P\bar{1}$, a = 6.5299(1), b = 7.5228(2), c = 12.1648(2) Å, α = 90.789(2), β = 96.847(2), γ = 103.795(3)°, V = 575.64(4) Å^3, Z = 1, R_{wp} = 0.088, R_I = 0.114.   X-Ray crystallographic data of $[Au(tmstfdt)_2]$: $C_{18}H_{12}S_8Se_4Au$, triclinic, space group $P\bar{1}$, a = 6.5220(7), b = 7.5032(3), c = 12.303(2) Å, α = 90.26(1), β = 96.85(2), γ = 103.43(2)°, V = 581.1(2) Å^3, Z = 1, R_{wp} = 0.017, R_I = 0.067.   X-Ray structure analyses of **1** and **2** revealed that there is a crystallographically independent half of the metal complex, and the central metal atoms is located on the lattice points, showing simple and dense molecular packing motif.   A large number of short $S(Se)\cdots S(Se)$ contacts less than the sum of van der Waals radii were observed in **1** and **2**, but we could not confirm such a contact along the molecular stacking direction in **1**.   The 3D and two-dimensional (2D) chalcogen networks are developed in **1** and **2**, respectively.

2.3 Electrical and Magnetic Properties

The electrical resistivities of the compressed pellet samples of **1** and **2** were measured by a four-probe method down to 4.2 and 10 K, respectively.   Complex **1** showed a fairly high room temperature conductivity of ca. 100 $S \cdot cm^{-1}$ even for the measurement on the compressed pellet sample. As seen from Figure 1, the resistivity of **1** decreased down to ca. 50 K, while its temperature dependence was fairly small, then the resistivity increased very slowly down to 4.2 K. The metallic temperature dependence of the resistivity down to 50 K and a fairly large conductivity even at to the liquid helium temperature ($\sigma_{4.2\ K} \approx \sigma_{r.t.}$) suggest the system to be essentially metallic down to 4.2 K. On the other hand, complex **2** showed a relatively high room temperature conductivity ($\sigma_{r.t.} = 11$ S·cm^{-1}) and a semiconducting behaviour with a small activation energy ($E_{a\ (200\text{-}300\ K)} = 16$ meV). The magnetic susceptibilities of **1** were measured by a SQUID magnetometer down to 2 K at 10 kOe. As shown in Figure 1, the temperature dependence of magnetic susceptibilities of **1** after subtracting a diamagnetic contribution and a small low-temperature Curie term indicate the Pauli-paramagnetism of the metal electrons, which is consistent with the conducting behaviour. The magnitude of the susceptibilities of **1** ($1.7\sim2.0\times10^{-4}$ emu·mol^{-1}) are about a half of that of [Ni(tmdt)$_2$],[1] suggesting that this system has larger intermolecular interactions than those in [Ni(tmdt)$_2$].[1] As indicated by both the conducting behaviour and magnetic properties, nickel complex **1** is a new single-component molecular metal with the STF skeleton.

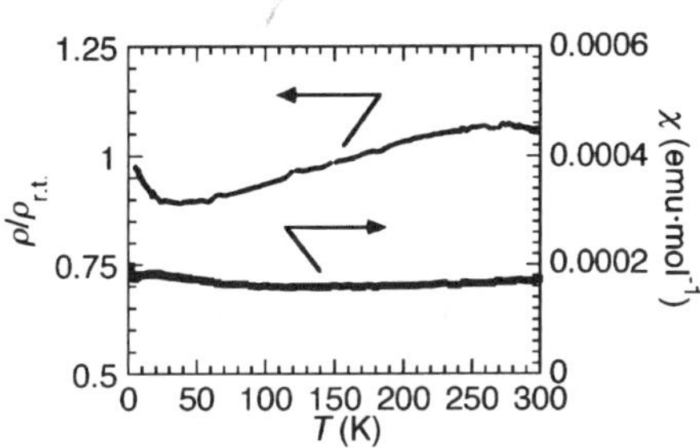

Figure 1 *Normalized resistivity ($\rho/\rho_{r.t.}$) and magnetic susceptibility (χ) as a function of temperature (T) for 1.*

2.4 Band Structure Calculation

The band structure calculation of **1** was carried out by a tight-binding method based on an extended Hückel approximation using Slater-type atomic orbitals. The metal complex has an approximately D_{2h} symmetry. The MO calculation of **1** gave the HOMO and the LUMO with b_{1u} and b_{2g} symmetry, respectively. The HOMO has the same sign on every chalcogen atoms, but the LUMO has a nodal plane on the central metal atom. The estimated HOMO-LUMO energy gap (ΔE_{HL}) is a small value of 0.15 eV. The calculated

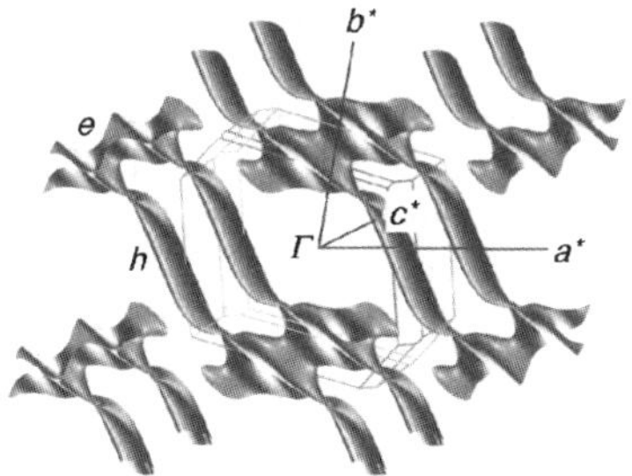

Figure 2 *Calculated electron (e) and hole (h) Fermi surfaces for 1. ΔE_{HL} is assumed to be 0.15 eV.*

Fermi surfaces for ΔE_{HL} = 0.15 eV are shown in Figure 2. Electron and hole Fermi surfaces are arranged alternately in the plane approximately parallel to the a^*-b^* and c^* axes. The configuration and arrangement of electron and hole pockets are similar to those of the single-component 3D molecular metal [Ni(tmdt)$_2$],[1] suggesting the 3D nature of metal electrons in **1**.

3 CONCLUSION

We have reported the synthesis, structure analysis and physical properties of the single-component molecular conductors with STF skeletons **1** and **2**, which showed the high room temperature conductivities. Especially, the band structure analysis and physical properties of **1** revealed that a 3D metallic state was also realized in the selenium analogue of [Ni(tmdt)$_2$]. To our knowledge, there is no 3D molecular metal based on metal complexes with STF skeletons. This discovery of the single-component molecular metal involving selenium atoms contributes to the development of new single-component molecular metals with enhanced intermolecular interactions. Investigations on the physical properties of the gold complex **2** are now in progress to confirm the existence of magnetic phase transitions.

References

1   (*a*) H. Tanaka, Y. Okano, H. Kobayashi, W. Suzuki and A. Kobayashi, *Science*, 2001, **291**, 285; (*b*) A. Kobayashi, H. Tanaka and H. Kobayashi, *J. Mater. Chem.*, 2001, **11**, 2078; (*c*) A. Kobayashi, E. Fujiwara and H. Kobayashi, *Chem. Rev.*, 2004, **104**, 5243.

2   H. Tanaka, M. Tokumoto, S. Ishibashi, D. Graf, E. S. Choi, J. S. Brooks, S. Yasuzuka, Y. Okano, H. Kobayashi and A. Kobayashi, *J. Am. Chem. Soc.*, 2004, **126**, 10518.

3   (*a*) A. J. Schultz, H. H. Wang, L. C. Soderholm, T. L. Sifter, J. M. Williams, K. Bechgaard and M. -H. Whangbo, *Inorg. Chem.*, 1986, **26**, 3757; (*b*) N. C. Schiødt, T. Bjørnholm, K. Bechgaard, J. J. Neumeier, C. Allgeier, C. S. Jacobsen and N. Thorup, *Phys. Rev. B, Condens. Mater.*, 1996, **53**, 1773; (*c*) D. Belo, H. Alves, E. B. Lopes, M. T. Duarte, V. Gama, R. T. Henriques, M. Almeida, A. Pérez-Benítez, C. Rovira and J. Veciana, *Chem. Eur. J.*, 2001, **7**, 511.

4   W. Suzuki, E. Fujiwara, A. Kobayashi, Y. Fujishiro, E. Nishibori, M. Takata, M. Sakata, H. Fujiwara and H. Kobayashi, *J. Am. Chem. Soc.*, 2003, **125**, 1486.

5   S. Ishibashi, H. Tanaka, M. Kohyama, M. Tokumoto, A. Kobayashi, H. Kobayashi and K. Terakura, *J. Phys. Soc. Jpn.*, 2005, **74**, 843.

6   (*a*) Y. Okano, H. Sawa, S. Aonuma and R. Kato, *Chem. Lett.*, 1993, 1851; (*b*) Y. Okano, H. Sawa, S. Aonuma and R. Kato, *Synth. Metals*, 1995, **70**, 1161.

7   (*a*) N. Svenstrup, K. M. Rasmussen, T. K. Hansen and J. Becher, *Synthesis*, 1994, 809; (*b*) L. Binet, J. M. Febre, C. Montginoul, K. B. Simonsen and J. Becher, *J. Chem. Soc., Perkin Trans. 1*, 1996, 783.

8   (*a*) M. Takata, B. Umeda, E. Nishibori, M. Sakata, Y. Saito, M. Ohno and H. Shinohara, *Nature*, 1995, **377**, 46; (*b*) M. Takata, E. Nishibori and M. Sakata, *Z. Kristallogr.*, 2001, **216**, 71.

CHARGE TRANSFER COMPLEXES OF CYTOSINE AND TCNQ DERIVATIVES

K. Nishimura, T. Murata, Y. Enomoto, G. Honda, Y. Shimizu, M. Sakata, and G. Saito
Graduate School of Science, Kyoto University, Kyoto, 606-8502, Japan.

1 INTRODUCTION

Many controversial reports have been published concerning with the conductivity of DNA molecules. Some reported metallic and even superconducting natures, on the other hand, others observed insulating features.[1-3] While, variety of multifunctional molecular conductors based on TTF derivatives having a nucleobase moiety have been reported to show exotic conducting and structural properties.[4-7] However, so far, no conducting charge transfer (CT) complexes have been prepared using nucleobase as a component molecule. Here we focus on cytosine (Figure 1), which is one of the nucleobases, and report several conductive complexes together with the crystal structure of a Mott insulator, 2:1 complex of cytosine·TCNQ.

Cytosine (C) is a Brönsted base ($pK_a = 4.55$) to form readily protonated cation of cytosine (CH^+, Fig. 1) in an acidic condition. In addition to C and CH^+ species, a pair of a neutral cytosine and a protonated one, namely hemiprotonated cytosine (CHC^+) has been observed in polycytidylic acid[8] and several salts of 1-methylcytosine.[9] Cytosine is a weak electron donor ($E_{ox}= 1.90$ V vs SCE in acetonitrile,[10] ionization potential = 8.68 eV[11]) compared to molecules in the TTF system. Cation radicals of cytosine were observed in X-ray irradiation[12] or as a photo-induced CT intermediate from excited triplet state.[13] In spite of a weak donor strength of cytosine, the formation of an anion radical salt of TCNQ from methanol/acetonitrile solution was reported.[14] Also an annealing at 350 K of neutral cytosine·TCNQ complex, which was obtained by co-sublimation method under vacuum, gave anion radical of TCNQ.[15] In both cases, however, cytosine radicals were not detected and no structural and conductive properties have been reported.

Cytosine has a strong tendency to form hydrogen bonds, such as the complementary hydrogen bonds in double strand DNA. Thus hydrogen bonds would play an important role in crystal structure formation, which has close relation to physical properties.

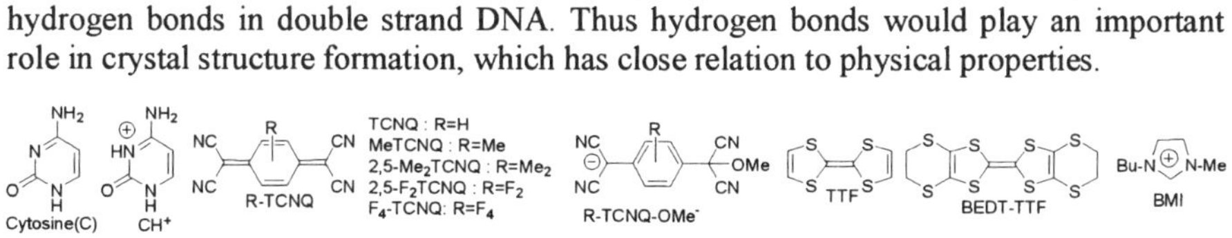

Figure 1   *Chemicals in text*

2 EXPERIMENTAL

2.1 Preparation of charge transfer complexes

The complexes of cytosine and TCNQ derivatives (R-TCNQ, Figure 1) were prepared by direct mixing of cytosine in methanol and R-TCNQ in acetonitrile. For TCNQ complexes, a 2:1 complex (**1**) was obtained as single crystals by slow diffusion of cytosine and TCNQ in methanol/acetonitrile (2:1) solution. **1** was also obtained as black powder by metathesis between equimolar portions of cytosine, cytosine hydrochloride and Li·TCNQ. Besides 2:1 complex, a 1:1 complex (**2**) was also obtained by mixing in different ratio. Some typical compounds between cytosine and R-TCNQ are summarized in Table 1.

Table 1 *Characteristics of complexes of cytosine and R-TCNQ*

R	Composition	Appearance	σ_{RT} / S cm^{-1}, (ε_a / eV)[a]	UV-Vis-NIR[d] / 10^{3}cm^{-1}			
				A	B	C	D
1 H	$(CHC^+)(TCNQ^{\cdot-})$	black plate	3×10^{-2} $(0.14)^b$	–	7.1	11.7,	16.5
2 H	$(CH^+)(TCNQ^{\cdot-})(H_2O)_{0.5}$	black	1×10^{-4} $(0.12)^c$	–	7.0	11.7,	16.4
3 Me	$(CHC^+)(MeTCNQ^{0.5-})_2$	black	2×10^{0} $(0.041)^c$	3.4	–	10.5,11.5	17.1
4 Me$_2$	$(C_4H^+)(Me_2TCNQ^{0.5-})_2(H_2O)_{0.2}$	dark blue	8×10^{-3} $(0.06)^c$	3.4	6.4	11.6,	16.8
5 F$_2$	$(C_3H^+)(F_2TCNQ\text{-}OMe^-)(H_2O)_{0.5}$	pale green	insulator	–	–	–	–
6 F$_2$	$(CHC^+)(F_2TCNQ^{\cdot-})$	blue-purple	$1\times10^{-3}(0.21)^c$	–	6.6	11.6	16.0
7 F$_4$	$(CHC^+)(F_4TCNQ\text{-}OMe^-)(H_2O)$	pale yellow	insulator	–	–	–	–

a) σ_{RT}: conductivity at room temperature (RT), ε_a: activation energy for conduction. b) //*a* direction. c) compaction pellet. d) below 20×10^3 cm^{-1}

2.2 Measurements

Absorption spectra were measured with a KBr disk on Perkin-Elmer PARAGON 1000 Series FT-IR in IR region (400-7800 cm^{-1}) and on SHIMADZU UV-3100 spectrometer in near-infrared, visible and ultraviolet (UV-Vis-NIR) region (3800-42000 cm^{-1}). D.c. conductivities were measured based on a standard two- or four-probe technique attaching gold wires (15-25 μmϕ) on samples by gold paste (Tokuriki 8560-1A). The stoichiometries of complexes were determined by elemental analysis or crystal structure. The intensity data of the structural analysis were collected in oscillator type X-ray imaging plate (DIP-2020K) with a monochromated MoK_α radiation at 200 and 9 K. The structures were solved by direct methods using SHELXS-97.[16] The refinements of structures were performed by full matrix least squares method (SHELXL-97).[16] The positions of hydrogen atoms were determined by D-synthesis. The parameters were refined adopting anisotropic and isotropic temperature factors for non-hydrogen atoms and hydrogen ones, respectively. EPR spectra were measured by JEOL JES-RE2 X Band EPR spectrometer.

3 RESULTS AND DISCUSSION

3.1 Composition of the products

In spite of very weak electron donating ability of cytosine molecule, it afforded only ionic compounds even with the weakest acceptor in Figure 1 (R = Me$_2$), which cannot be ionized by BEDT-TTF (E_{ox}= 0.53 V vs SCE in acetonitrile). The solid products are classified into three categories based on the UV-Vis-NIR and IR spectra: I) Anion radical salts with fully ionic TCNQ moieties (**1**, **2** and **6**). II) Anion radical salts with partially ionic TCNQ moieties (**3** and **4**). III) Ionic salts with destructed TCNQ anions (**5** and **7**). In all cases, cytosine molecules are protonated and play a role of counter cation in these salts as CH^+ (**2**), CHC^+ (**1**, **3**, **6** and **7**), C_3H^+ (**5**) and C_4H^+ (**4**).

Compounds in Class **I** exhibited very similar UV-Vis-NIR absorption bands to those of the Li or K salt of R-TCNQ$^{\cdot-}$ (Figure 2a) showing characteristic bands; band-B (6-8×10^3 cm^{-1}), band-C (~11×10^3 cm^{-1}), band-D (15-17×10^3 cm^{-1}) and band-E (27-28×10^3 cm^{-1}). Since bands-C, -D and -E are essentially assignable to intramolecular transitions of a R-TCNQ anion radical, they appeared commonly in Class **I** and **II** compounds (Table 1). Band-B is known to be associated with the optical transition in a segregated column or layer of fully or highly ionized TCNQ molecules ($\gamma > 0.5$; γ. degree of CT) and is related with the on-site Coulomb repulsion.[17]

While compounds in Class **II** (**3** and **4**) exhibited the band-A below 5×10^3 cm^{-1} (Figure 2b,c) indicating that the salts have a segregated column or layer of partially ionized or charge separated radical species. The spectrum of **3** resemble to that of BMI(TCNQ)$_2$ ($\gamma = 0.5$; BMI = 1-butyl-3-methylimidazolium) without any trace of B-band (Figure 2b), while that of **4** exhibited B-band faintly suggesting a charge separation or a dimerization (Figure 2c).

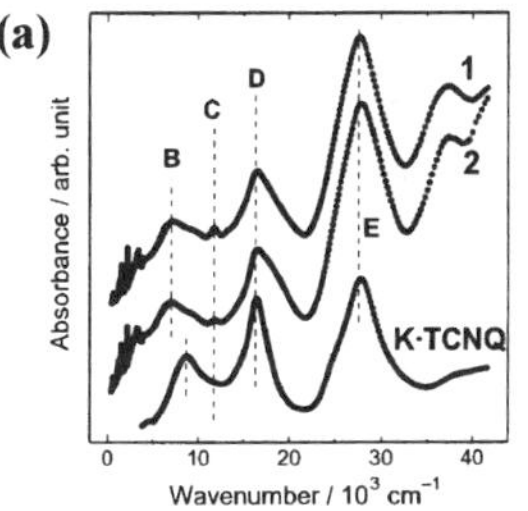

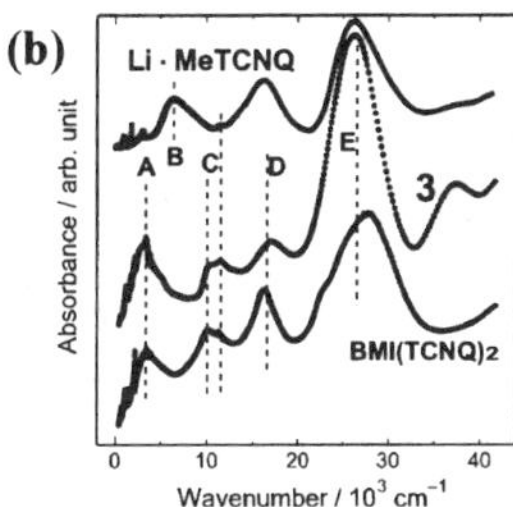

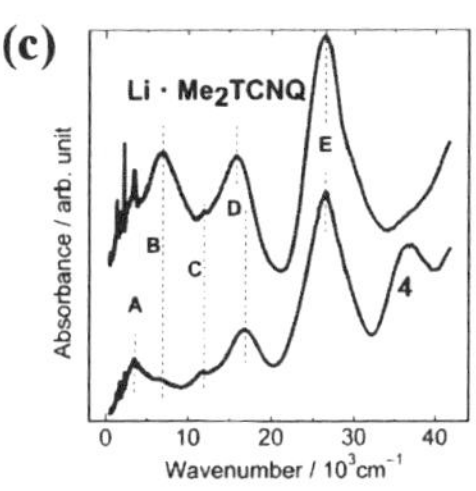

Figure 2 *UV-Vis-NIR spectra of (a) 1, 2 and K·TCNQ, (b) 3, Li•MeTCNQ and (BMI)(TCNQ)$_2$ and (c) 4 and Li•Me$_2$TCNQ in KBr. The optical bands A-E, see text. BMI: 1-butyl-3-methylimidazolium*

Compounds in Class **III** (**5** and **7**) showed common UV-Vis-NIR and IR spectra to each other but differed completely from those in other classes. There are no absorption peaks below 25×10^3 cm^{-1} and the first low-energy peak appeared at 27-28×10^3 cm^{-1} in solid. They exhibited two strong C≡N stretching peaks with a peak separation of 43-52 cm^{-1}, which is considerably large compared with that for R-TCNQ anion radicals (<30 cm^{-1}). Based on the elemental and optical analyses of them and especially the structural analysis of **7** indicated that the anion species of class **III** compounds are methoxy substituted R-TCNQs (R-TCNQ-OMe$^-$) (Figure 1).

EPR measurements indicated that there were no radical species of cytosine molecules. The inspection of IR spectra in the region of 1600-1800 cm^{-1} afforded information concerning the C=O and C=N stretching modes of cytosine molecules, and hence the kinds of protonated cytosine species. It has been known that C=O stretching modes strongly appeared in the range of 1710-1730 cm^{-1} for protonated cytosine salts,[18,19] while such modes were weak and red-shifted to 1701 cm^{-1} for neutral cytosine. For the compounds **1-7**, strong peaks appeared in the range of 1720-1735 cm^{-1} confirming the presence of protonated cytosine molecules. The compositions of **1-7** are deduced based on the optical, structural and elemental analyses as shown in Table 1.

3.2 Crystal structure of 1

1 crystallized in triclinic, and the crystallographic data at 200 K were listed in Ref. 20. One cytosine molecule and a half of TCNQ molecule were independent. No drastic change of crystal structure was found between the structures at 200 K and 9 K. The details of crystal structure will be discussed based on the data at 200 K. The γ value was deduced as 1.01-1.04 based on bond lengths of TCNQ molecule.[21] Figures 3a illustrates the cytosine and TCNQ segregated stacks along the *a*-axis.

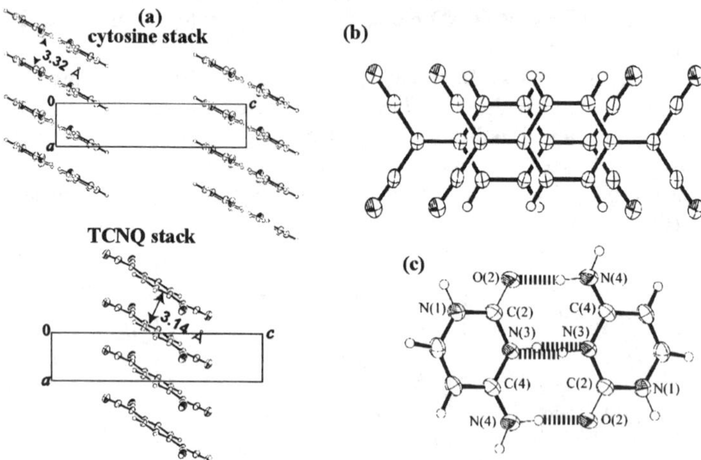

Figure 3   *Crystal structure of 1 at 200 K. (a) The b-axis projection to show cytosine and TCNQ stacks. (b) Overlap mode of TCNQ molecules. (c) Hemiprotonated cytosine pair. Dotted lines are short hydrogen bonds.*

Uniform stack of one-dimensional (1D) TCNQ columns was formed along the *a*-axis. Such a uniform 1D column was reported in the high-temperature phase of monovalent TCNQ radical salts of alkali metals.[22-23] Amazingly, the uniform stack of the TCNQ column was retained down to at least 9 K, though the TCNQ columns in Na·TCNQ, K·TCNQ and Rb·TCNQ(phase II) were dimerized below 348, 395 and 220 K, respectively.[22-24] This non-observation of monomer-dimer transition is the first case in fully ionized TCNQ complexes. The overlap mode of neighboring TCNQ molecules was ring-over-bond type (Figure 3b). The interplanar distance between TCNQ molecules was 3.14Å, which was shorter than any other TCNQ salts with uniform column (3.22-3.48Å). This distance was reduced to 3.10 Å at 9 K due to contraction of a crystal.

Cytosine molecules formed a ribbon along the *b*-axis with the width (9.10 Å) of two cytosine molecules, namely hemiprotonated cytosine pair CHC⁺ (Figures 3c and 4). Short triple complementary hydrogen bonds (two N-H···O, one N⁺-H···N) were observed in a cytosine pair. Since an inversion center laid at the center of a cytosine pair, the hydrogen atom on N(3) was located at two sites with site occupancy factor of 0.5. Figure 4 illustrates the crystal structure of **1** viewed along the *a*-axis. A flat ribbon-like structure along the *b*-axis was formed by double complementary hydrogen bonds (N-H···O) between hemiprotonated cytosine pairs. Such a ribbon-like structure was reported in cytosine hydrogen pentaiodide trihydrate,[25] while 2D sheet structure was observed in cytosine hydrochloride[26] and cytosinium nitrate.[27] The 1D cytosine ribbon is connected to the neighboring TCNQ molecules with strong N-H···N and weak C-H···N hydrogen bonds. Furthermore, a weak C-H···N hydrogen bond exists between TCNQ molecules, thereby

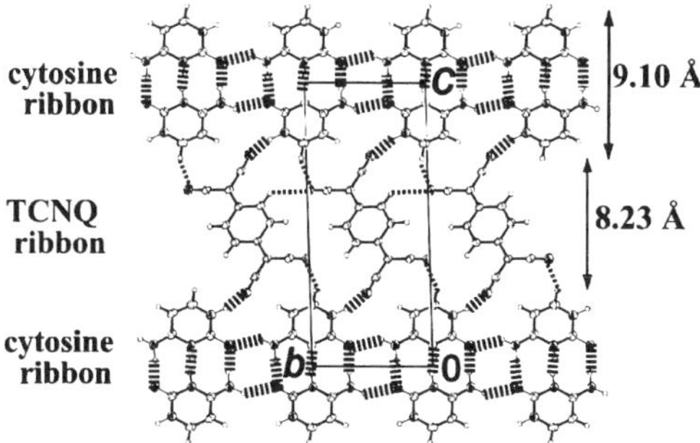

Figure 4 *Crystal structure of 1 at 200 K along the a-axis showing cytosine ribbons and TCNQ ribbon connected by hydrogen bonds (dotted lines).*

forming a uniform and flat TCNQ ribbon along the *b*-axis. Thick and thin dashed lines in Figure 4 represent strong N-H$\cdots$X (X = O, N) hydrogen bonds and weak C-H$\cdots$N ones, respectively. The cytosine ribbons composed of hemiprotonated cytosine pairs stack uniformly along the *a*-axis with interplanar separation of 3.32 Å. It is noteworthy that the ribbons of cytosine molecules in **1** forced to keep uniform TCNQ stacking along the *a*-axis down to 9 K by robust, numerous and uniform hydrogen bonds.

3.3 Conductivity and U_{eff}

The transport properties are summarised in Table 1. The conductivity of **1** at room temperature ($\sigma_{\mathrm{RT}} = 3 \times 10^{-2}$ S cm^{-1}, $\varepsilon_a = 0.14$ eV, measured on single crystal, //*a*) is higher than any of fully ionized TCNQ salts and in the same order as that of Rb$\cdot$TCNQ (phase II, 2×10^{-2} S cm^{-1}).[24] Since **1** has uniform column of fully ionized TCNQ radicals, it is classified to a Mott insulator, where the origin of energy gap is on-site Coulomb energy U. The effective U (U_{eff}) is evaluated as 1.08 eV according to an equation, $\varepsilon_g \approx U_{\mathrm{eff}} - 4t$, where transfer integral t is 0.20 eV based on band calculation and band gap ε_g is assumed twice of ε_a. The evaluated U_{eff} value is in good accordance with the proposed one for a TCNQ molecule (1.0-1.1 eV).[28] Other fully ionized complexes (**2** and **6**) have similar transport properties ($\sigma_{\mathrm{RT}} = 10^{-3}$-$10^{-4}$ S cm^{-1}, $\varepsilon_a = 0.12$-0.21 eV) even measured on compaction pellet samples. For partially ionized complexes, the conductivity of **3** was 2 S cm^{-1} at RT ($\varepsilon_a = 0.04$ eV), while that of **4** was 8×10^{-3} S cm^{-1} ($\varepsilon_a = 0.06$ eV) on compaction pellet samples. Even though the σ_{RT} value of **4** is a little less than that of **1** (single crystal), partial ionicity of **4** affected the transport resulting in a decreased ε_a value. For **4**, some structural factor such as lattice deformation may give rise to a decrease in the conductivity. The destructed R-TCNQ compounds (**5** and **7**) are good insulators.

4 CONCLUSION

A series of cytosine-TCNQ complexes were prepared by mixing cytosine in methanol and R-TCNQ in acetonitrile. Those compounds are classified into I) anion radical salts with fully ionized R-TCNQ, II) anion radical salts with partially ionized R-TCNQ, and III) ionic salts with methoxy adduct of R-TCNQ anion, and in all cases protonated cytosine molecules (CH$^+$, CHC$^+$, C$_3$H$^+$ or C$_4$H$^+$) act as counter cations. In the crystal of **1**, uniform

1D column of TCNQ molecules was found down to at least 9 K. Cytosine molecules formed a ribbon-like structure consisted of hemiprotonated cytosine pair. Strong triple and double complementary hydrogen bonds formed the cytosine ribbon. Hydrogen bonds were also observed between cytosine ribbon and TCNQ molecules and between TCNQ molecules. As a result of the robust, numerous, and uniform hydrogen bonds, the interplanar distance between TCNQ molecules is very short, **1** is highly conductive as a TCNQ Mott insulator, and a dimerization of TCNQ anion radicals is prevented down to low temperatures. The partial CT salt with MeTCNQ is highly conductive.

Acknowledgement

This work was in part supported by a Grant-in-Aid (15205019) from the Ministry of Education, Culture, Sports, Science and Technology, Japan.

References

1   B. Giese, *Acc. Chem. Res.* 2000, **33**, 631 and references therein.
2   A. Y. Kasumov et al., *Science*, 2001, **291**, 280.
3   Y. Zhang, R.H. Austin, J. Kraeft, E.C. Cox and N.P. Ong, *Phys. Rev. Lett.*, 2002, **89**, 198102.
4   O. Neilands, *Mol. Cryst. Liq. Cryst.*, 2001, **335**, 331.
5   G. Saito, A. Ota, Y. Yoshida, M. Maesato, H. Yamochi, K. Balodis, O. Neilands, S. Khasanov and M. Tanatar, *Mol. Cryst. Liq. Cryst.* 2004, **423**, 99.
6   Y. Morita, S. Maki, M. Ohmoto, H. Kitagawa, T. Okubo, T. Mitani and K. Nakasuji, *Synth. Met.*, 2003, **135-136**, 541.
7   T. Murata, Y. Morita, K. Fukui, K. Sato, D. Shiomi, T. Takui, M. Maesato, H. Yamochi, G. Saito and K. Nakasuji, *Angew. Chem. Int. Ed.*, 2004, **43**, 6343.
8   R. Langridge and A. Rich, *Nature*, 1963, **198**, 725.
9   R.E. Marsh, R. Bierstedt and E.L. Eichhorn, *Acta Cryst.*, 1962, **15**, 310.
10 C.A.M. Seidel, A. Schulz and M.H.M. Sauer, *J. Phys. Chem.*, 1996, **100**, 5541.
11 V.M. Orlov, A.N. Smirnov and Ya.M. Varshavsky, *Tetrahedron Lett.*, 1976, **17**, 4377.
12 E.O. Hole, W.H. Nelson, E. Sagstuen and D.M. Close, *Radiat. Res.*, 1998, **149**, 109.
13 J. Geimer, K. Hildenbrand, S. Naumov and D. Beckert, *Phys. Chem. Chem. Phys.*, 2000, **2**, 4199.
14 G.G. Sheina, E.D. Radchenko, Yu.P. Blagoii and B.I. Berkin, *Doklady Akademii Nauk SSSR*, 1978, **240**, 463.
15 G.G. Sheina, E.D. Radchenko, S.A. Egupoii, Yu.P. Blagoy and V.M. Orlov, *Int. J. Quant. Chem.*, 1979, **16**, 387.
16 G. M. Sheldrick, University of Göttingen, 1997.
17 C. S. Jacobsen, *Semiconductors and Semimetals*, ed. E. Conwell, Academic Press, Boston, 1988; vol. 27, Chapter 5.
18 Y. Nishimura and M. Tsuboi, *Chem. Phys.*, 1985, **98**, 71.
19 G. Ślósarek and R. Zamboni, *Spectrochim. Acta*, 1991, **47A**, 863.
20 Triclinic, P$\bar{1}$, a=3.764(1), b=7.408(2), c=17.010(5) Å, α=88.25(2), β=89.74(2), γ=84.65(2)°, V = 472.0(2) Å^3, Z=1, D_{calc}=1.504 g cm^{-3}, R_1=0.067, $_wR_2$=0.198, GOF=1.032. T= 200 K.
21 P.S. Flandrois and D. Chasseau, *Acta Cryst., B*, 1977, **33**, 2744.
22 M. Konno and Y. Saito, *Acta Cryst., B*, 1975, **31**, 2007.
23 M. Konno, T. Ishii and Y. Saito, *Acta Cryst. B*, 1977, **33**, 763.
24 H. Kobayashi, *Bull. Chem. Soc. Jpn.*, 1981, **54**, 3669.
25 Z. Wang, Y. Cheng, C. Liao and C. Yan, *CrystEngComm*, 2001, **50**, 1.
26 N.S. Mandel, *Acta Cryst., B*, 1977, **33**, 1079.
27 A. Cherouana, K. Bouchouit, L. Bendjeddou and N. Benali-Cherif, *Acta Cryst., E*, 2003, **59**, 983.
28 K.D. Cummings, D.B. Tanner and J.S. Miller, *Phys. Rev. B*, 1981, **24**, 4142.

ELECTRON DONORS FROM NATURAL PRODUCTS: DONATING AND LIGATING ABILITIES AND CHARGE-TRANSFER COMPLEX FORMATION OF IMIDAZO[1,2-a]PYRAZIN-3(7*H*)-ONES

Mai Suzuki,[1] Takashi Fujii,[1] Takashi Nogami,[1] Takashi Hirano[1] and Takayuki Ishida[1,2]

[1]Department of Applied Physics and Chemistry, The University of Electro-Communications, Chofu, Tokyo 182-8585, Japan; E-mail: ishi@pc.uec.ac.jp
[2]Course of Coherent Optical Science, The University of Electro-Communications

1 INTRODUCTION

There have been several examples of molecule-based electron conducting and/or magnetic materials inspired by ingenious biochemical systems.[1,2] Recently, cytochrome c and chlorophyll a have been available for organic light-emitting materials using the intrinsic electron-transferable nature.[3] We focused on the bioluminescent substrates, luciferins, because they are readily oxidized with atmospheric oxygen at the initial step of bioluminescence schemes,[4,5] and accordingly we can regard them as a good candidate for electron donor molecules. The imidazo[1,2-a]pyrazin-3(7*H*)-one structure (abbreviated as IP; Scheme 1) is the essential of various marine luciferins, which are provided from the crustacea *Cypridina* (*Vargula*) for example. We studied the electron-donating properties of IPs and their ability of formation of charge-transfer complexes[6] and transition-metal complexes. An oxidative dimerization reaction has been found in the course of our study.

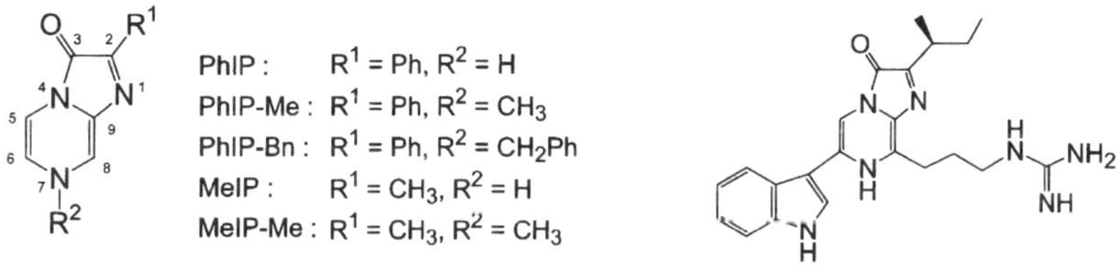

Scheme 1 *Molecular structures of imidazo[1,2-a]pyrazin-3(7H)-ones (IPs) and the Cypridina luciferin.*

2 RESULTS AND DISCUSSION

We have reported the first oxidation potential of PhIP measured by means of differential pusled voltammetry (DPV).[6] In this study, the first oxidation potentials of MeIP,[7-9] *N*-methyl,[9] and *N*–benzyl derivatives[10] were measured in acetonitrile, and summarized in Table 1. This result implies that the IP derivatives are good electron donors comparable to ferrocene. The donor ability was weakened by *N*-alkylation, and the potential values almost agree with the data previously reported on PhIP-Me.[11] A semiempirical calculation indicated that the HOMO levels did not change irrespective of *N*-methylation of PhIP.[9]

Table 1. *First oxidation potentials of IP derivatives in acetonitrile*

Compound	$E_{ox}^{(1)}$ / V(vs. Fc/Fc$^+$)
PhIP	0.095
PhIP-Me	0.17
PhIP-Bn	0.20
MeIP	-0.053
MeIP-Me	0.10

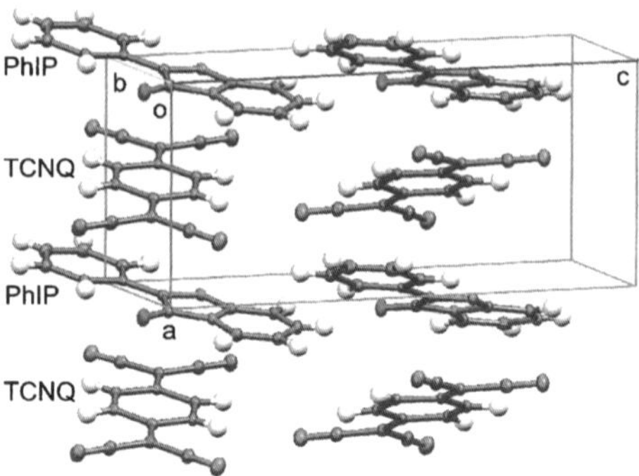

Figure 1 *Crystal structure of PhIP•TCNQ.*

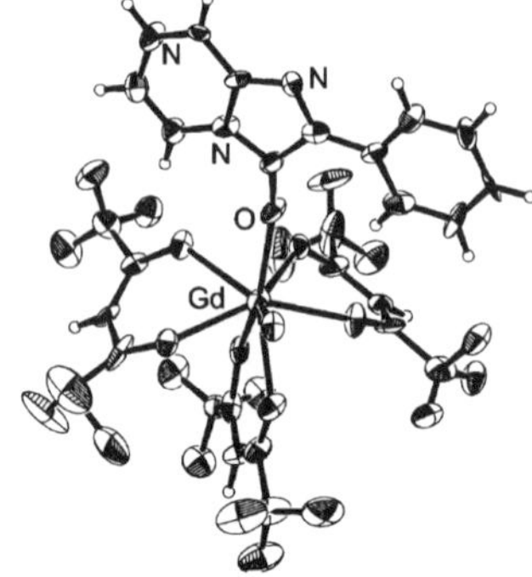

Figure 2 *Molecular structure of [Gd(hfac)₃(PhIP)(H₂O)].*

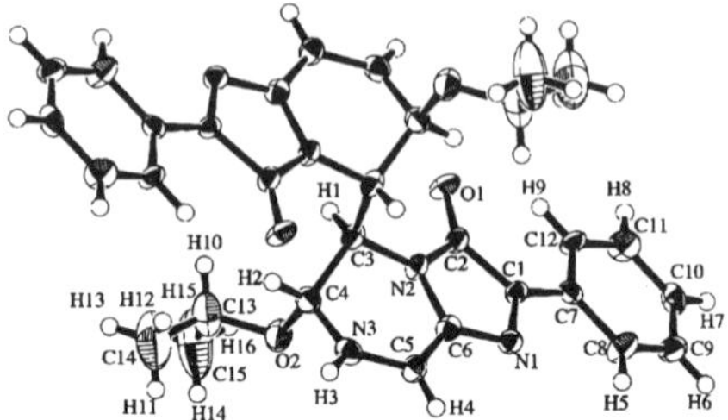

Figure 3 *Molecular structure of meso-(i-PrOPhIP)₂.*

The reaction of PhIP with TCNQ in acetonitrile gave brown crystals of a 1/1 molecular complex. Since the details have been reported elsewhere,[6] we summarize the results briefly. X-Ray crystallographic analysis revealed that they formed a mixed stack (Fig. 1). The PhIP portion was practically planar, as found in a neutral form of PhIP,[8,9] and arranged in a parallel manner with the planar TCNQ counter part. Interplane distances between PhIP and TCNQ were 3.30 and 3.37 Å. The IR absorption analysis suggested the charge-transfer degree of ca. 16%. Compressed pellet specimens of PhIP•TCNQ were electronically insulating. A mixed-stack structure with the small charge transfer degree on the present complex gives rise to low conductivity like BEDT-TTF-TCNQ.[12]

Lanthanide complexes having organic chromophores are of interest in development of photo-luminescent materials using an antenna effect.[13] The complexation of PhIP with Gd(hfac)₃ gave [Gd(hfac)₃(PhIP)(H₂O)] (hfac = 1,1,1,5,5,5-hexafluoropentane-2,4-dionate) Figure 2 shows the structure of one of two crystallographically independent molecules.[14] We safely concluded that the oxygen atom of PhIP was coordinated to the Gd ion for each unit, despite of the unsatisfactorily large *R* values. The PhIP ligand plays a role of an *O*-donor rather than an *N*-donor, being consistent with the arguments from the spectroscopic study on the M^{n+}-PhIP-Bn and MeIP-Bn systems where M^{n+} is alkali, alkaline-earth, and rare-earth metal ions.[10]

A similar reaction of PhIP with Cu(hfac)₂ in ethanol, however, gave unexpected products (EtOPhIP)₂ via oxidative dimerization. There were two isomers with a ca. 1/1 molar ratio, which could be manually separated by crystal habits under a microscope.

Preliminary X-ray diffraction analysis clarified that the dark yellow block crystals consisted of a *dl*-mixture of (EtOPhIP)$_2$ in a tetragonal lattice. Another needle-shaped crystals belonged to a monoclinic crystal system, being assigned to be *meso*-(EtOPhIP)$_2$. Dimerization took place at the 5- and 5'-positions and ethoxy groups were attached at the 6- and 6'-positions (Scheme 2). The ^{1}H, ^{13}C, and 2D NMR spectra supported the molecular structure and symmetry.[15] A similar oxidation in 2-propanol gave well-defined hexagonal prismatic crystals. The molecular structure of *meso*-(*i*-PrOPhIP)$_2$ was successfully determined in a trigonal $R\bar{3}$ space group (Figure 3).[16] The NMR spectra of *meso*-(*i*-PrOPhIP)$_2$[17] are quite similar to that of *meso*-(EtOPhIP)$_2$ except for the alkoxy groups.

As shown in Scheme 2, a mechanism can be proposed involving a radical coupling process of an oxidized form of PhIP. The copper(II) ion works as an oxidant, and the resultant cationic center is trapped by alcohol followed by radical combination at the 5-position. A pathway including dimerization before alcohol-trapping also seems possible. The active site for radical coupling seems to be the 5-position under the present conditions, whereas radical coupling with $O_2^{-\bullet}$ at the 2-position has been proposed under basic conditions or in the presence of the luciferase.[5] Dimerization at the 2-position would be prohibited owing to steric hindrance if the PhIP oxygen atom is coordinated to Cu(hfac)$_2$.

Goto et al. reported that 6-(3-indolyl)MeIP underwent the oxidative dimerization at the 5,5'-positions in the presence of K$_3$[Fe(CN)$_6$], but did not discuss possible diastereoisomerism with respect to a pair of chiral centers at the 5 and 5'- positions.[18] The optically resolved products prepared from the oxidation of the *Cypridina* Luciferin exhibited remarkably large CD ellipticities,[19] suggesting that the *d*- and *l*- isomers had been isolated. However, solely monomeric molecular structures were considered in that report.[19] In the present study, the diastereomers could be separated by preferential crystallization but the enantiomers could not. Optical resolution of *dl*-(ROPhIP)$_2$ by means of chiral-column chromatography for example would afford an answer to such large ellipticities.

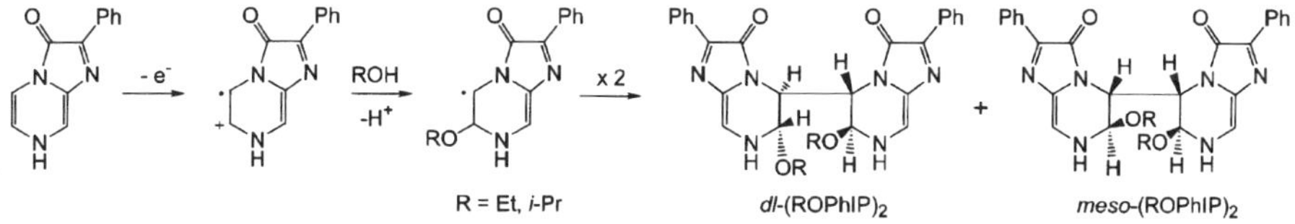

Scheme 2 *Plausible mechanism of the formation of (ROPhIP)$_2$.*

3 CONCLUSION

We have prepared five IP derivatives and their electron-donating properties were characterized by the oxidation potential measurements. The formation of the molecular PhIP•TCNQ complex and the dimeric products of (ROPhIP)$_2$ are explained in terms of the electron-donating properties of IP derivatives. The ligating ability of the carbonyl oxygen atom was unequivocally clarified. We are now studying the donor properties of other luciferins and their application to electronics-oriented materials chemistry.

References and Notes

1   Y. Morita, S. Maki, M. Ohmoto, H. Kitagawa, T. Okubo, T. Mitani and K. Nakasuji, *Org. Lett.*, 2002, **4**, 2185; O. Neilands, S. Belyakov, V. Tilika and A. Edzina, *Chem. Commun.*, **1995**, 325; K. Balodis, S. Khasanov, C. Chong, M. Maesato, H. Yamochi, G. Saito and O. Neilands, *Synth. Met.*, 2003, **133-134**, 353.

2   R. Feher, D. B. Amabilino, K. Wurst and J. Veciana, *Mol. Cryst. Liq. Cryst.*, 1999, **334**, 333; P. Taylor, P. R. Serwinski and P. M. Lahti, *Chem. Commun.*, **2003**, 1400; D. Shiomi, M. Nozaki, T. Ise, K. Sato and T. Takui, *J. Phys. Chem. B*, 2004, **108**, 16606.

3   H. Tajima, S. Ikeda, M. Matsuda, N. Hanasaki, J.-W. Oh and H. Akiyama, *Solid State Commn.*, 2003, **126**, 579; K. Shimatani, H. Tajima, T. Komino, S. Ikeda, M. Matsuda, Y. Ando and H. Akiyama, *Chem. Lett.*, 2005, **34**, 948.

4   Y. Ohmiya and T. Hirano, *Chem. Biol.*, 1996, **3**, 337; R. Saito, T. Hirano, H. Niwa and M. Ohashi, *J. Chem. Soc,, Perkin Trans. 2*, **1997**, 1711.

5   H. Kondo, T. Igarashi, S. Maki, H. Niwa, H. Ikeda and T. Hirano, *Tetrahedron Lett.*, 2005, **46**, 7701.

6   M. Suzuki, T. Fujii, T. Nogami, T. Hirano and T. Ishida, *Chem. Lett.*, 2006, **35**, 250.

7   G. B. Barlin, D. J. Brown, Z. Kadunc, A. Petric, B. Stanovnik and M. Tisler, *Aust. J. Chem.*, 1983, **36**, 1215.

8   I. Devillers, B. de Wergifosse, M.-P. Bruneau, B. Tinant, J.-P. Declercq, R. Touillaux, J.-F. Rees and J. Marchand-Brynaert, *J. Chem. Soc., Perkin Trans. 2*, **1999**, 1481.

9   S. Nakai, M. Yasui, M. Nakazato, F. Iwasaki, S. Maki, H. Niwa, M. Ohashi and T. Hirano, *Bull. Chem. Soc. Jpn.*, 2003, **76**, 2361.

10 T. Sekiguchi, S. Maki, H. Niwa, H. Ikeda and T. Hirano, *Tetrahedron Lett.*, 2004, **45**, 1065.

11 Y. Takamuki, S. Maki, H. Niwa, H. Ikeda and T. Hirano, *Tetrahedron*, 2005, **61**, 10073.

12 T. Mori and H. Inokuchi, *Solid State Commun.*, 1986, **59**, 355; T. Mori and H. Inokuchi, *Bull. Chem. Soc. Jpn.*, 1987, **60**, 402.

13 M. Kawa and J. M. J. Frechet, *Chem. Mater.*, 1998, **10**, 286; B. Alpha, R. Ballardini, V. Balzani, J.-M. Lehn, S. Perathoner and N. Sabbatini, *Photochem. Photobiol.*, 1990, **52**, 299.

14 Preliminary results. Triclinic, $P\bar{1}$, a = 13.761(8), b = 17.80(1), c = 18.69(1) Å, α = 64.73(4), β = 70.20(4), γ = 67.47(5)°, V = 3740(3) Å^3, Z = 4, and R_1 = 0.137.

15 *dl*-(EtOPhIP)$_2$: ^{1}H NMR (500 MHz, DMSO-d_6) δ 1.08 (t, 6H, J = 7 Hz), 3.46 – 3.52 (m, 4H), 3.56 – 3.62 (m, 2H), 4.44 (pseudo-s, 2H), 5.15 (d, 2H, J = 4 Hz), 6.92 (d, 2H, J = 6 Hz), 7.27 – 7.29 (m, 6H), 8.04 – 8.06 (m, 4H), 8.61 (dd, 2H, J = 4, 6 Hz). ^{13}C NMR (126 MHz, DMSO-d_6) δ 14.86, 53.22, 61.93, 78.16, 122.11, 125.96, 127.95, 128.38, 129.03, 132.39, 144.66, 160.46. *meso*-(EtOPhIP)$_2$: ^{1}H NMR (270 MHz, DMSO) δ 1.09 (t, 6H, J = 7 Hz), 3.71 – 3.82 (m, 4H), 4.49 (pseudo-s, 2H), 4.95 (d, 2H, J = 3 Hz), 7.09 (d, 2H, J = 6 Hz), 7.37 – 7.48 (m, 6H), 8.26 (d, 4H, J = 7 Hz), 8.94 (m, 2H, J = 3, 6 Hz). ^{1}H signals due to N-*H* were missing in CDCl$_3$. The negligible coupling constants $J_{H5\text{-}H6}$ ($J_{H5'\text{-}H6'}$) are consistent with *trans*-configuration with respect to substitution at the C5 and C6 (C5' and C6') positions.

16 Trigonal, $R\bar{3}$, a = 25.36(5), c = 12.41(2) Å, V = 6915(21) Å^3, Z = 9, d_{calc} = 1.168 g cm^{-3}, μ(Mo Kα) = 0.080 mm^{-1}, T = 90 K, R_1 = 0.079. CCDC reference number 291389.

17 *meso*-(*i*-PrOPhIP)$_2$: ^{1}H NMR (500 MHz, DMSO-d_6) δ 1.03 (d, 12H, J = 6 Hz), 3.80 (septet, 2H, J = 6 Hz), 4.43 (pseudo-s, 2H), 5.04 (d, 2H, J = 3 Hz), 7.10 (d, 2H, J = 6 Hz), 7.37 (t, 2H, J = 8 Hz), 7.44 (t, 4H, J = 8 Hz), 8.26 (d, 4H, J = 8 Hz), 8.93 (dd, 2H, J = 3, 6 Hz).

18 Y. Toya, S.-i. Nakatsuka and T. Goto, *Tetrahedron Lett.*, 1985, **26**, 239.

19 Y. Toya, S.-i. Nakatsuka and T. Goto, *Tetrahedron Lett.*, 1983, **24**, 5753.

ONE-DIMENSIONAL ALTERNATING ANTIFERROMAGNETIC CHAIN ACCOMPANYING CHARGE ORDERING IN DODHT ORGANIC SUPERCONDUCTORS

H. Nishikawa,[1] A. Machida,[2] T. Kodama,[2] I. Ikemoto,[2] K. Kikuchi,[2] J. Yamada,[3] and H. Oshio[1]

[1]Graduate School of Pure and Applied Sciences, University of Tsukuba, 1-1-1 Tennoudai, Tsukuba, Ibaraki 305-8571, Japan. E-mail: nishikaw@chem.tsukuba.ac.jp
[2]Department of Chemistry, Graduate School of Science, Tokyo Metropolitan University, Hachiohji, Tokyo 192-0397, Japan
[3]Department of Material Science, Graduate School of Material Science, University of Hyogo, Ako-gun, Hyogo 678-1297, Japan

1 INTRODUCTION

The relationship between charge ordering (CO) and superconductivity is of intense current interest in a wide range of strongly correlated electron materials both in inorganic and organic systems. There are several theoretical works suggesting that charge fluctuation mediates the superconductivity in organic conductors with quarter-filled band structure.[1] We are continuously investigating organic conductors based on reduced π-electron donors, and found pressure-induced superconducting salts from DODHT [(1,4-dioxane-2,3-diyl-dithio)dihydrotetrathiafulvalene].[2] All superconduting salts (PF_6, AsF_6, and $BF_4 \cdot H_2O$ salts) crystallize in β''-type structure and possess electronic structure with one open quasi-one-dimensional Fermi surface and one two-dimensional Fermi pocket. In these salts, the strong electron correlation is expected because of the enhancement of on-site Coulomb interaction, U, and decrease of intermolecular transfer integrals, t. As a result, the ground state of β''-$(DODHT)_2PF_6$ at ambient pressure corresponds to CO insulator.[3] On the other hand, SbF_6 and $Au(CN)_2$ salts do not exhibit superconducting transition up to pressure of 17 kbar. In this paper, we describe the resistive behavior of the SbF_6 and $Au(CN)_2$ salts under pressure and the magnetic properties of the superconducting and SbF_6 salts.

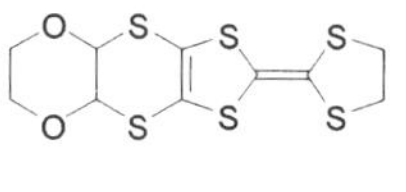

Figure 1 *Molecular Structure of DODHT*

2 EXPERIMENTS

Preparation of DODHT salts was carried out by the electrochemical oxidation at a constant current. Temperature dependence of resistivity within the two-dimensional conducting plane was measured by a four-probe dc method under various pressures of up to 17 kbar. Magnetic susceptibility was measured for a powder sample at ambient pressure in the temperature range of 2 – 300 K using a superconducting quantum interference device (SQUID) magnetometer with 10000 Oe field application. The diamagnetic contribution of DODHT was experimentally determined to be 1.52×10^{-4} emu mol^{-1} and those of counter anions were calculated using Pascal's law. Collection of diffraction data for X-ray crystallographic analyses was carried out using a Bruker SMART APEX three-circle diffractometer equipped with a CCD area detector and a full sphere of data was collected using graphite-monochromated Mo-Kα radiation, λ = 0.71073 Å. The data can be obtained free of charge from The Cambridge Cryatallographic Data Centre via www.ccdc.cam.ac.uk /data_request/cif.

3 RESULTS AND DISCUSSION

3.1 Electrical Conductivity of β''-(DODHT)$_2$X [X = SbF$_6$ and Au(CN)$_2$]

Temperature dependence of resistivity of β''-(DODHT)$_2$SbF$_6$ at ambient pressure was reported previously[4], which is fairly different from that of Au(CN)$_2$ salts; the temperature dependence of resistivity of SbF$_6$ salt is semiconductive without any distinct anomaly, on the other hand, Au(CN)$_2$ salt shows anomaly at 285 K as shown in Figure 2 (b), which corresponds to the insulator transition similarly to the superconducting DODHT salts. Figure 2 shows the resistivity as a function of temperature under various pressures of up to 17.6 kbar for the SbF$_6$ salt and 15.3 kbar for the Au(CN)$_2$ salt, respectively. The resistivity of SbF$_6$ salt at ambient pressure can be analyzed by activation-type conduction in the temperature range between 300 and 240 K to be 0.18 eV. The electrical conductivity at 300 K is 11 Scm^{-1}, which increases with applying pressure. The activation energy decreases with increasing pressure and the temperature dependence is very weak down to 100 K at 17.6 kbar. No metallic behaviour, however, was observed at least up to 17.6 kbar. In contrast to the resistive behavior of SbF$_6$ salt, temperature dependence of the resistivity of Au(CN)$_2$ salt at ambient pressure can not described by activation-type plot. With applying pressure, the resistive behavior turns to be activation-type conduction and the resistivtiy is fitted by Arrhenius plot in temperature range below 190 K. The activation energy decreases from 0.28 eV at 4.6 kbar to 0.19 eV at 15.3 kbar, however, no metallic reagion was observed up to the pressure of 15.3 kbar in this salt, either.

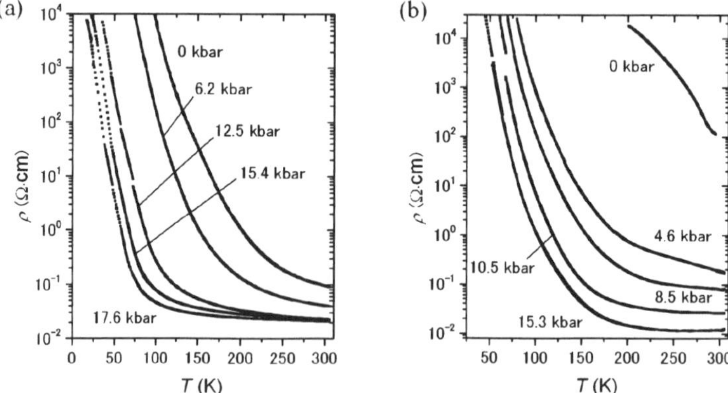

Figure 2   *Temperature dependence of resistivity of (a) SbF$_6$ salt and (b) Au(CN)$_2$ salt under various pressures.*

3.2 Magnetic Susceptibility of β''-(DODHT)$_2$X [X = BF$_4$·H$_2$O, PF$_6$ and SbF$_6$]

The magnetic susceptibility as a function of temperature of BF$_4$·H$_2$O, PF$_6$, and SbF$_6$ salts is shown in Figure 3. The magnetic susceptibility data of superconducting salts, BF$_4$·H$_2$O and PF$_6$ salts, are reproduced by one-dimensional $S = 1/2$ alternating Heisenberg chain model as shown by the solid lines in Figure 3. X-ray diffraction experiment for PF$_6$ salt proves the emergence of twofold superlattice along a-axis below the insulator transition temperature. These results indicate the CO with the stripe-type charge pattern accompanied by one-dimensional antiferromagnetic chain as shown in Figure 3. On the other hand, the magnetic susceptibility of SbF$_6$ salt at 300 K is larger ($\chi_m = 1.8 \times 10^{-3}$ emu mol^{-1}) than those of superconducting salts and increases slightly and monotonically down to approximately 30 K with decreasing temperature. This result suggests that the insulating state of SbF$_6$ salt does not seem to be the CO insulator unlike the case of superconducting salts. This fact corresponds to the different resistive behavior of SbF$_6$ salt as shown in Figure 1.

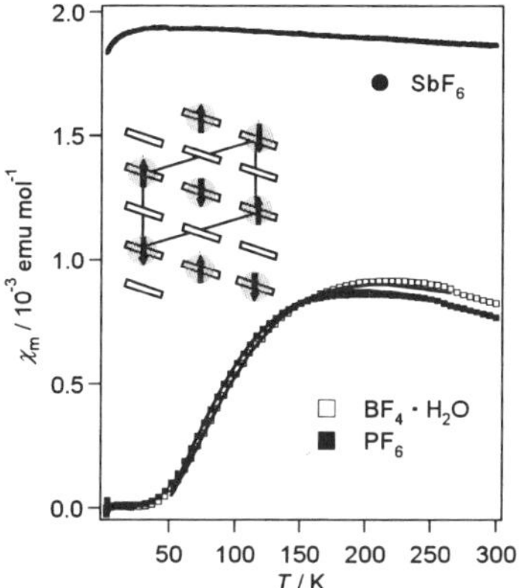

Figure 3   *Temperature dependence of magnetic susceptibility of BF$_4$·H$_2$O, PF$_6$, and SbF$_6$ salts, and the schematic diagram of CO pattern and spin interaction for BF$_4$·H$_2$O and PF$_6$ salts. The parallelogram, gray circles, and arrows represent the unit cell, the charge rich site, and the magnetic moments, respectively.*

3.3 Crystal structure of β''-(DODHT)$_2$Au(CN)$_2$

While the crystal structure of SbF$_6$ salt is isostructural to the superconducting salts, that of Au(CN)$_2$ salt is slightly different in terms of the space group although the donor arrangement is also β''-type; the space group of Au(CN)$_2$ salt is $P\bar{1}$ while all the other salts crystallize in $P1$ and two independent molecules exist in the unit cell unlike the DODHT salts with octahedral anions. Figure 4 shows the crystal structure, band structure and Fermi surface[5] of β''-(DODHT)$_2$Au(CN)$_2$. Similarly to the other DODHT salts, intermolecular sulfur contacts are observed only between the columns, and the intermolecular overlap integrals along molecular side-by-side direction are larger than those within the stack. However interactions along c direction are larger than those along p-q direction in contrast to the superconducting salts where the interaction along c direction is comparable to those along p-q direction. As a result, the band structure and Fermi surface of Au(CN)$_2$ salt are fairly different from those of superconducting salts; the Fermi surfaces are composed of two quasi-one-dimensional open Fermi surfaces. This type of electronic structure usually provides nesting of the Fermi surfaces, thus the origin of the insulating state of $Au(CN)_2$ salt may be the nesting rather than CO in spite of the large U of DODHT or consequently large inter-site Coulomb energy, V, in the DODHT salts.

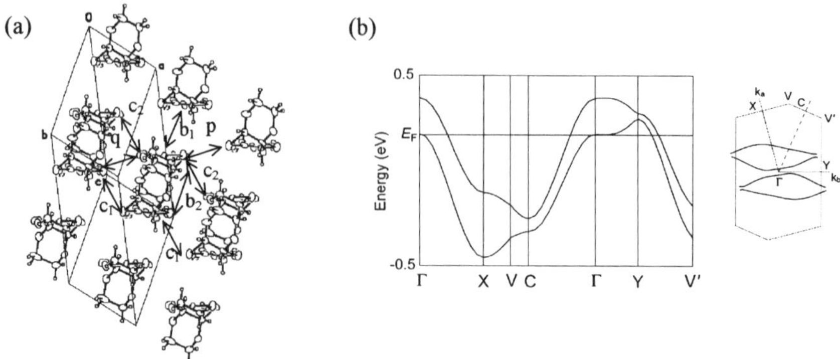

Figure 4   *(a) Crystal structure and intermolecular overlap integrals of Au(CN)$_2$ salt ($\times 10^{-3}$) b_1, b_2, p, q, c_1, and c_2 are -2.42, -1.39, 5.44, 7.81, -14.12, and -14.57, respecrively. (b) Band structure and Fermi surface of β''-(DODHT)$_2$Au(CN)$_2$.*

4 CONCLUSION

The resistivities under various pressures were investigated for β''-(DODHT)$_2$X [X = SbF$_6$ and Au(CN)$_2$]. Both the salts did not exhibit superconducting transition under the pressure of up to15 - 17 kbar. Magnetic susceptibility measurement for SbF$_6$ salt did not indicate the CO insulating state in contrast to the ground state at ambient pressure of the superconducting salts which corresponds to CO insulator with alternating antiferromagnetic chain. Although the crystal structure of Au(CN)$_2$ salt is also β''-type, this salt has a different Fermi surface which generally has a tendency to the nesting of Fermi surface. In conclusion, the CO state seems unlikely to realize in both the salts. Thus, the CO should play an important role in the emergence of superconductivity of the DODHT salts.

References

1   J. Merino and R. H. McKenzie, *Phys. Rev. Lett.*, 2001, **87**, 237002; A. Kobayashi, Y. Tanaka, M. Ogata and Y. Suzumura, *J. Phys. Soc. Jpn.*, 2004, **73**, 1115; Y. Tanaka, Y. Yanase and M. Ogata, *J. Phys. Soc. Jpn.*, 2004, **73**, 2053.

2   H. Nishiakwa, T. Morimoto, T. Kodama, I. Ikemoto, K. Kikuchi, J. Yamada, H. Yoshino and K. Murata, *J. Am. Chem. Soc.*, 2002, **124**, 730; H. Nishiakwa, A. Machida, T. Morimoto, K. Kikuchi, T. Kodama, I. Ikemoto, J. Yamada, H. Yoshino and K. Murata, *Chem. Commun.*, 2003, 494.

3   H. Nishiakwa, Y. Sato, K. Kikuch, T. Kodama, I. Ikemoto, J. Yamada, H. Oshio, R. Kondo and S. Kagoshima, *Phys. Rev. B*, 2005, **72**, 052510-1.

4   H. Nishiakwa, T. Morimoto, T. Kodama, I. Ikemoto, K. Kikuchi, J. Yamada, H. Yoshino and K. Murata, *Synth. Met.*, 2003, **133-134**, 193.

5   T. Mori, A. Kobayshi, Y. Sasaki, H. Kobayashi, G. Saito and H. Inokkuchi, *Bull. Chem. Soc. Jpn.*, 1984, **57**, 627.

DIMETHYL-SUBSTITUTED ANALOGUE OF BDH-TTP, DMDH-TTP AND ITS METALLIC SALTS

J. Yamada,[1] H. Song,[1] H. Akutsu,[1] S. Nakatsuji[1] and K. Kikuchi[2]

[1]Department of Material Science, Graduate School of Material Science, University of Hyogo, 3-2-1 Kouto, Kamigori-cho, Ako-gun, Hyogo 678-1297, Japan. E-mail: yamada@sci.u-hyogo.ac.jp
[2]Department of Chemistry, Graduate School of Science, Tokyo Metropolitan University, Hachioji, Tokyo 192-0397, Japan

1 INTRODUCTION

The achievement of organic superconductivity is undoubtedly a major topic in the field of molecular conductors.[1] We have found that the π-electron donor BDA-TTP [2,5-bis(1,3-dithian-2-ylidene)-1,3,4,6-tetrathiapentalene] can produce the ambient-pressure and pressure-induced superconducting salts.[2] The molecular design of BDA-TTP is based on the destabilization of the stable metallic state that occurs with the use of BDH-TTP [2,5-bis(1,3-dithian-2-ylidene)-1,3,4,6-tetrathiapentalene] by extending the σ-framework.[3] Actually, compared to rigid outer dithiolane rings of BDH-TTP, BDA-TTP has more expanded outer dithiane rings capable of adopting various kinds of chair conformations. Therefore, the replacement of two outer dithiolane rings in BDH-TTP with dithiane rings would be regarded as an increase of steric bulk on the periphery of BDH-TTP.

As another structural modification to BDH-TTP in this respect, we focused our next attention on the attachment of two methyl groups to one outer dithiolane ring of BDH-TTP, and accomplished the synthesis of DMDH-TTP [2-(4,5-dimethyl-1,3-dithiolan-2-ylidene)-5-(1,3-dithiolan-2-ylidene)-1,3,4,6-tetrathiapentalene] as a diastereomeric mixture of *meso*- and *trans*-forms.[4] In addition, we have revealed that despite the presence of two methyl groups, DMDH-TTP with the AsF_6 and BF_4 anions forms metallic salts stable down to liquid helium temperature (4.2 K), and that the metallic (DMDH-TTP)$_2$AsF$_6$ salt contains only *meso*-DMDH-TTP molecule as donor components, although a mixture of *meso*- and *trans*-forms of DMDH-TTP was used for preparation of the AsF$_6$ salt by electrocrystallization. Herein we report on the physical properties of other DMDH-TTP salts and also on the structural aspects of the BF$_4$ and PF$_6$ salts.

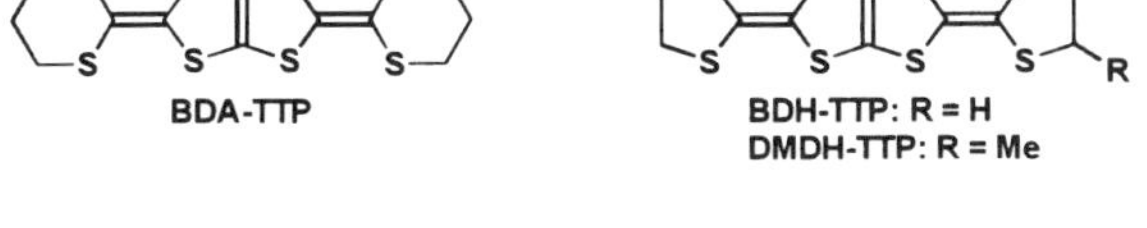

Figure 1 *Structures of BDA-TTP, BDH-TTP and DMDH-TTP*

2 METHOD AND RESULTS

2.1 Preparation and Physical Properties

The synthesis of DMDH-TTP was carried out as previously reported.[4] Controlled-current electrocrystallization[5] of a diastereomeric mixture of DMDH-TTP with n-Bu$_4$NX (X = I$_3$, AuI$_2$ and PF$_6$) in 1,1,2-trichloroethane or with Et$_4$NX (X = GaCl$_4$ and FeCl$_4$) in 5% acetone/PhCl gave the corresponding DMDH-TTP salts. Variable-temperature conductivity measurements were performed on single crystals by a four-probe technique. Magnetic susceptibility measurement was made on non-oriented crystals from 2 to 300 K.

The I$_3^-$ anion with DMDH-TTP formed two kinds of salts: a semiconducting salt with a room-temperature conductivity (σ_{rt}) of 7.7×10^{-5} S cm^{-1} and a metallic salt with $\sigma_{rt} = 88$ S cm^{-1} (Figure 2a). The resistivities of the AuI$_2$ and GaCl$_4$ salts ($\sigma_{rt} = 15$ and 36 S cm^{-1}, respectively) exhibited metallic behaviour from room temperature and slight increases below about 10 and 12 K, respectively (Figure 2b). As shown in Figure 2c, the FeCl$_4$ and PF$_6$ salts ($\sigma_{rt} = 26$ and 32 S cm^{-1}, respectively) retained the metallic state down to 4.2 K. The susceptibility of the FeCl$_4$ salt obeyed the Curie-Weiss law, giving $C = 4.35$ emu K mol^{-1} and $\theta = 0.65$ K. The small positive θ implies a very weak ferromagnetic interaction.

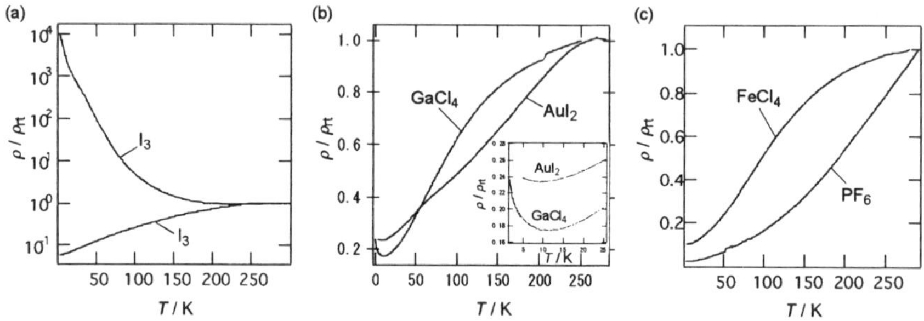

Figure 2 *Temperature dependence of the relative resistivities for the DMDH-TTP salts; (a) two I$_3$ salts, (b) the AuI$_2$ and GaCl$_4$ salts and (c) the FeCl$_4$ and PF$_6$ salts*

2.2 Crystal and Electronic Structures

While we have been unable to obtain good single crystals of the I$_3$, AuI$_2$, GaCl$_4$ and FeCl$_4$ salts of DMDH-TTP suitable for an X-ray diffraction study, the structures of the BF$_4$ and PF$_6$ salts were determined by X-ray analyses.[6,7]

The BF$_4$ salt consists of alternating layers containing only *meso*-DMDH-TTP donor molecule and only BF$_4^-$ anion (Figure 3a). The BF$_4^-$ anion has a large thermal motion around the boron atom and exhibits positional disorder. There are two crystallographically discrete *meso*-DMDH-TTP molecules in the asymmetric unit, which are stacked along the [01$\bar{1}$] direction in a head-to-tail manner with a four-folded period being 3.87, 3.75 and 3.61 Å apart. Accordingly, the donor molecule packing of this salt belongs to the λ-type category. Intermolecular S$\cdots$S contacts shorter than the van der Waals distance (3.70 Å) are observed among three consecutive donor molecules in the four-fold period as well as between donor stacks (Figure 3b). The overlap integrals between the HOMO's of neighbouring donor molecules were calculated on the basis of the extended Hückel method.[8] Compared to the absolute values of the interstack overlap integrals, larger absolute values are estimated for all the intrastack overlap integrals. The tight-binding band calculation led to the energy dispersion relation and the Fermi surfaces shown in Figures 3c,d, respectively. The two highest bands are partially filled, and the Fermi surfaces associated with these bands consist of a pair of wave-like lines and a closed rhombus-like loop centred at Γ. Thus, this salt has both one-dimensional and two-dimensional (2D) Fermi surfaces, and the 2D hole surface would result in the metallic conductivity down to 4.2 K.

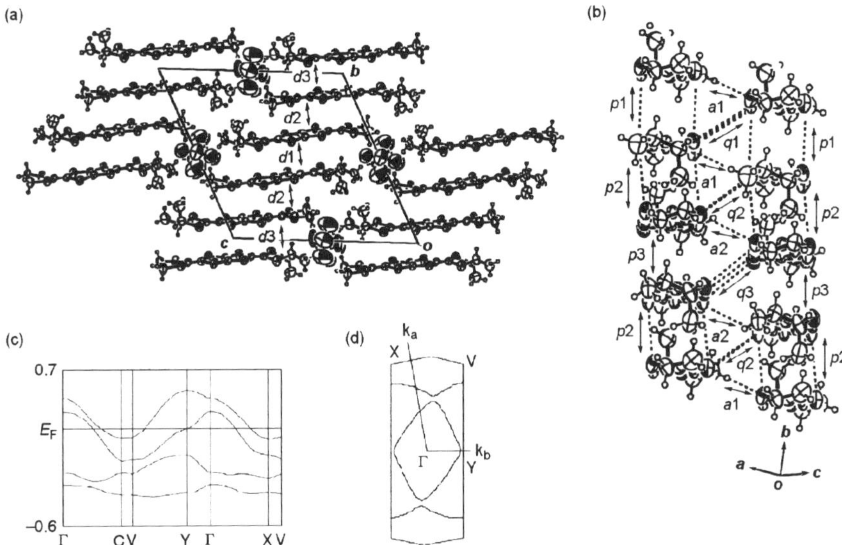

Figure 3 *(a) Crystal structure of λ-(meso-DMDH-TTP)$_2$BF$_4$ viewed along the a-axis. Interplanar distances in the donor column are 3.87 (d1), 3.75 (d2) and 3.61 (d3) Å (b) Donor arrangement in λ-(meso-DMDH-TTP)$_2$BF$_4$. Short S···S contacts [3.583(5)–3.695(4) Å] are shown by broken lines. Intermolecular overlap integrals p1, p2, p3, a1, a2, q1, q2 and q3 are 15.5, –16.5, 14.7, –4.93, –5.35, 5.17, –9.66 and –6.63 × 10^{-3}, respectively. (c) Energy band structure of λ-(meso-DMDH-TTP)$_2$BF$_4$. (d) Fermi surfaces of λ-(meso-DMDH-TTP)$_2$BF$_4$. The wave-like lines are electron surfaces, whereas the closed rhombus-like loop is a hole surface.*

According to structural information on the PF$_6$ salt, the salt also involves only *meso*-DMDH-TTP donor molecule and is isostructural to (*meso*-DMDH-TTP)$_2$AsF$_6$. Therefore, the structural features as found in the AsF$_6$ salt[4] hold good for the PF$_6$ salt.

3 CONCLUSION

As a structural modification to BDH-TTP, the introduction of two methyl groups into the periphery of BDH-TTP has not yet led to the achievement of superconductivity. In the present paper, we described only the utilization of a diastereomeric mixture of DMDH-TTP, and our successful X-ray studies suggested that *meso*-DMDH-TTP, like BDH-TTP, has the ability to form stable metallic salts. On the other hand, the intriguing question of what electronic state is brought about by the use of *trans* [(*S,S*) or (*R,R*)]-DMDH-TTP remains unsolved. Work currently in progress is addressing this issue.

This work was supported by MEXT (No. 15073102) and JSPS (No. 17550135).

References and Notes

1   J.M. Williams, J.R. Ferraro, R.J. Thorn, K.D. Carlson, U. Geiser, H.H. Wang, A.M. Kini and M.-H. Whangbo, *Organic Superconductors (Including Fullerenes) – Synthesis, Structure, Properties, and Theory*, Prentice Hall, Englewood Cliffs, NJ, 1992; T. Ishiguro, K. Yamaji and G. Saito, *Organic Superconductors*, 2nd edn., ed. P. Fulde, Springer, Berlin, 1998; *TTF Chemistry – Fundamentals and Applications of Tetrathiafulvalene*, eds. J. Yamada and T. Sugimoto, Kodansha & Springer, Tokyo, 2004; thematic issue on *Molecular Conductors*, *Chem. Rev.*, 2004, **104**.

2   J. Yamada, M. Watanabe, H. Akutsu, S. Nakatsuji, H. Nishikawa, I. Ikemoto and K. Kikuchi, *J. Am. Chem. Soc.*, 2001, **123**, 4174; J. Yamada, T. Toita, H. Akutsu, S. Nakatsuji, H. Nishikawa, I. Ikemoto, K. Kikuchi, E.S. Choi, D. Graf and J.S. Brooks, *Chem. Commun.*, 2003, 2230; E.S. Choi, D. Graf, J.S. Brooks, J. Yamada, H. Akutsu, K. Kikuchi and M. Tokumoto, *Phys. Rev. B*, 2004, **70**, 024517.

3   J. Yamada, H. Nishikawa and K. Kikuchi, in *TTF Chemistry – Fundamentals and Applications of Tetrathiafulvalene*, eds. J. Yamada and T. Sugimoto, Kodansha & Springer, Tokyo, 2004, Chapter 11, pp. 261–286; J. Yamada, *J. Mater. Chem.*, 2004, **14**, 2951; J. Yamada, H. Nishikawa, H. Akutsu and K. Kikuchi, *Chem. Rev.*, 2004, **104**, 5057.

4   J. Yamada, H. Song, H. Akutsu, S. Nakatsuji and K. Kikuchi, *Chem. Lett.*, 2005, **34**, 1404.

5   H. Anzai, J.M. Delrieu, S. Takasaki, S. Nakatsuji and J. Yamada, *J. Cryst. Growth*, 1995, **154**, 145; H. Nishikawa, T. Sato, T. Kodama, I. Ikemoto, K. Kikuchi, H. Anzai and J. Yamada, *J. Mater. Chem.*, 1999, **9**, 693.

6   *Crystal data* for (*meso*-DMDH-TTP)$_2$BF$_4$: $C_{24}H_{24}BF_4S_{16}$, M_r = 912.22, triclinic, space group $P\bar{1}$, a = 6.3678(4), b = 17.740(2), c = 17.827(4) Å, α = 64.246(6)°, β = 86.671(3)°, γ = 79.250(2)°, V = 1781.3(4) Å^3, Z = 2, ρ_{calcd} = 1.701 g cm^{-3}, μ(Mo Kα) = 1.0118 mm^{-1}, T = 296 K, R = 0.128 and R_w = 0.113 for 3654 observed reflections with $I > 3\sigma(I)$ from 7308 unique reflections.

7   *Crystal data* for (*meso*-DMDH-TTP)$_2$AsF$_6$: $C_{24}H_{24}PF_6S_{16}$, M_r = 970.47, triclinic, space group $P\bar{1}$, a = 6.4412(10), b = 8.8901(7), c = 32.213(6) Å, α = 93.700(4)°, β = 93.6419(14)°, γ = 97.6858(15)°, V = 1819.4(5) Å^3, Z = 2, ρ_{calcd} = 1.77 g cm^{-3}, μ(Mo Kα) = 1.047 mm^{-1}, T = 293 K, R = 0.126 and R_w = 0.108 for 3669 observed reflections with $I > 3\sigma(I)$ from 7314 unique reflections.

8   T. Mori, A. Kobayashi, Y. Sasaki, H. Kobayashi, G. Saito and H. Inokuchi, *Bull. Chem. Soc. Jpn.*, 1984, **57**, 627.

MONOOXYGEN-CONTAINING ANALOGUES OF DHDA-TTP, DHOTA-TTP AND OTDA-TTP AND THEIR CHARGE-TRANSFER SALTS

J. Yamada,[1] N. Hiratani,[1] K. Kunigita,[1] H. Akutsu,[1] S. Nakatsuji[1] and K. Kikuchi[2]

[1]Department of Material Science, Graduate School of Material Science, University of Hyogo, 3-2-1 Kouto, Kamigori-cho, Ako-gun, Hyogo 678-1297, Japan. E-mail: yamada@sci.u-hyogo.ac.jp
[2]Department of Chemistry, Graduate School of Science, Tokyo Metropolitan University, Hachioji, Tokyo 192-0397, Japan

1 INTRODUCTION

We have reported on the synthesis of DHOT-TTP [2-(1,3-dithiolan-2-ylidene)-5-(1,3-oxathiolan-2-ylidene)-1,3,4,6-tetrathiapentalene], in which one sulfur atom in the outer dithiolane ring of BDH-TTP [2,5-bis(1,3-dithiolan-2-ylidene)-1,3,4,6-tetrathiapentalene] is substituted by oxygen, and also on the physical and structural properties of the metallic α-(DHOT-TTP)$_2$(AuI$_2$)$_{0.87}$ and κ-(DHOT-TTP)$_2$X (X = PF$_6$, AsF$_6$ and FeCl$_4$) salts.[1,2] Through these studies, we have elucidated that the substitution of only one sulfur atom with oxygen holds the following two conflicting effects: (i) a decrease of interaction between donor molecules, which is caused by a very small HOMO coefficient on the oxygen atom,[1] and (ii) an increase of chemical pressure, which originates from the smaller size of the oxygen atom.[2] Additionally, our study of the DHDA-TTP donor, in which one outer dithiolane ring of BDH-TTP is replaced by a dithiane ring, has shown that despite of the introduction of a more expanded six-membered ring system, DHDA-TTP inherits to some extent the ability to generate the stable metallic state from BDH-TTP.[3]

Meanwhile, our design of π-donors leading to organic superconductors has evolved from the destabilization of the stable metallic state.[4] It was thus of continuous interest to reveal what electronic states occur with the use of π-donors that are designed on the basis of a combination of the aforementioned chalcogen substitution and expansion of σ-framework, because the settlement of this issue would provide further information on our molecular design strategy for destabilizing the stable metallic state. We describe here the study of two monooxygen-substituted derivatives of DHDA-TTP, DHOTA-TTP [2-(1,3-dithiolan-2-ylidene)-5-(1,3-oxathian-2-ylidene)-1,3,4,6-tetrathiapentalene] and OTDA-TTP [2-(1,3-oxathiolan-2-ylidene)-5-(1,3-dithian-2-ylidene)-1,3,4,6-tetrathiapentalene].

BDH-TTP: X = S
DHOT-TTP: X = O

DHDA-TTP: X = Y = S
DHOTA-TTP: X = O, Y = S
OTDA-TTP: X = S, Y = O

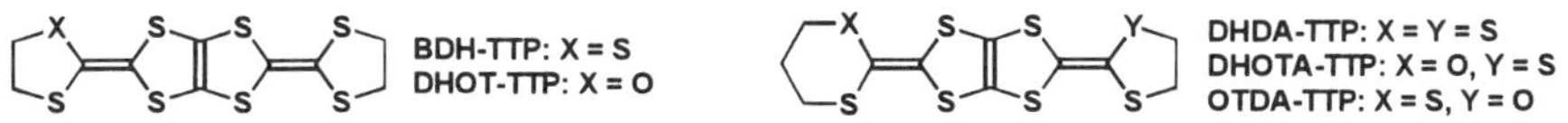

Figure 1 *Structures of BDH-TTP, DHOT-TTP, DHDA-TTP, DHOTA-TTP and OTDA-TTP*

2 METHOD AND RESULTS

2.1 Synthesis and Molecular Structures

Similarly to the synthetic route to DHOT-TTP,[1] the dimethyl acetal-appended oxone **1** was utilized as starting material for the synthesis of DHOTA-TTP (Figure 2a). Reaction of **1** with 3-mercaptopropanol in benzene containing a catalytic amount of *p*-TsOH at reflux gave oxone **2** in 78% yield. Subsequently, treatment of **2** with MeMgBr followed by trapping with Cl_2SnBu_2 in THF afforded tin dithiolate **3**, which was immediately coupled with 2-ethoxycarbonyl-1,3-dithiolane in the presence of Me_3Al (2 equiv.) in CH_2Cl_2 to give compound **4** as a precursor of DHDA-TTP in 48% overall yield. Finally, oxidation of **4** using DDQ in refluxing toluene furnished DHOTA-TTP in 86% yield.

The synthesis of OTDA-TTP began with the oxathiolane-attached oxone **5**[1] and proceeded via reactions similar to those used for the construction of DHOTA-TTP, as outlined in Figure 2b. Thus, conversion of **5** to tin dithiolate **6** via sequential treatment with MeMgBr and Cl_2SnBu_2 in THF followed by the Me_3Al (2 equiv.)-promoted reaction with 2-ethoxycarbonyl-1,3-dithiane in CH_2Cl_2 gave the coupling product **7** in 41% overall yield. DDQ oxidation of **7** in toluene at reflux led to OTDA-TTP in 74% yield.

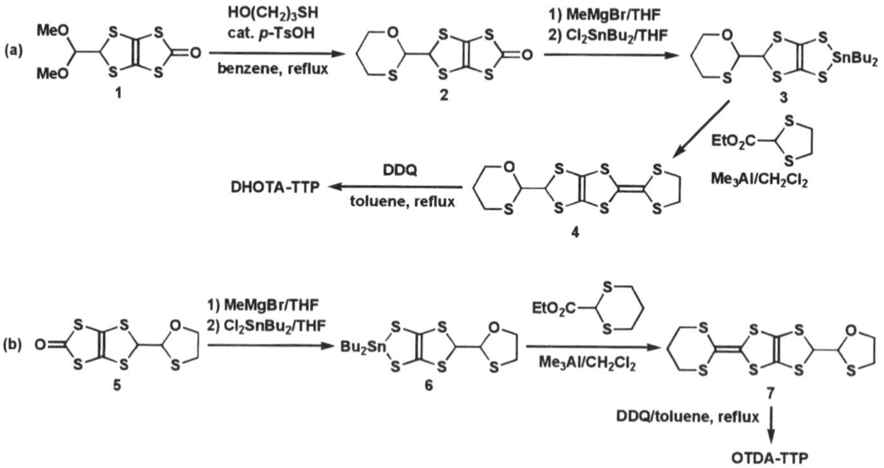

Figure 2 *Synthetic routes to (a) DHOTA-TTP and (b) OTDA-TTP*

In order to gain an insight into the conformational change induced by the introduced oxygen atom, we carried out X-ray diffraction studies of DHDA-TTP, DHOTA-TTP and OTDA-TTP in the neutral state.[5–7] Surprisingly, in contrast to the conformational change from the planar structure of BDH-TTP to the nonplanar structure of DHOT-TTP,[1] the molecular structures of DHDA-TTP, DHOTA-TTP and OTDA-TTP are all isomorphous. The DHOTA-TTP and OTDA-TTP molecules have orientational disorder between the oxygen and sulfur atoms in the oxathiane and oxathiolane rings, respectively. Population analyses suggested that the respective occupancy factors for the oxygen atoms in DHOTA-TTP and OTDA-TTP are 0.65 and 0.76. As shown in Figure 3, the structural features common to the present three donor molecules are that (i) the outer dichalcogenane ring adopts a chair conformation, (ii) the central tetrathioethylene moiety and the external tetrachalcogenoethylene moiety linking the trimethylene group form a common plane and (iii) the trimethlene end group and another external tetrachalcogenoethylene moiety linking the ethylene group are far out of the common plane in opposite directions. Electrochemical properties of DHDA-TTP and OTDA-TTP will be reported separately.

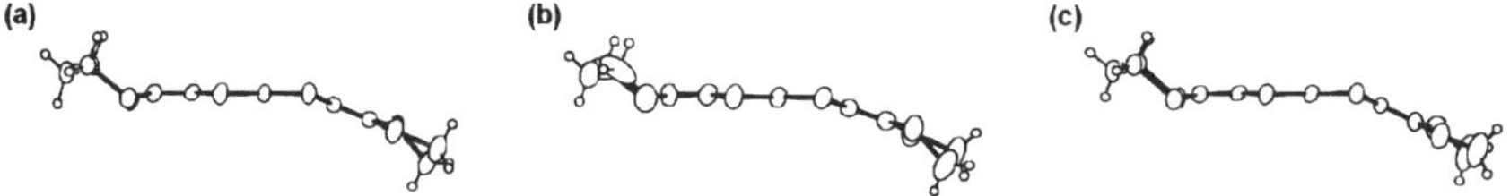

Figure 3 *Side views of the molecular structures of (a) DHDA-TTP, (b) DHOTA-TTP and (c) OTDA-TTP*

2.2 Electrical Conducting Behaviour

As reported by us earlier,[3] β-(DHDA-TTP)$_2$X (X = PF$_6$ and AsF$_6$) undergo metal-insulator transitions at about 30 and 60 K, respectively (Figure 4a). Thus, to compare with the electrical transport and structural properties of these DHDA-TTP salts, we prepared single crystals of the corresponding DHOTA-TTP and OTDA-TTP salts by controlled-current electrocrystallization[8] with *n*-Bu$_4$NX (X = PF$_6$ and AsF$_6$) in 1,1,2-trichloroethane. Variable-temperature conductivity measurements were made by a four-probe technique.

The PF$_6$ and AsF$_6$ salts of DHOTA-TTP exhibited room-temperature conductivities (σ_{rt}'s) of 7.5 and 1.7 S cm^{-1} and weak metallic behaviour down to around 250 and 270 K, respectively, below which temperatures both salts behaved as semiconductors (Figure 4b). X-ray diffraction analyses of these salts indicated that they crystallize isostructurally and have the κ-type donor arrangement.[9] On the other hand, the PF$_6$ [σ_{rt} = 2.5 S cm^{-1}, E_a (activation energy) = 26 meV (291–200 K)] and AsF$_6$ [σ_{rt} = 1.9 S cm^{-1}, E_a = 9 meV (290–167 K)] salts of OTDA-TTP were semiconductors as illustrated in Figure 4c by the Arrhenius plots of the resistivity as a function of temperature. By X-ray analyses, these salts were found to have isomorphous structures with the β″-type donor arrangements.[9]

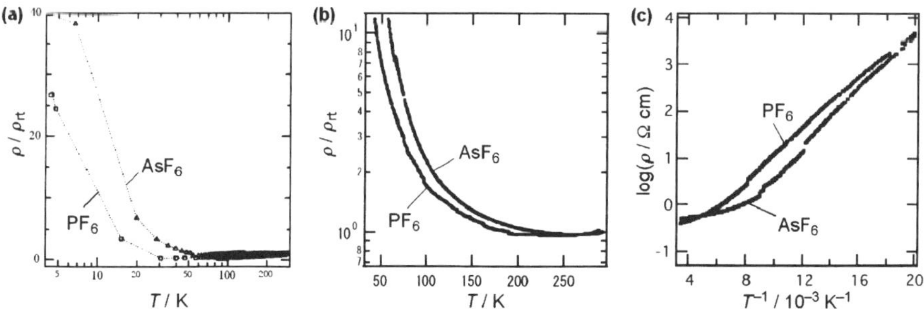

Figure 4 *Temperature dependence of the relative resistivities for (a) β-(DHDA-TTP)$_2$X and (b) κ-(DHOTA-TTP)$_2$X, where X = PF$_6$ and AsF$_6$. (c) Arrhenius plots for β″-(OTDA-TTP)$_2$X (X = PF$_6$ and AsF$_6$).*

3 CONCLUSION

We have accomplished the synthesis of DHOTA-TTP and OTDA-TTP and revealed the effects of the introduced oxygen atom on the molecular structure in the neutral state and on the transport properties and structural aspects of derived charge-transfer (CT) salts. Although the incorporation of oxygen into either outer ring of DHDA-TTP led to almost no change in the conformation of the prototype DHDA-TTP molecule, the difference in the position of the introduced oxygen atom allowed the control of the electrical property and the molecular packing mode in the CT state. Therefore, the chalcogen substitution described herein would serve as a fine tuning of the intermolecular interaction. Such a fine tuning may play a key role in a destabilization of the stable metallic state required for realizing organic superconductivity. Further work on this is ongoing in our laboratories.

This work was supported by MEXT (No. 15073102) and JSPS (No. 17550135).

References and Notes

1   J. Yamada, K. Kunigita, H. Akutsu, S. Nakatsuji and K. Kikuchi, *Chem. Lett.*, 2005, **34**, 32.

2   J. Yamada, K. Kunigita, H. Akutsu, S. Nakatsuji and K. Kikuchi, *Chem. Lett.*, 2005, **34**, 1126.

3   J. Yamada, M. Watanabe, T. Toita, H. Akutsu, S. Nakatsuji, H. Nishikawa, I. Ikemoto and K. Kikuchi, *Chem. Commun.*, 2002, 1118.

4   J. Yamada, H. Nishikawa and K. Kikuchi, in *TTF Chemistry – Fundamentals and Applications of Tetrathiafulvalene*, eds. J. Yamada and T. Sugimoto, Kodansha & Springer, Tokyo, 2004, Chapter 11, pp. 261–286; J. Yamada, *J. Mater. Chem.*, 2004, **14**, 2951; J. Yamada, H. Nishikawa, H. Akutsu and K. Kikuchi, *Chem. Rev.*, 2004, **104**, 5057.

5   *Crystal data* for DHDA-TTP: $C_{11}H_{10}S_8$, $M_r = 398.68$, monoclinic, space group $P2_1/n$, $a = 6.416(2)$, $b = 14.231(2)$, $c = 16.933(2)$ Å, $\beta = 93.59(2)°$, $V = 1543.2(5)$ Å^3, $Z = 4$, $\rho_{calcd} = 1.716$ g cm^{-3}, μ(Mo Kα) $= 1.1371$ mm^{-1}, $T = 298$ K, 4018 reflections, of which 3701 were independent and 2156 were included in the refinement $[I > 3\sigma(I)]$, 202 parameters, $R = 0.032$, $R_w = 0.023$, GOF $= 1.490$.

6   *Crystal data* for DHOTA-TTP: $C_{11}H_{10}OS_7$, $M_r = 382.66$, monoclinic, space group $P2_1/n$, $a = 6.353(6)$, $b = 14.460(5)$, $c = 16.564(4)$ Å, $\beta = 93.90(4)°$, $V = 1518.1(16)$ Å^3, $Z = 4$, $\rho_{calcd} = 1.67$ g cm^{-3}, μ(Mo Kα) $= 1.025$ mm^{-1}, $T = 296$ K, 3954 reflections, of which 3489 were independent and 2448 were included in the refinement $[I > 3\sigma(I)]$, 173 parameters, $R = 0.066$, $R_w = 0.045$, GOF $= 1.245$.

7   *Crystal data* for OTDA-TTP: $C_{11}H_{10}OS_7$, $M_r = 382.62$, monoclinic, space group $P2_1/n$, $a = 6.380(3)$, $b = 14.278(2)$, $c = 16.610(2)$ Å, $\beta = 93.30(2)°$, $V = 1510.7(5)$ Å^3, $Z = 4$, $\rho_{calcd} = 1.682$ g cm^{-3}, μ(Mo Kα) $= 1.0294$ mm^{-1}, $T = 294$ K, 4094 reflections, of which 3631 were independent and 2531 were included in the refinement $[I > 3\sigma(I)]$, 173 parameters, $R = 0.040$, $R_w = 0.050$, GOF $= 1.800$.

8   H. Anzai, J.M. Delrieu, S. Takasaki, S. Nakatsuji and J. Yamada, *J. Cryst. Growth*, 1995, **154**, 145; H. Nishikawa, T. Sato, T. Kodama, I. Ikemoto, K. Kikuchi, H. Anzai and J. Yamada, *J. Mater. Chem.*, 1999, **9**, 693.

9   Structural details will be reported and discussed elsewhere.

TTF DERIVATIVES LINKING A DIMETHYLDIOXOLANE RING AND THEIR
CHARGE-TRANSFER SALTS

J. Yamada,[1] Y. Kuri,[1] R. Oka,[1] H. Akutsu,[1] S. Nakatsuji,[1] H. Nishikawa[2] and K. Kikuchi[2]

[1]Department of Material Science, Graduate School of Material Science, University of
Hyogo, 3-2-1 Kouto, Kamigori-cho, Ako-gun, Hyogo 678-1297, Japan. E-mail:
yamada@sci.u-hyogo.ac.jp
[2]Department of Chemistry, Graduate School of Science, Tokyo Metropolitan University,
Hachioji, Tokyo 192-0397, Japan

1 INTRODUCTION

Aiming at achieving metallic conductivity and superconductivity, extensive studies on
structural modifications of the TTF molecule have been made.[1] In this regard, our
molecular design strategy for the synthesis of new π-electron donors, including new
derivatives of TTF, has evolved from the extension of σ-bond framework.[2] For instance,
we have reported that metallic charge-transfer (CT) materials are obtainable from TTF-
based donor molecules appended with a dioxolane ring, such as DO-MET[3] and DO-
MMT.[4,5] These results contradict the preconception that the dioxolane ring almost
perpendicularly bonded to the MET and MT molecules hinders the face-to-face interaction
necessary for constructing metallic electronic structures. It is especially noteworthy that
DO-MET displays its ability to form stable metallic salts down to about 2 K.[5]

On the other hand, we have succeeded in finding a series of organic superconductors
from π-donors BDA-TTP and DODHT that are designed on the basis of a concept of
destabilizing the stable metallic state.[2] Therefore, the intriguing question of what
electronic states occur with the use of derivatives of DO-MET with a more expanded σ-
framework arose quite naturally. As a synthetic approach to this issue, we focused on the
attachment of two methyl groups to the dioxolane ring of DO-MET and accomplished the
synthesis of (R,R)-DMDO-MET.[6] In this paper, we report on the structural aspects of the
(R,R)-DMDO-MET-based salts and also on the synthesis and study of *meso*-DMDO-MET.

Figure 1 *Structures of DO-MET, DO-MMT, (R,R)-DMDO-MET and meso-DMDO-MET*

2 METHOD AND RESULTS

2.1 Crystal Structures of (*R,R*)-DMDO-MET Salts

As we reported briefly,[6] the CT salts of (*R,R*)-DMDO-MET, so far prepared by the controlled-current electrocrystallization method[7] in 1,1,2-trichloroethane (TCE) or PhCl, are all semiconductors (Table 1). Preliminary X-ray diffraction analyses of the BF_4, ClO_4 and AsF_6 salts suggested that these salts possess θ-type donor arrangements very similar to one another.[8] In the three salts, all the anions have considerable positional disorder.[9] Figures 1a,b show the donor layer and the donor column in the ClO_4 salt, respectively. (*R,R*)-DMDO-MET donor molecules are stacked in a head-to-head manner with a constant interplanar spacing of 3.54 Å (3.59 Å in the BF_4 salt, 3.51 Å in the AsF_6 salt): a similar donor stacking mode can be found in $(DO\text{-}MET)_2AsF_6$.[5] The dihedral angle between the planes formed by donor molecules on neighbouring stacks is 123.6° (124.2° in the BF_4 salt, 122.9° in the AsF_6 salt). Several intermolecular S···S contacts shorter than the van der Waals distance (3.70 Å) are observed between donor stacks, whereas there is only one short S···S contact within a donor stack. This is probably because two (*R,R*)-DMDO-MET donor molecules are mutually shifted along the molecular short axis as well as the molecular long axis so as to avoid the steric hindrance of the dimethyl-substituted dioxolane ring (Figure 1c), which would be responsible for the semiconducting behaviour.

Table 1 *Electrical conducting behaviour of the (R,R)-DMDO-TTP salts*

Anion	Solvent	$\sigma_{rt}/S\ cm^{-1a}$
I_3^-	TCE	3.0×10^{-2} (E_a = 79 meV)[b]
AuI_2^-	TCE	6.3×10^{-1} (E_a = 130 meV)
BF_4^-	TCE	2.2×10^{-4} (E_a = 380 meV)
ClO_4^-	TCE	3.3×10^{-2} (E_a = 110 meV)
PF_6^-	TCE	1.2 (E_a = 28 meV)
PF_6^-	PhCl	1.6×10^{-3} (E_a = 240 meV)
AsF_6^-	TCE	3.2×10^{-3} (E_a = 200 meV)

[b]Room-temperature conductivity measured on a single crystal by a four-probe technique unless otherwise noted. [d]Measured by a two-probe technique.

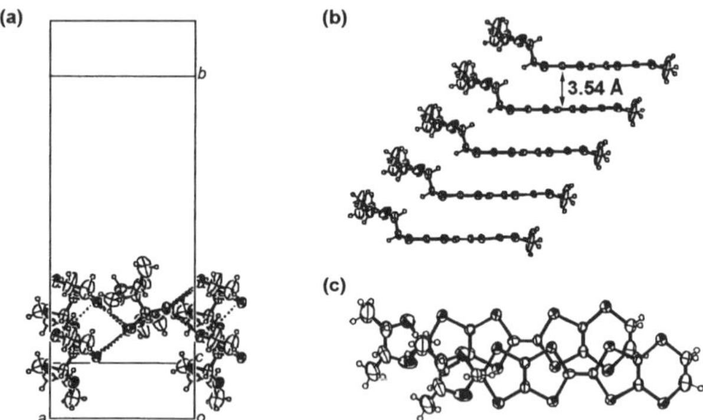

Figure 1 *(a) Donor arrangement in [(R,R)-DMDO-MET]ClO₄. Dotted lines indicate short S···S contacts (< 3.70 Å). (b) (R,R)-DMDO-MET donor column. (c) Overlap mode of two (R,R)-DMDO-MET donor molecules.*

2.2 Synthesis and Electrochemical Properties of *meso*-DMDO-MET

Preparation of oxone **3** endowed with a *meso*-dimethyldioxolane ring as a building block leading to *meso*-DMDO-MET began with the dioxane-fused thione **1**[10] (Figure 2). Thus, the $BF_3 \cdot OEt_2$ (10 equiv.)-promoted reaction of **1** with 2,3-butanediol as a mixture of *meso*- and *trans* [(*R,R*) and (*S,S*)]-forms in $CHCl_3$ for 3 days, after sequential purification by silica gel chromatography and recrystallization, gave a 7:1 mixture of *meso*- and *trans*-dimethyl isomers of thione **2** in 22% yield. The ratio was determined by comparing with the ^1H NMR spectrum of (*R,R*)-**2** obtained via a similar $BF_3 \cdot OEt_2$-promoted reaction of **1** with (*R,R*)-2,3-butanediol. Treatment of **2** with $Hg(OAc)_2$ in THF-HOAc followed by purification using silica gel chromatography and recrystallization afforded **3** as a pure *meso*-form in 47% yield. Subsequent $(MeO)_3P$-mediated cross-coupling of **3** with thione **4** in toluene at reflux allowed the construction of *meso*-DMDO-MET in 58% yield. Cyclic voltammetry of *meso*-DMDO-MET [0.1 M *n*-Bu_4NClO_4 in PhCN, Pt electrode, scan rate 50 mV s^{-1}] showed three reversible oxidation waves [$E_1 = 0.52$, $E_2 = 0.81$, $E_3 = 1.48$, ΔE ($E_2 - E_1$) = 0.29 V (*vs.* SCE)]. The E_1 and ΔE values are almost comparable to those of (*R,R*)-DMDO-MET ($E_1 = 0.53$, $\Delta E = 0.27$ V)[6] measured under identical conditions.

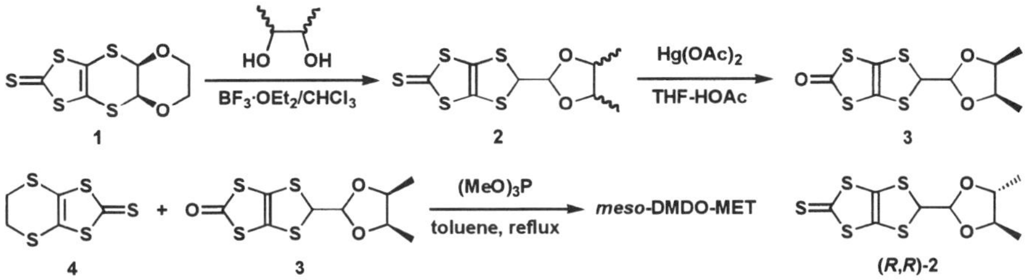

Figure 4 *Synthesis of meso-DMDO-MET*

2.3 Electrical Conducting Behaviour of *meso*-DMDO-MET Salts

Controlled-current electrocrystallization[7] of *meso*-DMDO-MET with *n*-Bu_4NX (X = AuI_2 and BF_4) in PhCl gave single crystals of the corresponding salts. Variable-temperature conductivity measurements indicated that the AuI_2 ($\sigma_{rt} = 2.3 \times 10^{-1}$ S cm^{-1}, $E_a = 270$ meV) and BF_4 ($\sigma_{rt} = 7.8 \times 10^{-2}$ S cm^{-1}, $E_a = 88$ meV) salts are semiconductive. It should be noted that *meso*-DMDO-MET with the BF_4 anion forms a smaller gap simiconductor compared to the BF_4 salt of (*R,R*)-DMDO-MET. However, the quality of the resulting single crystals of the AuI_2 and BF_4 salts was not suitable for an X-ray diffraction study.

3 CONCLUSION

Although, in contrast to DO-MET, (*R,R*)-DMDO-MET and *meso*-DMDO-MET have failed to achieve metallic conductivity, a subtle variation in the stereochemistry of two methyl groups added to the dioxolane ring of DO-MET has been found to alter the transport properties of derived CT salts. Useful applications of this result will be found in the design and synthesis of π-donors used for the control of solid-state properties.

This work was supported by MEXT (No. 15073102) and JSPS (No. 17550135).

References and Notes

1   G. Schukat and E. Fanghänel, *Sulfur Rep.*, 2003, **24**, 1; *TTF Chemistry – Fundamentals and Applications of Tetrathiafulvalene*, eds. J. Yamada and T. Sugimoto, Kodansha & Springer, Tokyo, 2004; thematic issue on *Molecular Conductors*, *Chem. Rev.*, 2004, **104**.

2   J. Yamada, H. Nishikawa and K. Kikuchi, in *TTF Chemistry – Fundamentals and Applications of Tetrathiafulvalene*, eds. J. Yamada and T. Sugimoto, Kodansha & Springer, Tokyo, 2004, Chapter 11, pp. 261–286; J. Yamada, *J. Mater. Chem.*, 2004, **14**, 2951; J. Yamada, H. Nishikawa, H. Akutsu and K. Kikuchi, *Chem. Rev.*, 2004, **104**, 5057.

3   Abbreviations of donor molecules: DO-MET, (1,3-dioxolan-2-yl)methylenedithio-(ethylenedithio)tetrathiafulvalene; DO-MMT, (1,3-dioxolan-2-yl)methylenedithio-(methylenedithio)tetrathiafulvalene; MET, methylenedithio(ethylenedithio)tetrathiafulvalene; MT, bis(methylenedithio)tetrathiafulvalene; BDA-TTP, 2,5-bis(1,3-dithian-2-ylidene)-1,3,4,6-tetrathiapentalene; DODHT, (1,4-dioxane-2,3-diyldithio)dihydrotetrathiafulvalene; DMDO-MET, (4,5-dimethyl-1,3-dioxolan-2-yl)methylenedithio(ethylenedithio)tetrathiafulvalene.

4   J. Yamada, M. Hamasaki, O. Jinih, S. Tanaka, K. Hagiya and H. Anzai, *Tetrahedron Lett.*, 1997, **38**, 3439.

5   J. Yamada, R. Oka, T. Mangetsu, H. Akutsu, S. Nakatsuji, H. Nishikawa, I. Ikemoto and K. Kikuchi, *Chem. Mater.*, 2001, **13**, 1770.

6   J. Yamada, *Trends Org. Chem.*, 2001, **9**, 115.

7   H. Anzai, J.M. Delrieu, S. Takasaki, S. Nakatsuji and J. Yamada, *J. Cryst. Growth*, 1995, **154**, 145; H. Nishikawa, T. Sato, T. Kodama, I. Ikemoto, K. Kikuchi, H. Anzai and J. Yamada, *J. Mater. Chem.*, 1999, **9**, 693.

8   *Crystal data* for [(*R,R*)-DMDO-MET]BF$_4$: C$_{14}$H$_{14}$O$_2$BF$_4$S$_8$, M_r = 557.55, orthorhombic, space group $P2_12_12_1$, a = 10.862(7), b = 41.497(9), c = 5.057(9) Å, V = 2280(3) Å^3, Z = 4, ρ_{calcd} = 1.624 g cm^{-3}, μ(Mo Kα) = 0.8249 mm^{-1}, T = 297 K, R = 0.050 and R_w = 0.046 for 1463 observed reflections with $I > 3\sigma(I)$ from 3125 unique reflections. *Crystal data* for [(*R,R*)-DMDO-MET]ClO$_4$: C$_{14}$H$_{14}$O$_6$ClS$_8$, M_r = 570.19, orthorhombic, space group $P2_12_12_1$, a = 10.796(3), b = 41.221(4), c = 5.055(3) Å, V = 2250(1) Å^3, Z = 4, ρ_{calcd} = 1.683 g cm^{-3}, μ(Mo Kα) = 0.9415 mm^{-1}, T = 299 K, R = 0.061 and R_w = 0.051 for 2006 observed reflections with $I > 3\sigma(I)$ from 3082 unique reflections. *Crystal data* for [(*R,R*)-DMDO-MET]$_2$AsF$_6$: C$_{28}$H$_{28}$O$_4$AsF$_6$S$_{16}$, M_r = 2262(4), orthorhombic, space group $P2_12_12_1$, a = 10.79(2), b = 41.67(1), c = 5.034(9) Å, V = 2262(4) Å^3, Z = 4, ρ_{calcd} = 1.659 g cm^{-3}, μ(Mo Kα) = 1.5479 mm^{-1}, T = 298 K, R = 0.112 and R_w = 0.090 for 1141 observed reflections with $I > 3\sigma(I)$ from 3111 unique reflections.

9   X-ray study of the ClO$_4$ salt at 110 K revealed that this disorder is not suppressed at a lower temperature, thereby suggesting that it is static in origin. The respective refinements for X-ray data of the present three salts were based on the assumption that the BF$_4$ and ClO$_4$ salts have a 1:1 (donor:anion) stoichiometry while the AsF$_6$ salt has a 2:1 (donor:anion) stoichiometry.

10 J. Yamada, S. Tanaka, J. Segawa, M. Hamasaki, K. Hagiya, H. Anzai, H. Nishikawa, I. Ikemoto and K. Kikuchi, *J. Org. Chem.*, 1998, **63**, 3952.

STRUCTURES AND ELECTRICAL PROPERTIES OF NEW PDT-TTP
CONDUCTORS

M. Daini[1], Y. Misaki[2] and K. Tanaka[1, 3]

[1]Department of Molecular Engineering, Kyoto University, Kyoto 615-8510, Japan.
[2]Department of Applied Chemistry, Ehime University, Matsuyama 790-8577, Japan.
[3]JST-CREST

1 INTRODUCTION

Development of conducting materials with multi-dimensional electronic structure has received considerable attention in the search for molecular metals stable down to low temperature and superconductors as a result.[1] We have reported that bis-fused TTF, 2,5-bis(1,3-dithiol-2-ylidene)-1,3,4,6-tetrathiapentalene (BDT-TTP, or simply TTP) and its derivatives have afforded a large number of molecular metals stable down to low temperatures.[2] On the other hand, examples of molecular conducting materials with three-dimensional electronic structure is very rare.[3] In all cases, three-dimensional (3D) electronic structure has been realized based on metal dithiolate complexes or by participation of inorganic counter ions, and no example of 3D metal based on purely organic molecules. In order to introduce the third directional interaction along the molecular long axis, we have studied on synthesis and electrical properties of the cation radical salts based on the derivatives of 2-(1,3-dithiol-2-ylidene)-5-(pyran-4-ylidene)-1,3,4,6-tetrathiapentalene (PDT-TTP) and its thiopyran analog (TPDT-TTP), which has chalcogen atoms in their edges.[4,5] Among them, TM-TPDS gave radical the AsF_6^- salt possessing three-dimensional "windmill-type" donor array through side-to-edge sulfur-sulfur contacts between tetrathiapentalene moiety and thiopyran rings as well as stacking of the donors, although calculated Fermi surface is quasi-1D due to weak side-to-edge interactions.[6] In contrast, a selenomethyl analog SM-PDT gave a PF_6^- salt with two-dimensional molecular packing through edge-to-edge selenium-selenium contacts between methylseleno groups.[7] In this connection, systematic investigation of the substituent effect on the donor arrangement of (T)PDT-TTP conductors is of considerable interest to elucidate correlation between molecular structure and packing motif and to control the molecular arrangement of an organic conductor as a result. In these Proceedings, we report the synthesis of ethylthio derivative of TPDT-TTP (TE-TPDT), and structures and conductivity of its cation radical salt.

2 RESULTS AND DISCUSSION

2.1 Synthesis and electrochemistry

Synthesized of TE-TPDT was carried out according to Scheme 1. By the cross-coupling reaction of a ketone **1**[5] with 4,5-bis(ethylthio)-1,3-dithiol-2-thione (**2**) in neat triethylphosphite at 110 °C, **3** was obtained in 93% yield. Dehydrogenation of **3** with a 2.5equiv of 2, 3-dichloro-5,6-dicyano-*p*-benzoquinone (DDQ) in refluxing xylene gave TE-TPDT in 60% yield. Cyclic votammogram of TE-TPDT measured in benzonitrile consist of four pairs of one-electron redox waves at 0.03, 0.25, 0.54, 0.73 V (vs. Fc/Fc$^+$), respectively. No significant change of redox potentials was observed compared with bis(methylthio) derivative of TPDT-TTP, indicating the length of alkyl groups little affects the donating ability of the donor molecules.[5]

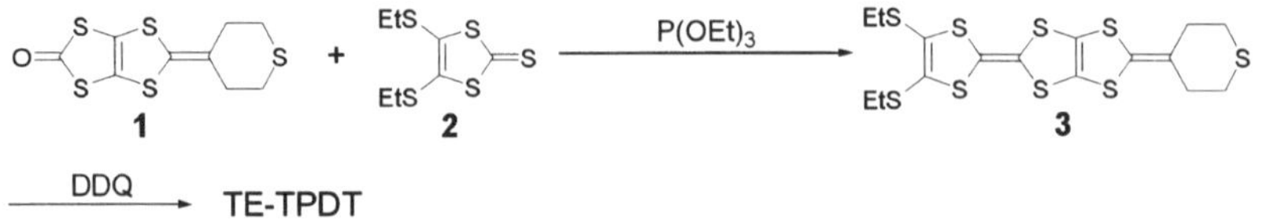

Scheme 1   *Synthesis of TE-TPDT*

2.2 Crystal structure and conducting property of (TE-TPDT)AsF$_6$(THF)

Single crystals of (TE-TPDT)AsF$_6$(THF) suitable for X-ray structure analysis[8] were prepared by electrochemical oxidation of TETPDT in THF containing ethanol (5%, v/v) in the presence of Bu$_4$NAsF$_6$ at 25 °C. The current was stepwisely changed from 0.2 to 0.7 μA during electrocrystallization for 2 weeks.[9] The molecular structure of TE-TPDT in the present salt is shown in Figure 1. The TPDT-TTP skeleton is almost planar. On of the ethyl groups is bent upward from the molecular plane, while the other projects outside of the molecular long axis. Figure 2 shows the crystal structure of (TE-TPDT)AsF$_6$(THF). The ratio of the donor to anion has been determined to be 1 : 1, because both the donor

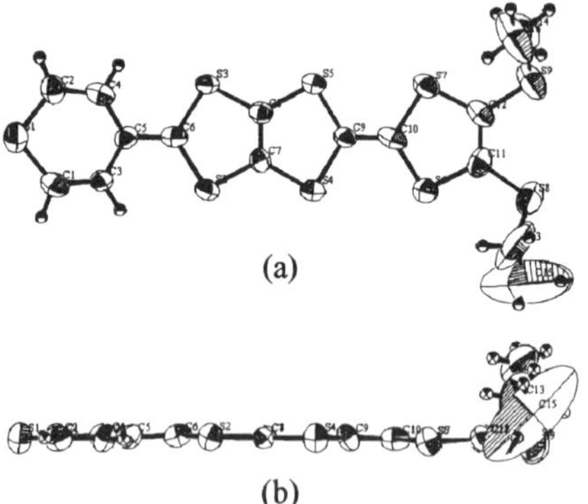

(a)

(b)

Figure 1   *(a) ORTEP drawing of (TE-TPDT)AsF$_6$(THF) and (b) the side view.*

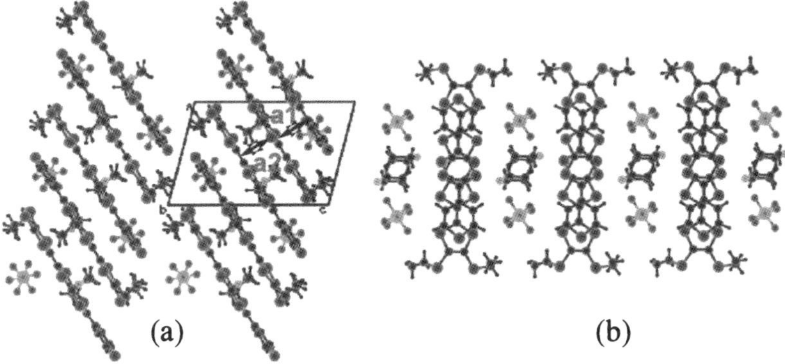

Figure 2   *(a) Crystal structure of (TE-TPDT)AsF₆(THF) viewed along the a-axis, and (b) viewed along c-axis.*

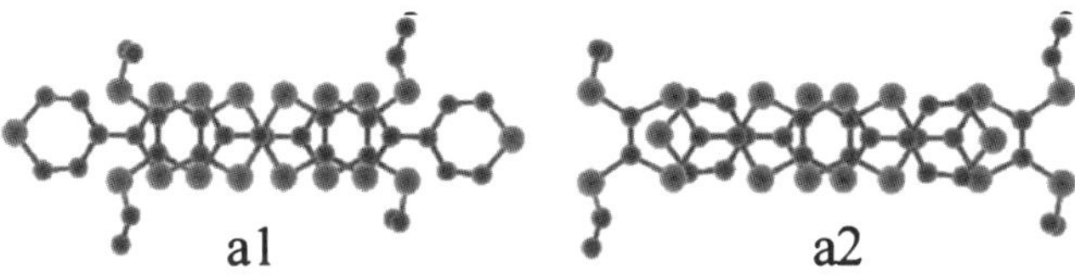

Figure 3   *The overlap mode of donor molecules in (TE-TPDT)AsF₆(THF).*

molecule and the AsF_6 anion are located on a general position. The unsymmetrical donors are stacked in a head-to-tail manner along the b axis. There is no side-by-side interaction, because the anion and the solvent molecules exist between columns along this direction. On the other hand, no significant edge-to-edge interaction along the donor long axis is observed for this salt in contrast to $(SM\text{-}PDT)PF_6(PhCl)_x$[7] with relatively large edge-to-edge interaction through selenium-selenium contact. Thus, the present salt should have one-dimensional (1D) electronic structure. The donors are dimerized in the stack. Thus, there are two kinds of intrastack overlaps a1 and a2. The interplanar distances are 3.38 Å and 3.50 Å for a1 and a2, respectively. On the other hand, the slip distances along the donor long axis are 4.8 and 1.6 Å for a1 and a2, respectively (see Figure 3). As a result, the calculated overlap integral[9] of a2 (21.9×10^{-3}) is quite larger than that of a1 (14.2×10^{-3}). Figure 4 shows the calculated band structure. A unit cell contains two donor molecules, resulting in the formation of two bands from the HOMO of TE-TPDT. The upper and lower bands are separated from each other with an energy gap of 0.16 eV owing to dimerization in 1D column of the donors. Thus, the present salt is indicated to be a band insulator with no Fermi surface, because of 1 : 1 composition of the donor and the anion. This salt showed conductivity of $\sigma_{rt} = 1.1 \times 10^{-3}$ S cm^{-1} on a single crystal, and exhibited semiconductive temperature dependence with the activation energy of 0.11 eV.

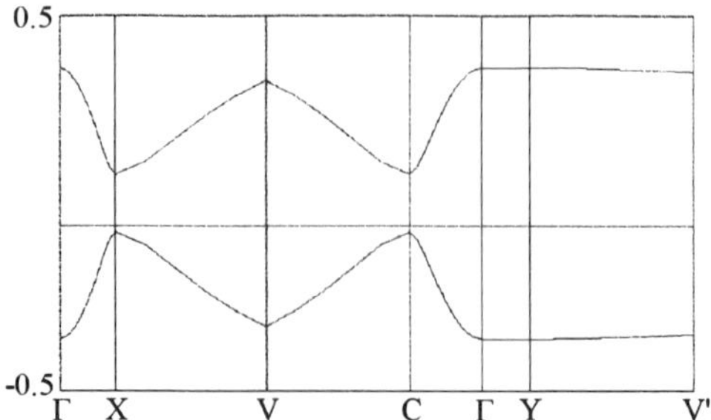

Figure 4   *Energy band structure of (TE-TPDT)AsF₆(THF).*

Acknowledgement

The present research was partially supported by Grant-in-Aid for Scientific Research No.15073216 and 16350100 from the Ministry of Education, Science, Sports and Culture.

References

1   J. M. Williams, J. R. Ferraro, R. J. Thorn, K. D. Carlson, U. Geiser, H. H. Wang, A. M. Kini, and M. H. Whangbo, "Organic Superconductors" (Prentice Hall, New Jersey, 1992); T. Ishiguro, K. Yamaji, and G. Saito, "Organic Superconductors, 2nd Ed." (Springer-Verlag Berlin, Heidelberg, 1998).

2   Y. Misaki, H. Nishikawa, K. Kawakami, S. koyanagi, T. Yamabe, and M.Shiro, *Chem.Lett.*, 1992, 2321; Y. Misaki, H. Fujiwara, T. Yamabe, T.Mori, and S. Tanaka, *ibid.*, 1994, 1653

3   S. Uji, T. Terashima, H. Aoki, J. S. Brooks, R. Kato, H. Sawa, S. Aonuma, M. Tamura, and M. Kinoshita, *Phys. Rev.*, 1994, **B**, **50**, 15597; T. Terashima, S. Uji, H. Aoki, M. Tamura, M. Kinoshita, and M. Tokumoto, *Solid State Commun.*, 1994, **91**, 595.

4   Y. Misaki, H. Fujiwara, and T. Yamabe, *Synth. Met.*, 1995, **70**, 1147.; Y. Misaki, H. Fujiwara, T. Maruyama, M. Taniguchi, T. Yamabe, T. Mori, H. Mori, and S. Tanaka, *Chem. Mater.*, 1999, **11**, 2360.

5   Y. Misaki, H. Fujiwara, and T. Yamabe, *J. Org. Chem.*, 1996, **61**, 3650.

6   Y. Misaki, T. Kaibuki, M. Taniguchi, K. Tanaka, T. Kawamoto, T. Mori, and T. Nakamura, *Chem. Lett.*, 2000, 1274

7   K. Takahashi, T. Nakayashiki, M. Taniguchi, Y. Misaki, and K. Tanaka, *Chem. Lett.*, 2001, 162

8   Crystal Data for (TE-TPDT)AsF₆(THF): triclinic, space group $P\bar{1}$, a = 9.4209(5) Å, b = 11.000(1) Å, c = 15.1490(7) Å, α = 72.332(4)°, β = 74.025(6)°, γ = 83.443(6)°, V = 1437.3(2) Å^3, Z = 2, D_c =1.737 g cm^{-3}, R = 0.086, R_w = 0.200.

9   H. Anzai, J. M. Delrieu, S. Takasaki, S Nakatsuji, J. Yamada, *J. Cryst. Growth*, 1995, **154**, 145.

10   T. Mori, A. Kobayashi, Y. Sasaki, H. Kobayashi, G. Saito, H. Inokuchi, *Bull. Chem. Soc. Jpn.* 1984, **57**, 627.

CRYSTAL STRUCTURES AND PHYSICAL PROPERTIES OF $(Mo_6X_{14})^{2-}$ SALTS (X = Br , I)

H. Hosoda[1] , K. Nishimura[1], H. Yamochi[2], G. Saito[1], K. Kirakci[3], S. Cordier[3], C. Perrin[3]

[1]Division of Chemistry, Graduate School of Sciences, Kyoto University, Sakyo-ku, Kyoto 606-8502, Japan
[2] Research Center for Low Temperature and Materials Science, Kyoto University, Sakyo-ku, Kyoto 606-8502, Japan & ERATO/JST
[3]Institut de Chimie de Rennes, Laboratoire de Chimie du Solide et Inorganique Moléculaire, UMR 6511 CNRS-Université de Rennes 1, Avenue du Général Leclerc, 35042 Rennes Cedex, France

1 INTRODUCTION

Low oxidation state molybdenum halides are well known to form $Mo_6X^i_8X^a_6$ (X = Cl, Br and I) cluster in which the octahedral Mo_6 unit is face-capped by eight inner ligands (X^i) to form a $Mo_6X^i_8$ core, to which six terminal apical ligands (X^a) are bonded additionally (^{i} = inner, ^{a} = apical, according to the Schäfer and Schnering notation[1], Figure 1a). Within the solid, these units can be discrete or interconnected by shared inner and/or apical ligands leading to strong electronic interactions. The organic salts with tetrathiafulvalene monocation radical ($TTF^{+\bullet}$) showed antiferromagnetic fluctuations at low temperatures[2]. In this work, we report the crystal structures and physical properties of new salts with TTF derivatives; HMTTF, TTC$_1$-TTF, TTT, and BO (Figure 1b).

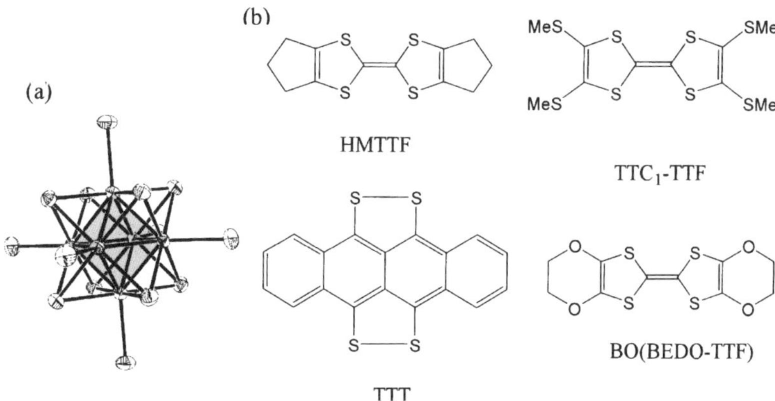

Figure 1 *(a): Structure of $Mo_6X^i_8X^a_6$ (central octahedron is shaded) , (b): chemical structures of donor molecules*

2 SYNTHESIS

$(HMTTF)_2Mo_6I_{14}$·4PhCN(**1**), $(TTC_1\text{-}TTF)_2Mo_6Br_{14}$(**2**), $(TTT)Mo_6Br_{14}$·3PhCN(**3**)   and
$(BO)_2Mo_6Br_{14}$·4PhCN(**4**)   (PhCN: benzonitrile) salts were prepared by conventional
electrooxidation method. $(TBA)_2Mo_6Br_{14}$ was prepared according to the procedure
described in Ref 3. For example, a solution of BO (10.0 mg) and $(TBA)_2Mo_6Br_{14}$ (100 mg)
in 18 ml PhCN (TBA: tetrabutylammonium) in a glass cell was subjected to constant
current of 0.5 µA for 4 weeks to afford single crystals of **4**.

3 RESULTS AND DISCUSSION

3.1 $(HMTTF)_2Mo_6I_{14}$·4PhCN(1) and $(TTC_1\text{-}TTF)_2Mo_6Br_{14}$(2)

In **1** and **2**, monocationic donor molecules are dimerized (Figure 2). Their UV-Vis-NIR
absorption spectra in KBr clearly exhibited the bands which are attributed to the intradimer
transition at 14 and 11×10^3 cm^{-1}, respectively. Temperature dependence of static magnetic
susceptibility of **1** in the range of 2 - 300 K was well fitted with the singlet-triplet model (J
$/k_B$ = 190 K). On the other hand, **2** showed a singlet ground state and no triplet excitation
was observed even at room temperature. These behaviors are in good agreement with the
calculated intradimer overlap integrals (5.9×10^{-3} for **1** and 38.6×10^{-3} for **2**).

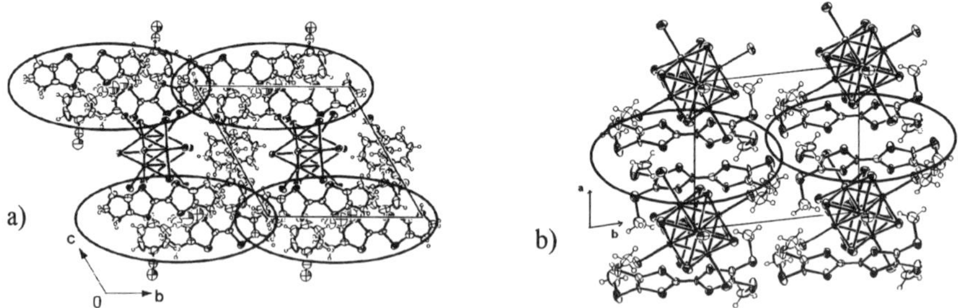

Figure 2 *a) : Crystal structure of $(HMTTF)_2Mo_6I_{14}$·4PhCN (1, a-axis projection), b) :
Crystal structure of $(TTC_1\text{-}TTF)_2Mo_6Br_{14}$ (2, c-axis projection), ellipses indicate the
donor dimers*

3.2 $(TTT)Mo_6Br_{14}$·3PhCN(3)

In the crystal of **3**, dicationic TTT molecules were separated to each other and also from
the anion clusters (Figure 3). The UV-Vis-NIR absorption spectrum in KBr exihibited a
band ascribable to the intramolecular transition from second HOMO to HOMO at 22×10^3
cm^{-1}, that is common to those observed in the dicationic TTT salts reported previously[4].

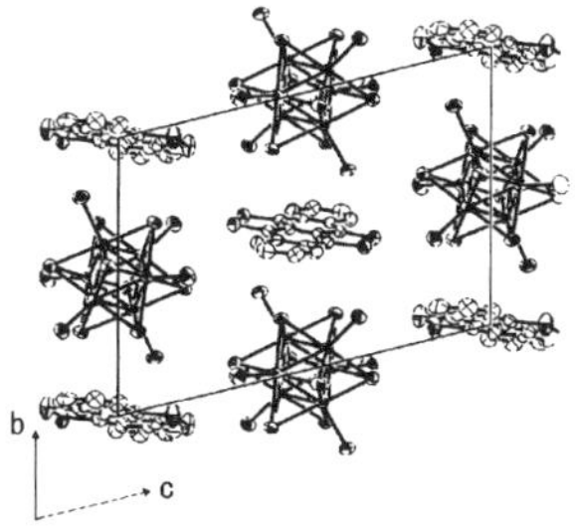

Figure 3 *Crystal structure of (TTT)Mo$_6$Br$_{14}$·3PhCN (3, a-axis projection)(PhCN is omitted for clearity)*

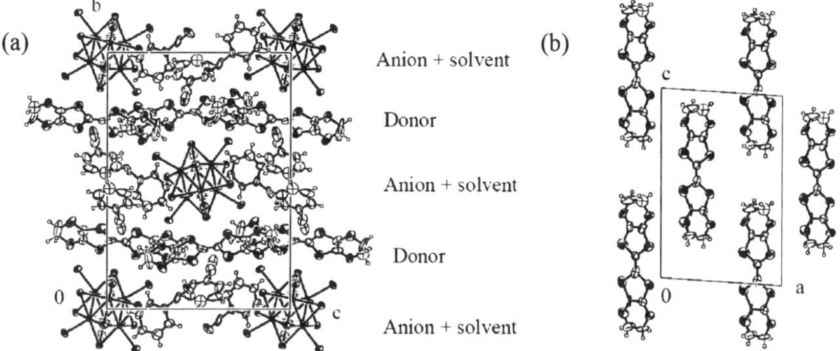

Figure 4 *Crystal structure of (BO)$_2$Mo$_6$Br$_{14}$·4PhCN(4), (a) a-axis projection. (b) b-axis projection of BO molecules located on y = 1/4 plane*

3.3 (BO)$_2$Mo$_6$Br$_{14}$·4PhCN(4)

In the crystals of **4**, monocationic BO molecules form a layer separated by the layer composed of Mo$_6$Br$_{14}$ dianion and PhCN molecules (Figure 4). The S$\cdots$S, S$\cdots$O and O$\cdots$O intermolecular distances (5.75(7), 3.86(5) and 3.07(2) Å, respectively) are too long to enable the significant overlap between HOMO of BO. This feature was confirmed by the calculated overlap integrals between neighboring BO molecules ($<10^{-4}$). The UV-Vis-NIR absorption spectrum of **4** clearly showed a band at 10.4×10^3 cm^{-1}(band-C of curve a in Figure 5). The spectrum is in excellent agreement with that observed for the monomeric form of BO cation radical (curve b, 10.6, 16.6, 20.4 and 21.7×10^3 cm^{-1}) [5], but differs considerably from that of BO cation radical salts having segregated column, (BO$^+$)I$_3$, that displays the intra- and interdimer transitions at 7.8, 9.8, respectively in addition to the intramolecular one at $\sim12 \times 10^3$ cm^{-1} (curve c). The temperature dependence of magnetic susceptibility of **4** in the range of 2.0 - 300 K was well fitted by Curie-Weiss law($\chi = C$ /(T- Θ)) with $C = 0.38$ emu / (spin of BO)· K and $\Theta = - 4.2$ K. The Curie constant is close to that of independent $S = 1/2$ spins ($C = 0.375$ emu / spin·K), indicating an isolated feature of BO cation molecules in this salt.

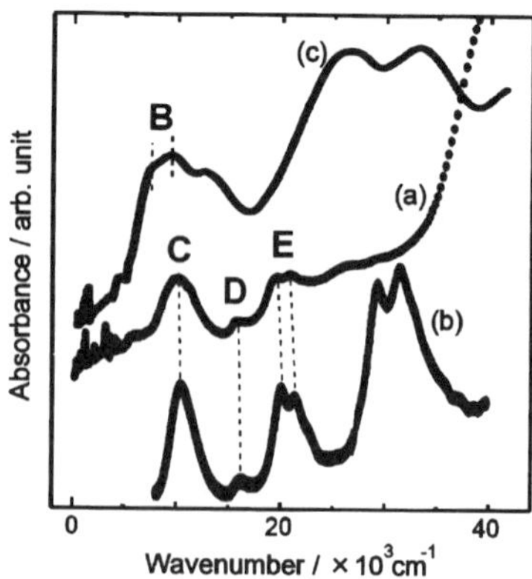

Figure 5 *UV-Vis-NIR absorption spectra (a: (BO)$_2$Mo$_6$Br$_{14}$·4PhCN(**4**) in KBr, b: monocationic BO in methanol, c: (BO)I$_3$ in KBr)*

4 CONCLUSION

In this study, we obtained single crystals of $(Mo_6X_{14})^{2-}$ salts (X = Br , I), **1** - **4**, determined their crystal structures and measured physical properties. The face-to-face dimer of cation molecules (**1**, **2**) or monomeric dication (**3**) or cation (**4**) molecules are stabilized in the crystals by being surrounded by large cluster anions. As a result, the donor molecules are not able to have segregated columns or layers with large overlap integrals. The optical and magnetic properties are in good agreement with the structural features.

Acknoledgement

This work was supported by a Grant-in-Aid (21st Century COE program on Kyoto University Alliance for Chemistry) from the Ministry of Education, Culture, Sports, Science and Technology, Japan.

References

1 H. Schäfer, H.-G. v. Schnering, *Angew. Chem. Int. Ed. Engl.* 76 (1964) 833.
2 P. Batail, C. Livage, S. Parkin, C. Coulon, J. Martin. E. Canadell, *Angew. Chem. Int. Ed. Engl.* **1991**, 11(30), 1498.
3 K. Kirakci, S. Cordier, C. Perrin, *Z. Anorg. Allg. Chem.* 631 (2005) 411.
4 S. Sekizaki, C. Tada, H. Yamochi, G. Saito, *J. Mater. Chem*, 11, (2001) 2293.
5 S. Horiuchi, H. Yamochi, G. Saito, K. Sakaguchi, M. Kusunoki, *J. Am. Chem. Soc.* 118 (1996) 8604.

PRONOUNCED ENHANCEMENT OF CHARGE ORDERING TRANSITION TEMPERATURES IN TMTTF SALTS WITH DEUTERATION

T. Nakamura[1,2], T. Hara[1] and K. Furukawa[1,2]

[1]Institute for Molecular Science, Okazaki 444-8585, Japan
[2]The Graduate University for Advanced Studies, Okazaki 444-8585, Japan

1 INTRODUCTION

The charge-ordering phenomena in TMTTF (tetramethyltetrathiafulvalene)-based organic salts are attracting significant attention.[1-3] The following ESR investigation for a series of TMTTF salts clarified that their low-temperature behavior can roughly be divided into three groups, (Type I) ReO_4, ClO_4 (Td-symmetry), (Type II) SbF_6, AsF_6, PF_6 (Oh-symmetry) and (Type III) Br and SCN salts.[4] The variation in ESR behavior is considered to be originated from the variation of the charge ordering patterns at low temperatures, such as -O-O-o-o- and -O-o-O-o- along the stacking axes.[4] This model is also consistent with X-ray results.[5] It is well known that deuteration often has a great influence on the electronic states of organic conductors. Moreover, ferroelectric transition temperatures are strongly affected by the deuteration.[6] It is also advantageous to compare pristine and deuterated salts since they have most parts in common. Hence, in order to investigate the effect of deuteration on the charge ordering phenomena of $(TMTTF)_2X$, we synthesized perdeuterio-TMTTF, TMTTF-d_{12}, and performed magnetic resonance and X-ray measurements to clarify their low-temperature electronic and crystal structures. We observed an enhancement of the charge-ordering transition temperature, T_{CO}, by deuteration, and also found a relationship between the T_{CO}'s and crystallographical parameters.[7] We proposed a possible origin of the charge ordering phenomena in one-dimensional 1/4-filling systems.[7,8] In this paper, we introduce our recent progress for the investigation of the charge-ordering phenomena for a series of TMTTF salts.

2 EXPERIMENTAL

Detailed experimental conditions are shown in the previous literatures.[4,7,8] TMTTF molecules were synthesized following the synthesis method described in the previous reports.[9-11] Rectangular plate-like crystals were prepared by the electrochemical oxidation in a pristine or deuterated acetonitrile solution under a constant current of $1.0\mu A$. ESR measurements were carried out for a single crystals using the X-band spectrometer Bruker E500 with the new-type Oxford cryostat ESR900. Broad-line NMR measurements were performed using a pulsed-NMR spectrometer operated at 85.071MHz. X-ray measurements were carried out using the single crystal X-ray diffractmeter Rigaku R-AXIS IV with MERCURY CCD. Low-temperature X-ray measurements were achieved by a N_2 gas flow cryostat, at temperatures between 300K and 100K.

3 RESULTS AND DISCUSSION

The temperature dependences of the ESR linewidth, ΔH_{pp}, of the pristine and deuterated $(TMTTF)_2SbF_6$ salts, $(TMTTF-h_{12})_2SbF_6$ and $(TMTTF-d_{12})_2SbF_6$, are shown in Fig 1. A clear hump of the ΔH_{pp} is observed at around 155K for pristine $(TMTTF-h_{12})_2SbF_6$. The hump of ΔH_{pp} shows the charge-ordering phase transition.[4] The hump temperature of $(TMTTF-d_{12})_2SbF_6$ was about 163 K as shown in Fig. 1 (b). The 8K of enhancement is clearly beyond the experimental error. Significant enhancement of T_{CO} about 16K was also observed in $(TMTTF-d_{12})_2AsF_6$.[7]

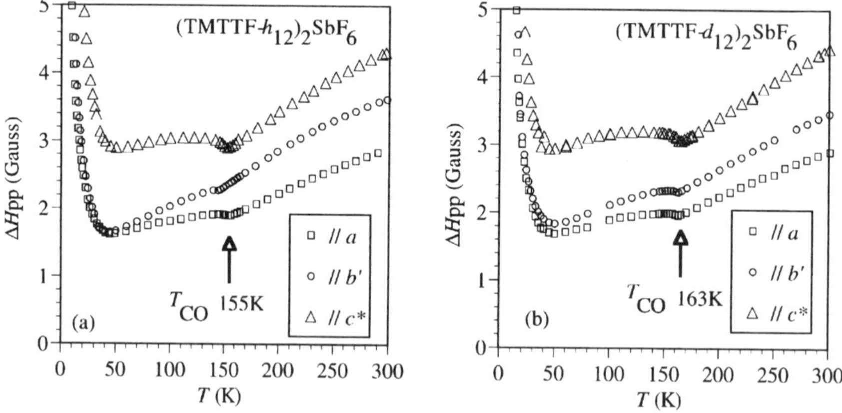

Figure 1 *Temperature dependence of the ESR linewidth, ΔH_{pp}, of (a) $(TMTTF-h_{12})_2SbF_6$ and (b) $(TMTTF-d_{12})_2SbF_6$.*

As for the type II salts, -O-o-O-o- type charge orderings are expected below T_{CO}. These behaviors are different from *the ground states* of the Type I salts. $(TMTTF)_2ReO_4$ undergoes a spin-singlet transition followed by anion ordering (T_{AO}). Below T_{AO}, the donor molecules show -O-O-o-o- type charge-ordering. This charge-ordering is considered to originate from an additional strong anion potential below T_{AO}. On the other hand, we recently found another -O-o-O-o- type charge ordering in the intermediate paramagnetic phase at around 225 K besides the -O-O-o-o- type charge-ordering for $(TMTTF)_2ReO_4$.[8,12] The -O-o-O-o- charge-ordering of $(TMTTF)_2ReO_4$ is also sensitive to the deuteration, while there is no observable change with deuteration for T_{AO}.[7,12]

We clarified that there is no direct relation between the SbF_6 anion motion and the TMTTF charge-order by ^{19}F NMR of $(TMTTF)_2SbF_6$. Hence, we performed temperature-dependence analysis on single crystal X-ray measurements for pristine and deuterated salts to understand the deuteration effects on the crystal structure. One possible reason for the enhancement of T_{CO} with deuteration is a chemical pressure effect. In general, a 'positive' pressure effect is expected for deuteration, since the C-D bond length is shorter than that of C-H. However, the T_{CO} of $(TMTTF)_2MF_6$ (M = As, Sb) rapidly decreases with applied pressure, according to ^{13}C NMR by the UCLA group.[3] Thus, a simple chemical pressure scenario can be ruled out. Hence we considered the dimerization effect. Noted that charge-order phases are not realized in simple Mott-Hubbard systems with large on-site Coulomb interactions but prefer 1/4-filling systems with uniform stacks. In order to discuss the deuteration effect in the above scenario, we estimated the deuteration- and temperature-dependence of the inter-molecular distances. The crystal structure and definition of the inter-molecular distances, A and B, of $(TMTTF)_2SbF_6$ are shown in Fig. 2 (a). Here we adopt a parameter, B/A, to discuss the dimerization degree. As the temperature decreases, both A and B decrease. However, the quantity B/A increases as the temperature decreases as seen in Fig. 2(b). This indicates that the donor stacking becomes more uniform at lower temperatures. As for the deuterated $(TMTTF\text{-}d_{12})_2SbF_6$ salts, both the inter-molecular distances of A and B are shorter than those of $(TMTTF\text{-}h_{12})_2SbF_6$ for each temperature. On the other hand, the ratio B/A, for each temperature is large and is close to unity in comparison with that for the pristine salt. In other words, dimerization is suppressed by deuteration, and T_{CO} increases.

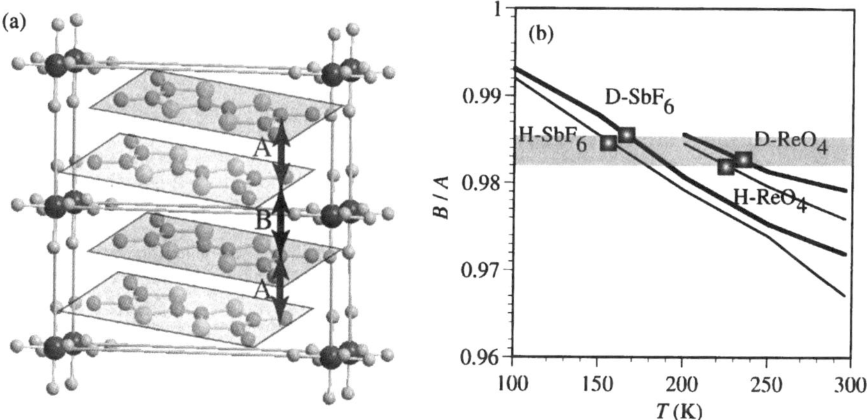

Figure 2 *(a) Crystal structure and definition of the inter-molecular distances, A and B, (see text). (b) Temperature dependence of inter-molecular distances and their ratio of (TMTTF-h$_{12}$)$_2$SbF$_6$, (TMTTF-d$_{12}$)$_2$SbF$_6$ (TMTTF-h$_{12}$)$_2$ReO$_4$ and (TMTTF-d$_{12}$)$_2$ReO$_4$. The measured points (every 50 K) are interpolated with a straight line. Solid squares correspond to T$_{CO}$'s for each of salts.*

It is interesting to compare the X-ray parameters of $(TMTTF)_2ReO_4$ and $(TMTTF)_2SbF_6$ salts in the aspect of the charge-ordering phenomena. The T_{CO}'s of these two salts are different while the lattice parameters along the stacking axis, a, are similar. The deuteration suppresses the dimerization also for $(TMTTF)_2ReO_4$ salts as seen in Fig. 2 (b). We can understand the enhancement of T_{CO} by deuteration within this scenario. To compare other salts, we tentatively interpolated the measured points with a straight line. Solid squares correspond to the T_{CO}'s for each of salts. The hatched area is only a guide, but it seems likely that there is a threshold of B/A to stabilize the charge-ordering phase. In other words, the 1/4-filling plays an important role in stabilizing the charge-ordering phenomena.

4 CONCLUSION

In conclusion, an enhancement by deuteration of T_{CO}'s is found in ESR measurements for (TMTTF)$_2$X salts. There is a threshold value of B/A to stabilize the charge-ordering, and the enhancement of T_{CO}'s by deuteration can be explained within this framework. The intermediate -o-O-o-O- charge ordering phase is a universal phenomenon of TMTTF-based salts, and originates from long-range Coulomb interactions.

5 ACKNOWLEDGMENTS

The authors thank F. Nad, P. Monceau (CNRS), Y. Nogami (Okayama Univ.) and T. Takahashi (Gakushuin Univ.) for useful discussions. This study was supported by Grant-in-Aid for Scientific Research on Priority Areas of 'Molecular Conductors' (No. 15073223) and for Creative Scientific Research Collaboratory on Electron Correlation-Toward a New Research Network between Physics and Chemistry- (13NP0201) from the Ministry of Education, Culture, Sports, Science and Technology, Japan.

References

1   D. S. Chow, F. Zamborszky, B. Alavi, D. J. Tantillo, A. Baur, C. A. Merlic and S. E. Brown, *Phys. Rev. Lett.*, 2000, **85**, 1698.
2   P. Monceau, F. Ya. Nad and S. Brazovskii, *Phys. Rev. Lett.*, 2001, **86**, 4080.
3   F. Zamborszky, W. Yu, W.Raas, S.E. Brown, B. Alavi, C.A. Merlic and A. Baur, *Phys. Rev. B*, 2002, **66**, 081103(R).
4   T. Nakamura, *J. Phys. Soc. Jpn.*, 2003, **72**, 213.
5   Y. Nogami and T. Nakamura, *J. Phys. IV*, 2002, **12**, Pr9-145.
6   for example, R. Blinc, *J. Phys. Chem. Solids*, 1960, **13**, 204, and references therein.
7   K. Furukawa, T. Hara and T. Nakamura, to be published in *J. Phys. Soc. Jpn.* 2005, **74**, 3288.
8   T. Nakamura, K. Furukawa and T. Hara, to be published in *J. Phys. Soc. Jpn.*, 2006, **75** No.1
9   J. P. Ferraris, T. O. Poehler, A. N. Bloch and D. O. Cowan, *Tetrahedron Lett.*, 1973, **14**, 2553.
10   A. Mas, J.-M. Fabre, E. Torrilles, L. Giral and G. Brun, *Tetrahedron Lett.*, 1977, **18**, 2579.
11   F. Wudl, E. Aharon-Shalom and S. H. Bertz, *J. Org. Chem.*, 1981, **46**, 4612.
12   F. Nad, P. Monceau, T. Nakamura and K. Furukawa, *J. Phys.: Condens. Matter*, 2005, **17**, L399.

RAMAN SPECTROSCOPY OF κ-(BEDT-TTF)$_2$Cu[N(CN)$_2$]I

O. Drozdova[1], M. A. Tanatar[1,2], R. Swietlik[1,3], K. Yakushi[1], T. Nakamura[1], S. Kagoshima[4], V. S. Yefanov[4,5], N. D. Kushch[6,7], and E. B. Yagubskii[6]

[1]Institute for Molecular Science, 38 Nishigo-naka, Myodaiji, Okazaki 444-8585, Japan
[2]Institute of Surface Chemistry, National Academy of Sciences of Ukraine, 17 General Naumov Str., Kyiv 03164, Ukraine
[3]Institute of Molecular Physics, Polish Acad. Sci, 17 M. Smoluchowskiego, 601-79 Poznan, Poland
[4]Department of Pure and Applied Sciences, University of Tokyo, 3-8-1 Komaba, Meguro, Tokyo 153-8902, Japan
[5]Institute of Semiconductor Physics, National Academy of Sciences of Ukraine, 45 pr. Nauki, Kyiv, Ukraine
[6]Institute of Problems of Chemical Physics, Russian A.S., 142432 Chernogolovka, MD, Russia
[7]Walther-Meissner-Institut, D-85748 Garching, Germany

1 INTRODUCTION

The family of the highest T_c layered organic superconductors κ-(BEDT-TTF)$_2$Cu[N(CN)$_2$]X, where X stands for the halogen atom or pseudohalide group, attracts a vivid interest of researchers since its discovery at the beginning of the 90s.[1] The compounds are isostructural, but, contrary to the prediction of the band structure calculations, show variety of properties depending on X.[1-3] The origin of this diversity is discussed in terms of strong electronic correlations,[4] deviations from perfect stoichiometry,[5] and order/disorder transformation[6] with formation of superstructures.[7,8]

Full structure determination of the compounds shows that the conformational disorder of BEDT-TTF molecule, observed at room temperature in X=Cl, Br and I, tend to disappear on cooling, leading to structural distortions confined in the vicinity of the ethylene fragment and involving CN groups and halogen atoms in the polymeric anion chain.[9] Despite the observation of similar temperature dependence of the lattice thermal expansion with numerous anomalies in all compounds,[10] diffuse x-ray studies showed a link between the anomalies with definite structural changes only in X=I salt,[7,8] which is presumably caused by a difficulty to detect small lattice displacement of the elements with low atomic numbers, producing small charge density modulations. An insight in these distortions can be gained by studying temperature dependence of lattice vibrations involving respective groups. In this article we report characterization of main Raman vibrations of the salt with X=I, and contrast them with the results of diffuse x-ray. We found a clear correlation between the temperature dependence of the Raman frequency shift and of the intensities of superstructure spots. This observation forms a ground for application of Raman spectroscopy for the characterization of small structural distortions in organic compounds.

2 METHODS AND RESULTS

Single crystals of κ-(BEDT-TTF)$_2$Cu[N(CN)$_2$]I (I salt in the following) were grown as described elsewhere.[7,8] The technique and apparatus for Raman measurements were identical to those used in our previous studies.[11] All the spectra were taken in identical thermal cycle conditions: cooling at a rate 1 K/min to 4 K, measurements on warming at fixed temperatures with 10 K steps (smaller steps were used close to the known temperatures of structural anomalies).

In Figure 1 we show panorama Raman spectra of I salt, taken with laser excitation polarized along a- and c- axes and collection of scattered light of all polarizations. The dominant features of the spectra are group of C=C double bond vibrations of BEDT-TTF at around 1500 cm^{-1}, CN stretching vibration in the polymeric anion chain at around 2240 cm^{-1}, and broad bands of ethylene group vibrations at around 2900 cm^{-1}, clearly seen only at low temperatures. We concentrate in the following on characterization of the temperature dependence of the peak positions of v_2 and v_3 modes of BEDT-TTF (here main contributions come from vibrations of the ring and central carbon-carbon double bonds, respectively) and CN stretching mode.

The line due to CN stretching vibration is singular, and its shape is well described by a Lorenzian. Therefore the determination of its position is straightforward. The lines due to v_2 and v_3 modes of BEDT-TTF overlap, however, their characterization does not meet notable difficulty because they dominate the spectra in different polarizations. In this case the overlap with the minor bands was easily diminished by fitting the line with a sum of two Lorenzians.

The temperature dependence of the peak position for the 3 bands is shown in Figure 2. The peaks of all three bands shifts to higher frequencies with cooling. For v_2 and v_3 modes the variation of the peak position with temperature roughly follows T^2 dependence, as expected for crystalline compounds. On top of this smooth variation, notable discontinuities are observed at around 220 K, 155 K and 45 K. The position of the peak of CN vibration mode does not follow closely the T^2 dependence, which may indicate a strong anharmonicity of the vibration. The anomalies of the temperature dependence of the peak position are seen at the same temperatures as for the vibration modes of BEDT-TTF. Worth noting that anomalous shift variation with temperature accounts for almost a half of the total shift of the peaks with temperature variation in 300 K to 4 K range.

In Figure 2 (d) we show the temperature dependence of the intensities of the superstructure spots in diffuse x-ray measurements. In the crystal under study the superstructure reflexes are well defined, and 3 characteristic superstructure wavenumbers (and temperatures of transformations) were determined as 0.5c* (forming below $T_{1/2}$~200 K), 0.4c* ($T_{2/5}$=150 K and) and 0.33c* ($T_{1/3}$=110K).

It can be clearly seen from Fig.2 that the peak position jumps correlate well in temperature with formation of the superstructures. However, variation of the Raman shift for second order structural transitions is expected to follow predictions of Ginzburg-Landau theory of phase transitions $\Delta v \sim (T-T_0)^{1/2}$, rather than show a jump. The origin of this discrepancy is not clear at present.

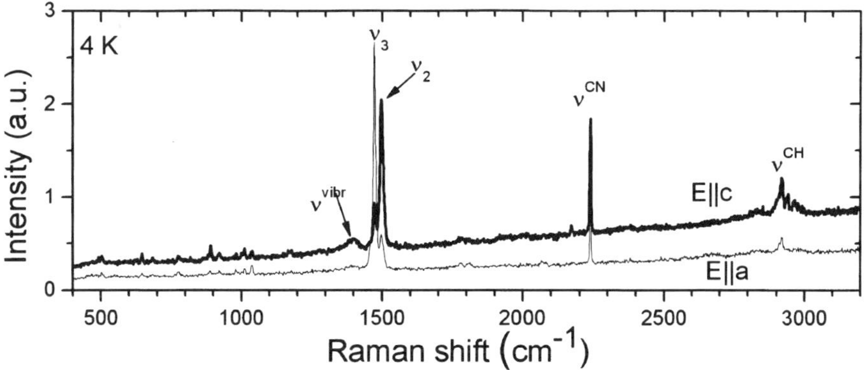

Figure1 Panorama Raman spectrum of κ-(BEDT-TTF)$_2$Cu[N(CN)$_2$]I with excitation light polarized along a- and c- crystallographic directions and detection of light of all polarizations.

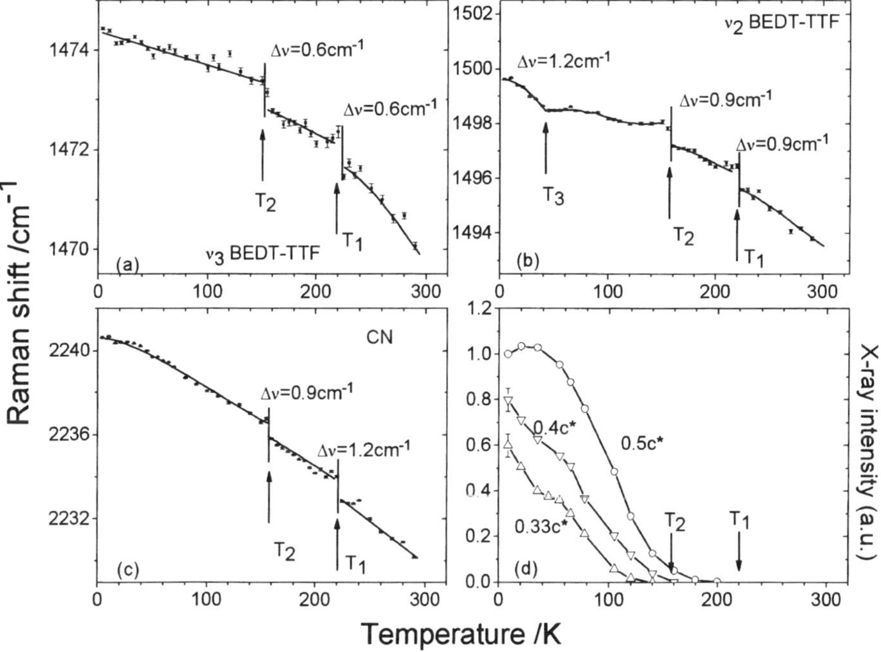

Figure 2 Temperature dependence of the peak position of ν_2 (a) and ν_3 (b) modes of BEDT-TTF and of anion CN stretching vibration (c). (d) Temperature dependence of the superstructure intensities.

3   CONCLUSION

Raman spectroscopy shows a clear correlation between the temperature dependence of the frequencies of CN vibrations in the anion layer and double C=C bond vibration modes ν_2 and ν_3 of BEDT-TTF molecule with the intensities of c*/2 and 0.4c* superstructures in organic superconductor κ-(BEDT-TTF)$_2$Cu[N(CN)]$_2$I. These studies provide the first evidence of intra-molecule distortion during structural transformations in the compound. They show that Raman spectroscopy provides a means for characterization of distortions in the parts of the crystals with low charge modulation, which in the most compounds are inaccessible by x-ray analysis.

ACKNOWLEDGEMENT

This work was partly supported by a Grant-in-Aid for Scientific Research on Priority Areas of Molecular Conductors (No. 15073223) from the Ministry of Education, Culture, Sports, Science and Technology, Japan and from grant RFBR-DFG 03-02-04023. V.S.Y. acknowledges JSPS Postdoctoral Fellowship for Foreign Researchers and Grant-in-Aid for Scientific Research No.13-01190 from Japan Society of Promotion of Sciences. M. A. T. and R. S. acknowledge IMS visiting fellowship.

References

1 J. M. Williams, A. J. Shultz, U. Geiser, K. D. Carlson, A. M. Kini, H. H. Wang, W. K. Kwok, M. H. Whangbo, and J. E. Schierber, *Science*, 1991, **252**, 1501.

2 T. Komatsu, T. Nakamura, H. Matsukawa, H. Yamochi, G. Saito, H. Ito, T. Ishiguro, M. Kusonoki, and K. Sakaguchi, *Solid State Communs.*, 1991, **80**, 843.

3 V. A. Bondarenko, R. A. Petrosov, M.A.Tanatar, V. S. Yefanov, V. M. Ogenko, N. D. Kushch, and E. B. Yagubskii, *Synth. Metals*, 1995, **70**, 955.

4 K. Kanoda, *Hyperfine Interactions*, 1997, **104**, 235.

5 T. Komatsu, H. Matsukawa, T. Inoue, and G. Saito, *J. Phys. Soc. Jpn.*, 1996, **65**, 1340

6 J. P. Pouget, *Mol. Cryst. Liq. Cryst.*, 1993, **230**, 101.

7 M.A.Tanatar, S. Kagoshima, T. Ishiguro, H. Ito, V. S. Yefanov, V. A. Bondarenko, N. D. Kushch, and E. B. Yagubskii, *Phys. Rev.*, 2000, **B62**, 15561.

8 M. A. Tanatar, T. Ishiguro, S. Kagoshima, N. D. Kushch, and E. B. Yagubskii, *Phys. Rev.* 2002, **B65**, 064516.

9 U. Geiser, A. J. Shultz, H. H. Wang, D. M. Watkins, D. L. Stupka, J. M. Williams, J. E. Schirber, D. L. Overmyer, D. Jung, J. J. Novoa, and M.-H. Whangbo, *Physica C*, 1991, **174**, 475.

10 M. Kund, H. Muller, N. D. Kushch, K. Andres, and G. Saito, *Synth. Metals*, 1995, **70**, 951.

11 O. O. Drozdova, H. Yamochi, K. Yakushi, M. Uruichi, M. Horiuchi, and G. Saito, *J. Amer. Chem. Soc.*, 2000, **122**, 4436.

CHARGE DISPROPORTIONATION WITH METALLIC CONDUCTIVITY IN NEW ET SALT, (ET)₃(AuBr₂)₂(AuBr₄)₂(solv)ₓ, HAVING A PECULIAR CRYSTAL STRUCTURE

Masafumi Sakata[1], Mitsuhiko Maesato[1], Gunzi Saito[1],
Takashi Yamamoto[2], Kyuya Yakushi[2],
Salavat S. Khasanov[1], Nataliya G. Spitsina[3], Alexander D. Dubrovskii[3],
Eduard B. Yagubskii[3], Paola Deplano[4], Maria L. Mercuri[4]

[1]Division of Chemistry, Graduate School of Science, Kyoto University, Sakyo-ku, Kyoto 606-8502, Japan
[2]Institute for Molecular Science, Myodaiji, Okazaki 444-8585, Japan
[3]Institute of Problems of Chemical Physics, RAS, Chernogolovka, Moscow District, 142 432, Russia
[4]Dipartimento di Chimica Inorganica ed Analitica, Università di Cagliari, I-09042 Monserrato-Cagliari, Italy

1 INTRODUCTION

The new salt, $(ET)_3(AuBr_2)_2(AuBr_4)_2(solv)_x$ (ET = bis(ethylenedithio)tetrathiafulvalene) (**I**), has a peculiar structure of the conducting layer which consists of ET and $AuBr_2$ molecules as shown in Figure 1.[1] It is a so-called τ-phase like structure.[2] The conducting layer and the insulating layer which is composed of $AuBr_4$ and unknown solvent molecules are alternately stacking along the c-axis. The average formal charge on ET is estimated as +4/3 from the composition. Although several salts containing the ET molecule with a charge exceeding +1 have been reported, most of them are insulators or contain ET dications which do not contribute to the conductivity.[3,4] One of the exceptions is $(ET)_6(AuBr_2)_6Br(TIE)_3$ (TIE = tetraiodoethylene), in which ET molecules have the average formal charge of +7/6.[5] To our knowledge, this salt is the first (weakly) metallic ET salt where the ET with a formal charge exceeding +1 contributes to conductivity. The structure of conducting layer of this salt is similar to that of **I**. However no detailed data such as reflectance and Raman spectra, magnetic properties have been reported. In this paper, we

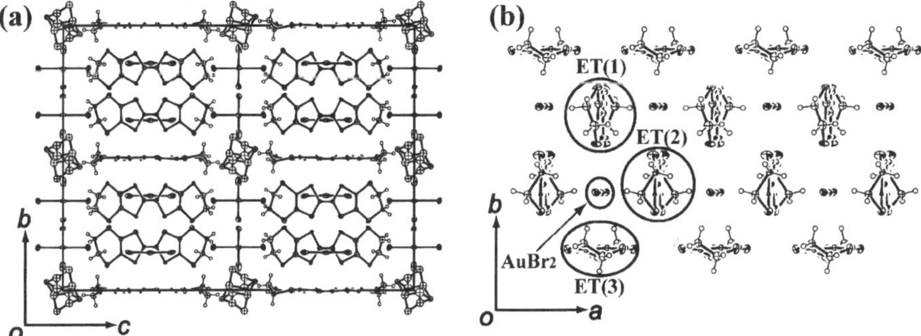

Figure 1 *Crystal structure of (ET)₃(AuBr₂)₂(AuBr₄)₂(solv)ₓ at 120 K. (a) View along the a-axis. (b) Structure of the conducting layer. Three crystallographically independent ET molecules (ET(1), ET(2), and ET(3)) and the AuBr₂ anion are indicated by circles.*

report the transport and optical properties of **I**. We discuss the charge disproportionation in the conducting layer based on the Raman and infrared conductivity spectra.

2 EXPERIMENTAL

The single crystals of **I** were obtained by the electrocrystallization described by Spitsina *et. al.*.[1] The d.c. resistivity measurements were performed by standard four probe method. Annealed gold leads (10 µm) were attached to the crystal with carbon paste (Ladd Research Ind.). Raman spectra were collected using a Renishaw Ramascope System 1000 composed of a notch filter, single monochromator, and charge-coupled device cooled by a thermoelectric device. Excitation light from the near-infrared laser (780 nm) was focused on approximately 10 µm diameter spot through a microscope equipped with an objective lens Mitsutoyo M Plan Apo 20x. The scattered light was collected with a backscattering geometry. Infrared reflectance spectra were observed from 600 to 10000 cm^{-1} using a microscopic measurement system composed of an IR microscope Spectratech IR plan and a Fourier-transformed IR spectrometer Nicolet Magna 760. Conductivity spectra were calculated through the Kramers-Kronig analysis from the reflectance spectra. The low-frequency side was extrapolated by Hagen-Rubens curve and the high-frequency side was extrapolated by the constant reflectivity.

3 RESULTS AND DISCUSSION

3.1 Transport property

Figure 2 shows the temperature dependence of resistivity (ρ) along the conducting layer. This salt shows the metallic behavior down to around 100 K with the room temperature conductivity (σ_{rt}) of 3×10 Scm^{-1}. The temperature dependence of resistivity is small in the metallic region, and the semiconducting behavior is observed below 100 K. The inset in Figure 2 shows the $d(\ln\rho)/d(1/T)$ which is negative above 125 K confirming metallic nature of **I**. The metal-insulator transition temperature (T_{MI}) is defined as the temperature

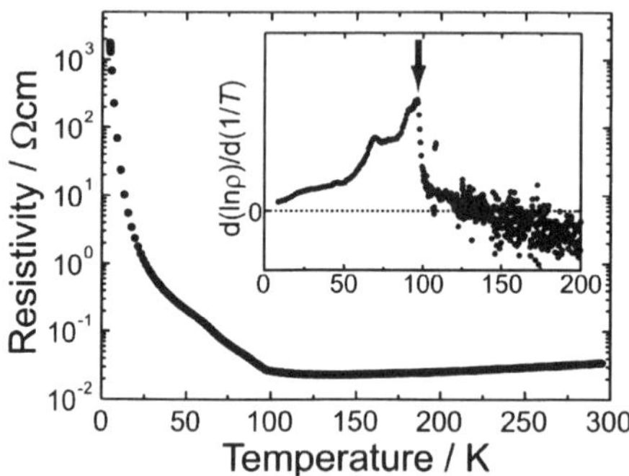

Figure 2   *Temperature dependence of resistivity of (ET)₃(AuBr₂)₂(AuBr₄)₂(solv)ₓ along the conducting layer. The arrow indicates the metal-insulator transition temperature (inset).*

where the $d(\ln\rho)/d(1/T)$ shows maximum[6], which appeared at around 100 K. These transport properties are quite similar to that of $(ET)_6(AuBr_2)_6Br(TIE)_3$.[5]

3.2 Reflectance spectra

The reflectance spectra polarized parallel to the *a*-axis within the conducting layer are shown in Figures 3a and 3c. At room temperature the spectrum exhibits the Drude-like feature regardless of direction of the polarization in the conducting *ab*-plane. On the other hand, no plasma-edge is observed in the spectrum polarized perpendicular to the conducting layer (*E//c*, Figure 3b). These results confirm the two-dimensional metallic nature of **I**. A small gap appears below 100 K as indicated by the suppression of reflectance below 1000 cm^{-1} (Figure 3c). It is consistent with the metal-insulator transition observed in the transport measurement.

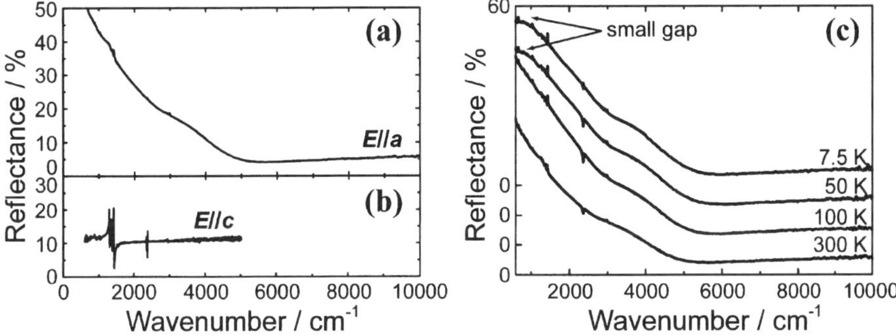

Figure 3 *Reflectance spectra of $(ET)_3(AuBr_2)_2(AuBr_4)_2(solv)_x$ at 300 K polarized parallel to the a-axis (a) and the c-axis (b). (c) Temperature dependence (E//a)*

3.3 Vibrational spectra

Figure 4 shows the vibrational spectra polarized perpendicular to the conducting layer. In the metallic phase, the peaks are observed at 1453, 1423, 1394, 1338, and 1318 cm^{-1} in the Raman spectrum. These peaks are assigned to v_2 and v_3 modes of ET molecule which are mainly contributed from the stretching vibrations of ring and central C=C bonds, respectively as shown in Figure 5. Table 1 summarizes the frequencies of such bands for typical ET species reported so far.[7,8] It is known that these modes show the frequency shift depending on the oxidation state of ET molecule.[7,9] We assigned the peak at 1453 cm^{-1} to $v_2(ET^+)$ mode and that at 1423 cm^{-1} to $v_3(ET^+)$ mode as shown in Figure 4a. The Raman peaks assignable to less ionized ET appear at higher frequencies; 1464 cm^{-1} for $v_3(ET^{0.5+})$, 1489 cm^{-1} for $v_2(ET^{0.5+})$ and 1493 cm^{-1} for $v_3(ET^0)$, and they are absent. Other peaks at lower frequencies in Figure 4a (at 1394, 1338, and 1318 cm^{-1}) are assigned to v_2 or v_3 modes of much ionized species of ET than +1. In the Raman spectrum of the dication ET salt, $ET(ClO_4)_2$, v_2 and v_3 modes are observed at 1375 and 1306 cm^{-1}, respectively.[8] Although it is hard to estimate the charge on ET exceeding +1 from Raman spectrum, because there is no systematic studies about the relationship between the frequency shift and the charge on ET exceeding +1, the present spectrum clearly indicates that ET molecules in **I** consist of plural and considerably more ionized species than +0.5 suggesting the charge disproportionation.

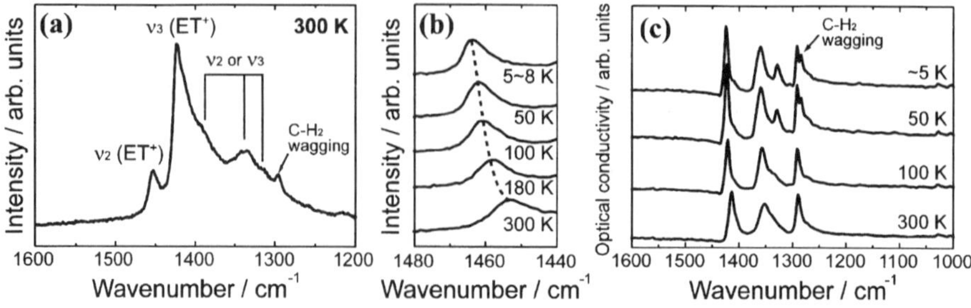

Figure 4   *(a) Raman spectrum of (ET)$_3$(AuBr$_2$)$_2$(AuBr$_4$)$_2$(solv)$_x$ at 300 K. (b) Temperature dependence of the v_2 peak assigned to ET$^+$ site in the Raman spectra. (c) Temperature dependence of the IR spectra (E//c).*

Figure 5   *Schematic views of C=C stretching modes*

Table 1   *The C=C stretching frequencies (cm^{-1}) in Raman and IR spectra of ET species*

ET	Measured			Calculated[10,11]		
	Raman[7,8]		IR[7,10]	Raman		IR
species	v_2	v_3	v_{27}	v_2	v_3	v_{27}
ET^0	1551	1493	1509	1570	1523	1538
ET$^{+0.5}$	1489	1464	1473	-	-	-
ET$^+$	1455	1431	1400	1449	1398	1396
ET^{2+}	1375	1306	-	1345	1305	1267

In IR spectrum (*E//c*), three peaks appear at 1414, 1352, and 1290 cm^{-1} in the metallic phase (Figure 4c). The v_2 and v_3 modes can be IR activated by coupling with charge transfer between ET molecules. However no charge transfer should be induced by electric field perpendicular to the conducting layer as seen in Figure 3b. Therefore we assigned all peaks observed at 300 K in the region from 1200 cm^{-1} to 1500 cm^{-1} to the v_{27} mode that is contributed from the stretching vibration of ring C=C bonds as shown in Figure 5. This mode is also charge-sensitive[7(d)]. To our knowledge, the frequency of v_{27} peak has not been reported on dication ET salt yet. The theoretical calculation suggests that the v_{27} peak on dication ET appears at 1267 cm^{-1} as shown in Table 1.[11] Using the theoretical values and assuming of the linearity in the charge dependence of frequency shift, the charges on each

ET sites are evaluated as +0.9, +1.4, and +1.8 at 300 K. The sum of them (+4.1) is in good agreement with the formal charge, +4 per $(ET)_3$. Three types of ET charge correspond to three crystallographically independent ET molecules as shown in the crystal structure in Figure 1b. However, we could not determine the spatial pattern of charge order. The accuracy of our X-ray crystal structure analysis is not enough to evaluate the charge on each ET molecule from the intramolecular bond lengths because of the large X-ray absorption coefficient of the sample.[12]

Figure 4b shows the temperature dependence of Raman ν_2 peak assigned to ET site having the nearly unit charge. The small blue shift is observed with decreasing temperature from 1453 cm^{-1} at 300 K to 1464 cm^{-1} at 5-8 K. It corresponds to the small charge variation toward more neutral direction by about 0.13 charge. However other Raman peaks show no remarkable change. On the other hand, noticeable temperature dependence is detected in IR spectra as shown in Figure 4c. The highest-frequency peak slightly shifts to higher frequency as observed in the Raman spectra. Additionally the central peak splits into two peaks below 100 K. The lowest peak observed below 100 K was assignable to ethylene wagging mode which is charge-insensitive. The shift and split of ν_{27} peaks suggest the charge redistribution from +0.9, +1.4, and +1.8 at 300 K to +0.8, +1.3, +1.5, and +1.8 below 100 K. This charge redistribution is probably related to the metal-insulator transition at 100 K.

4 CONCLUSION

We investigated the transport and optical properties of $(ET)_3(AuBr_2)_2(AuBr_4)_2(solv)_x$ which has the peculiar τ-like structure and the average formal charge on ET more than +1. This salt showed the metallic conductivity down to 100 K. In spite of the metallic behavior, the charge disproportionation was suggested by Raman and IR spectra. The metal-insulator transition was observed at 100K, possibly associated with charge redistribution.

Acknowledgements

This work was partly supported by 21st Century COE on Kyoto University Alliance for Chemistry, the COE Research on Elements Science No.12CE2005, Grant-in-Aid for Scientific Research on Priority Areas of Molecular Conductors No. 15073215, the Grants-in-Aid for Scientific Research No. 14740209 from the Ministry of Education, Culture, Sports, Science and Technology, Japan, the Grants-in-Aid for Scientific Research No. 15205019 and No. 14204033 by JSPS, Grant No. 05-03-33132 by RFBR, and the Joint Studies Program (2004-2005) of the Institute for Molecular Science.

References

1   N. G. Spitsina, A. D. Dubrovskii, E. B. Yagubskii, P. Deplano, M. L. Mercuri, M. Sakata, G. Saito and M. Maesato, to be submitted.
2   T. Mori, *Chem. Rev.*, 2004, **104**, 4947
3   L-K. Chou, M. A. Quijada, M. B. Clevenger, G. F. de Oliveira, K. A. Abboud, D. B. Tanner and D. R. Talham, *Chem. Mater.*, 1995, **7**, 530
4   a) T. Mori and H. Inokuchi, *Bull. Chem. Soc. Jpn.*, 1988, **61**, 591 b) H. Mori,

M. Kamiya, M. Haemori, H. Suzuki, S. Tanaka, Y. Nishio, K. Kajita and H. Moriyama, *J. Am. Chem. Soc.*, 2002, **124**, 1251

5   H. M. Yamamoto, J. Yamaura and R. Kato, *J. Am. Chem. Soc.*, 1998, **120**, 5909.

6   R. H. Friend, M. Miljak and D. Jérome, *Phys. Rev. Lett.*, 1978, **40**, 1048

7   a) M. E. Kozlov, K. I. Pokhodnia and A. A. Yurchenko, *Spectrochim. Acta*, 1987, **43A**, 323 b) M. E. Kozlov, K. I. Pokhodnia and A. A. Yurchenko, *Spectrochim. Acta*, 1989, **45A**, 437 c) R. Zamboni, D. Schweitzer and H. J. Keller, *Synth. Met.*, 1991, **42**, 91 d) J. Moldenhauer, C. Horn, K. I. Pokhodnia, D. Schweitzer, I. Heinen and H. J. Keller, *Synth. Met.*, 1993, **60**, 31

8   H. H. Wang, A. M. Kini and J. M. Williams, *Mol. Cryst. Liq. Cryst.*, 1996, **284**, 211.

9   a) H. H. Wang, J. R. Ferraro, J. M. Williams, U. Geiser and A. Schlueter, *J. Chem. Soc. Chem. Commun.*, **1994**, 1893 b) K. Yamamoto, K. Yakushi, K. Miyagawa, K. Kanoda and A. Kawamoto, *Phys. Rev. B*, 2002, **65**, 085110 c)T. Yamamoto, K. Yakushi, Y. Shimizu and G. Saito, *J. Phys. Soc. Jpn.*, 2004, **73**, 2326

10 T. Yamamoto, M. Uruichi, K. Yamamoto, K. Yakushi, A. Kawamoto and H. Taniguchi, *J. Phys. Chem. B*, 2005, **109**, 15226

11 T. Yamamoto, unpublished data; Quantum chemical calculation was performed with the Gaussian 03 program. Geometrical optimization for the ET dication was conducted using the DFT (B3LYP) method using standard 6-31 G** basis sets.

12 P. Guionneau, C. J. Kepert, G. Bravic, D. Chasseau, M. R. Truter, M. Kurmoo and P. Day, *Synth. Met.*, 1997, **86**, 1973

NANO-SIZE MOLECULAR CONDUCTORS DIRECTLY FORMED ON SILICON SUBSTRATES

H. M. Yamamoto,[1,2] H. Ito,[1,2] M. Ikeda,[1,2] R. Kato,[1,2] K. Shigeto,[1] K. Tsukagoshi[1]

[1]RIKEN, [2]JST-CREST, Hirosawa, Wako-shi, Saitama-ken 351-0198, JAPAN
E-mail: yhiroshi@riken.jp

1 INTRODUCTION

Recently, the physical properties of nano- and microcrystals of molecular conductors are attracting wide attention.[1-5] Because molecular conductors exhibit a variety of conduction properties, such as photo-induced insulator-metal transition,[6,7] humidity sensing behavior,[8] superconductivity,[9] ferroelectricity,[10] and giant negative magnetoresistance,[11] the size effect on these properties is interesting. For such studies, we have developed procedures that enable multi-probe measurement of nano-/microcrystals of molecular conductors on SiO_2/Si substrates at various temperatures.[1,2] These procedures are based on direct crystal growth on a substrate, and are applicable to the study of not only the size effect on nanocrystals but also the gate field effect on these materials. The field effect is especially interesting in a highly correlated electron system, which is often observed in molecular conductors, because the physical properties of the system are very sensitive to the carrier concentration. Indeed, the transformation of an insulator into a superconductor by an electric field was reported in an inorganic material.[12,13] Whereas such kind of study in molecular conductors has yet to be conducted, the field effect on bulk materials was studied in some molecular conductors.[14,15]

In our previous paper, we reported the two-probe measurement of $(DMe\text{-}DCNQI)_2Ag$, $(DMe\text{-}DCNQI\text{-}d7)_2Cu$, $(MeBr\text{-}DCNQI)_2Cu$, $(BEDT\text{-}TTF)_2I_3$, $(TMTSF)_2PF_6$, $(EDT\text{-}TTF)_4BrI_2(TIE)_5$, and $(EDO\text{-}TTF)_2PF_6$, and the four-probe measurement of only $(DMe\text{-}DCNQI\text{-}d7)_2Cu$ (DMe-DCNQI = 2,5-dimethyl-N,N'-dicyanoquinonediimine; MeBr-DCNQI = 2-methyl-5-bromo-N,N'-dicyanoquinonediimine; TIE = tetraiodoethylene; BEDT-TTF = bis(ethylenedithio)tetrathiafulvalene; TMTSF = tetramethyltetraselenafulvalene; EDT-TTF = ethylenedithiotetrathiafulvalene; EDO-TTF = ethylenedioxotetrathiafulvalene). Here we show the results of the four-probe measurement of $\alpha\text{-}(BEDT\text{-}TTF)_2I_3$ as well as the gate field effect on this material. α-$(BEDT\text{-}TTF)_2I_3$ is known to exhibit very high mobility exceeding 10^5 [cm^2 V^{-1} s^{-1}], and therefore its behavior under an electric field is interesting.

2 METHODS AND RESULTS

2.1   Crystal Growth

Single crystals of α-(BEDT-TTF)$_2$I$_3$ were grown by electrochemical reaction on a substrate. Ti (2 nm), Au (10 nm), and Pt (10 nm) were evaporated or sputtered onto SiO$_2$ (200 nm)/doped-Si substrate to form an electrode. The electrode pattern was drawn by electron beam lithography using PMMA/MMA resist. Then, the substrate was placed in a glass cell containing 5 ml chlorobenzene solution of BEDT-TTF (ca. 2 mg) and tetrabutylammonium triiodide. An anodic probe was attached to the Ti/Au/Pt electrode. The cathodic probe was positioned approximately 1 mm above the electrode on the substrate. When a voltage of approximately 2 V was applied to the probes with 5 MΩ protect resistance, tiny platelike crystals were formed on the electrode by the oxidation reaction on the anode. The crystal growth was observed with a microscope, and sufficient crystal size was achieved within 20 minutes. After the substrate was lifted out of the solution and dried, the electrodes were cut by laser ablation to form a circuit including the BEDT-TTF salt. The SEM image of α-(BEDT-TTF)$_2$I$_3$ on a substrate is shown in Figure 1.

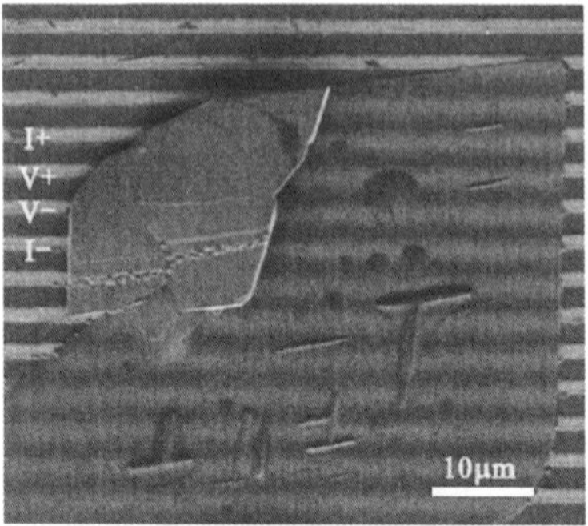

Figure 1   *Scanning electron microscope image of the α-(BEDT-TTF)$_2$I$_3$ crystal whose resistance was measured by the four-probe method. The configuration of the measurement is shown as 'I+', 'V+', 'V-', and 'I-'.*

2.2   Conductivity Measurements

Four-probe measurement was performed on the α-(BEDT-TTF)$_2$I$_3$ single crystal, and the temperature dependence of the resistance is shown in Figure 2. The four probes were selected as shown in Figure 1. The bulk crystal of α-(BEDT-TTF)$_2$I$_3$ is known to exhibit metal-insulator (M-I) transition at 135 K; however, the microcrystal showed M-I transition at approximately 150 K. This result was reproducible for several samples. Because the hard Si substrate tends to expand the soft organic crystal at low temperatures, this elevation of the transition temperature may have originated in the negative pressure produced by the difference in thermal expansion coefficient between the substrate and the microcrystal. However, the cell expansion seems to be not the only reason, because such elevation was not observed in (TMTSF)$_2$PF$_6$ in our previous study,[2] for example.

The I-V characteristics at gate voltages of $\pm$50 V were measured at 90 K. The source and drain electrodes for this measurement were 'V+' and 'V-' in the four-probe configuration. Because the insulator phase of α-(BEDT-TTF)$_2$I$_3$ is known to be a charge order state, it is plausible to consider this material as an intrinsic semiconductor in a band picture. The numbers of electrons and holes are the same at zero gate voltage. Figure 3 shows that the source-drain current is increased at the positive gate voltage. This means that the mobility of electrons is higher than that of holes in this system. Further study focusing on the influence of gate field in the temperature range of the M-I transition is under way.

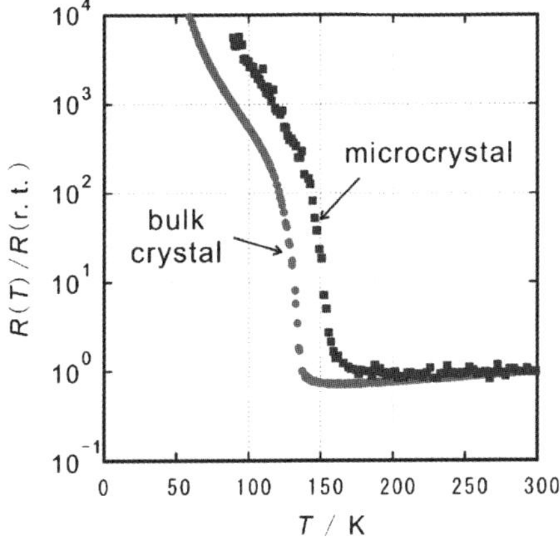

Figure 2   *Temperature dependence of resistance measured by four-probe method for a-(BEDT-TTF)$_2$I$_3$ bulk and microcrystal. The data were obtained with the same instrument.*

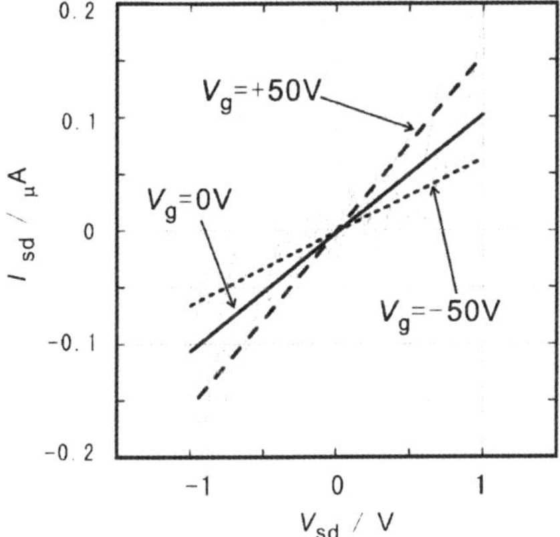

Figure 3   *I-V characteristics of a-(BEDT-TTF)$_2$I$_3$ measured at 90 K and gate voltages of +50 V and -50 V. The source-drain current was changed by a factor of more than two according to the gate voltage.*

3 CONCLUSIONS

The procedures we have developed to fabricate conductive molecular devices on silicon substrate are quite effective for the multi-probe measurement of micro-/nanocrystals of molecular conductors. In this study, the change of the transition temperature of α-(BEDT-TTF)$_2$I$_3$ was revealed by the four-probe measurement. The device was also proved to be effective in applying the gate electric field to the molecular crystal.

Acknowledgements

This work was partially supported by a Grant-in-Aid for Scientific Research (No. 16GS50219) from the Ministry of Education, Culture, Sports, Science and Technology of Japan.

References

1 H. M. Yamamoto, K. Shigeto, I. Yagi, K. Tsukagoshi, and R. Kato, PCT/JP2005/007945.

2 H. M. Yamamoto, H. Ito, R. Kato, K. Tsukagoshi, and R. Kato, *J. Low Temp. Phys.* in press

3 H. Hasegawa, T. Kubota, S. Mashiko, *Synth. Metals* 2003, **135-136**, 763.

4 M. Mas-Torrent and P. Hadley, *Small* 2005, **1**, 806.

5 J. H. Schott and M. D. Ward, *J. Amer. Chem. Soc.* 1994, **116**, 6806.

6 N. Tajima, J. Fujisawa, N. Naka, T. Ishihara, R. Kato, Y. Nishio, and K. Kajita, *J. Phys. Soc. Jpn.* 2005, **74**, 511.

7 M. Chollet, L. Guerin, N. Uchida, S. Fukaya, H. Shimoda, T. Ishikawa, K. Matsuda, T. Hasegawa, A. Ota, H. Yamochi, G. Saito, R. Tazaki, S. Adachi, and S. Koshihara, *Science* 2005, **307**, 86.

8 H. Yamochi, T. Haneda, A. Tracz, J. Ulanski, O. Drozdova, K. Yakushi, and G. Saito, *J. Phys. IV*, 2003, **114**, 591.

9 T. Ishiguro, K. Yamaji, and G. Saito, *Organic Superconductors*, Springer: Berlin, (1997)

10 P. Monceau, F. Nad, and S. Brazovskii, *Phys. Rev. Lett.* 2001, **86**, 4080.

11 N. Hanasaki, H. Tajima, M. Matsuda, T. Naito, and T. Inabe, *Mol. Cryst. Liq. Cryst.* 2000, **343**, 41.

12 K. A. Parendo, K. H. S. B. Tan, A. Bhattacharya, M. Eblen-Zayas, N. E. Staley, and A. M. Goldman, *Phys. Rev. Lett.* 2005, **94**, 197004.

13 C. H. Ahn, J.-M. Triscone, and J. Mannhart, *Nature* 2003, **424**, 1015.

14 J. S. Brooks, *Adv. Materials Opt. Elec.* 1998, **8**, 269.

15 T. Hasegawa, K. Mattenberger, J. Takeya, and B. Batlogg, *Phys. Rev. B*, **69**, 2004, 245115.

THEORETICAL STUDIES OF INTERDIMER PARAMETERS IN β'-Me$_4$P[Pd(dmit)$_2$]$_2$
CRYSTAL AND ITS SPIN FRUSTRATION.

T. Kawakami, M. Shoji, T. Taniguchi, Y. Nishimura, M. Takenaka, Y. Kitagawa, S.
Yamanaka, M. Okumura and K. Yamaguchi

Department of Chemistry, Graduate School of Science, Osaka University, Osaka 560-
0043, Japan

1 INTRODUCTION

The scientific interest in molecular magnets and superconductors with strong electron
correlation has been attracted.   In past decades, many types of important donor-based
organic superconductive salts have been reported.[1, 2]

Recently, some research groups reported the spin frustration behaviour in molecular
crystals.   Such spin frustration will be deduced from triangle geometry of
antiferromagnetic interactions.   First in the field of molecular magnetism, Awaga and his
co-workers reported that m-MPYNN$^+$(ClO$_4^-$)$_{0.72}$I$^-_{0.28}$ crystal had spin fluctuation, about a
decade ago[3, 4].   Very recently, Kanoda, Saito and their co-workers reported κ-(BEDT-
TTF)$_2$Cu$_2$(CN)$_3$ crystal[5, 6].   Tamura, Kato and their co-workers reported β'-
Me$_4$P[Pd(dmit)$_2$]$_2$ (**1**) crystal in earlier time[7-10].   In this paper, we employed the crystal
1 in order to investigate magnetic interactions.

In our previous studies[11-15], we studied many types of crystals in order to discuss
electronic and magnetic properties.   In our theoretical calculations, MO-based approaches
but extended Hückel band approaches[16-18] were employed .

2 CRYSTAL STRUCTURES

We treated the crystal **1** to study spin frustration, which was the most interesting
crystal by many scientists now.   For this crystal, Pd(dmit)$_2$ salt series (β-
Me$_4$P[Pd(dmit)$_2$]$_2$, β-MeAs[Pd(dmit)$_2$]$_2$ etc.) were also reported[7-10].   Here, we
employed only Me$_4$P salts for our purpose and one unit cell at the temperature of 295K is
illustrated in **Fig. 1** [9].   The view from c-axis is also illustrated in this figure and we can
find Pd(dmit)$_2$ layers.   The two molecules contact each other with close face-to-face type
stacking.   Thus, dimer structure is remarkable, which is commonly found among TTF-
based superconductive salts.

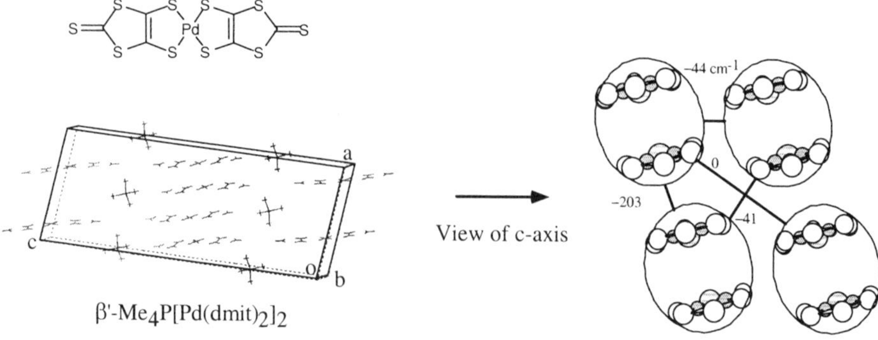

β'-Me$_4$P[Pd(dmit)$_2$]$_2$

Figure 1   *Crystal structure of β'- Me$_4$P[Pd(dmit)$_2$]$_2$ (**1**) and layers of Pd(dmit)$_2$ acceptors with the evaluated $J_{ab(M)}$ values.*

3 THEORETICAL CALCULATIONS

3.1   Evaluation of Inter-dimer $J_{ab(D)}$ Values

Effective exchange integrals (J_{ab}) successfully describe magnetic interaction. When we evaluate the inter-dimer magnetic interaction ($J_{ab(D)}$), one of our calculation schemes is direct evaluation in cluster of four molecules with two electrons ((M-M)$^{(-)}\bullet - \bullet$(M-M)$^{(-)}$).

Each dimer is constructed by nearest neighbour of two Pd(dmit)$_2$ molecules in the crystal. In such dimers electron transfer from counter ions occurs and one radical spin is created. Our calculations were carried out by the Gaussian 98 program package[19] and the simple 6-31G and LanL2DZ basis sets included in the program package were used.

For cluster systems of Pd(dmit)$_2$ dimers, the evaluated $J_{ab(D)}$ values by our scheme are also showed in **Fig. 1**. The spin structure is recognized as isosceles triangle-like of antiferromagnetic interactions (-203 cm^{-1}, -44 and -41). Face-to-face and side-by-side contact of [Pd(dmit)$_2$]$_2^{-}\bullet$ units in two-dimensional layer are repeated.

3.2   Noncollinear spin structure with classical Heisenberg spin model

The most stable spin orientation can be calculated easily with classical Heisenberg spin model. A simple model of isosceles triangle with only J_1 and J_2 parameters is illustrated in **Fig. 2A**. The angle parameter θ is defined as the twist angle between two spin vectors. When θ values are equal to 120 ° and 180, a regular triangle and collinear spin structure are construced, respectively. The classical Heisenberg spin Hamiltonian is

$$J_1 S^2 \cos(2\pi - 2\theta) + 2 J_2 S^2 \cos\theta \tag{1}$$

and S = 1/2. We calculated the suitable θ values, which minimized the total energy. In such condition, the optimized θ values are 96 °. In **Fig. 2B**, spin vectors with this θ value are illustrated. It is found that spin vectors are arranged in horizontal and vertical manner in each one-dimensional line, where opposite spin orientation occur in each line. Two spin vectors in neighboring lines make a right angle ($\theta = 96 \approx \angle 90°$), which is deduced from two magnitude of $J_{ab(D)}$ values ($J_1 = -203$ cm^{-1}, $J_2 = -44 \approx -41$).

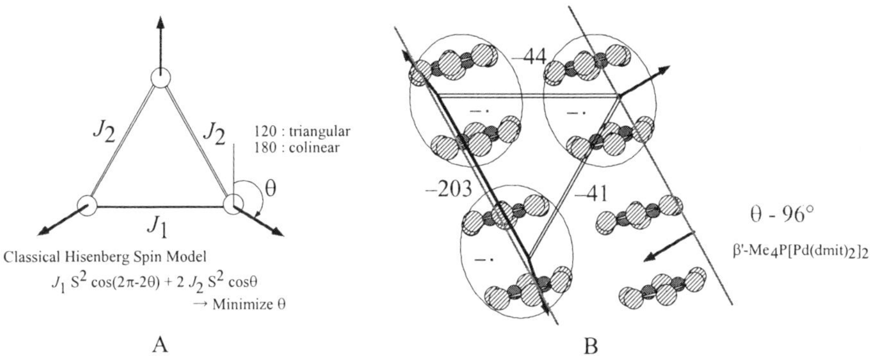

Figure 2   *Illustration of classical Heisenberg spin model (A) and minimization of θ value for the crystal 1 (B).*

4 CONCLUSION

In the crystal **1**, the $J_{ab(D)}$ values which describe spin coupling were evaluated by our theoretical calculations and gave us very important information. The spin frustration derived from competition of three antiferromagnetic interactions is expected. However, spin structure in layers of Pd(dmit)$_2$ dimers is not a regular triangle. We analysed the spin lattice with the classical Heisenberg spin model. If we study these systems by the generalized spin orbital (GSO) MO methods (i.e. G-HF and G-DFT) developed by Yamanaka, Yamaguchi and their co-workers[20], we can calculate non-collinear spins directly. In addition, numerical diagonalization for Heisenberg Hamiltonian for quantum spin lattice is very powerful approach[21, 22].

Acknowledgements

This work has been supported by a Grant-in-Aid for Scientific Research on Priority Areas (No. 15750120) from Ministry of Education, Culture, Sports, Science and Technology, Japan.

References

1   "Yuukidendoutai no Kagaku" (Ed. G Saito, Maruzen Inc., Tokyo, 2003).
2   T. Mori, *Chem. Rev.*, 2004, **104**, 4947.
3   Awaga et al., *Chem. Phys. Letters*, 1992, **195**, 21.
4   Awaga et al., *Phys. Rev.*, 1994, **B49**, 3975.
5   Y. Shimizu, K. Miyagawa, K. Kanoda, M. Maesato and G. Saito, *Phy. Rev. Lett.*, 2003, **91**, 107001.

6　H. Yamochi, T. Komatsu, N. Matsukawa, G. Saito, T. Mori, M. Kusunoki and K. Sakaguchi, *J. Am. Chem. Soc.*, 1993, **115**, 11319.

7　R. Kato, *Chem. Rev.*, 2004, **104**, 5319.

8　M. Tamura and R. Kato, *J. Phys: Cond. Matt.*, 2002, **14**, L729.

9　C. Faulmann, J. -P. Legros, P. Cassoux, J. Cornelissen, L. Brossard, M. Inokuchi, H. Tajima and M. Tokumoto, *J. Chem. Soc., Dalton Trans.*, 1994, 249.

10　A. Kobayashi, H. Kim, Y. Sasaki, K. Murata, R. Kato and H. Kobayashi, *J. Chem. Soc., Faraday Trans.*, 1990, **86(2)**, 361.

11　T. Kawakami, T. Taniguchi, Y. Kitagawa, Y. Takano, H. Nagao and K. Yamaguchi, *Molecular Physics*, 2002, **100**, 2641.

12　T. Kawakami, Y. Kitagawa, T. Taniguchi, S. Nakano and K. Yamaguchi, *Mol. Cryst. Liq. Cryst.*, 2002, **379**, 489.

13　T. Kawakami, Y. Kitagawa, T. Taniguchi, S. Nakano, R. Takeda and K. Yamaguchi, *Synthetic Metals*, 2003, **133-134**, 565.

14　T. Kawakami, T. Taniguchi, S. Nakano, Y. Kitagawa, K. Yamaguchi, *Synthetic Metals*, 2003, **137**, 1259.

15　T. Kawakami, T. Taniguchi, S. Nakano, Y. Kitagawa and K. Yamaguchi, *Polyhedron*, 2003, **22**, 2051.

16　G. Saito and S. Kagoshima Eds., The Physics and Chemistry of Organic Superconductors (Springer, Berlin, 1990).

17　T. Mori, *Bull. Chem. Soc. Jpn.*, 1998, **71**, 2509.

18　T. Mori, H. Mori and S. Tanaka, *Bull. Chem. Soc. Jpn.*, 1999, **72**, 179.

19　Gaussian 98, M. J. Frisch et al. and J. A. Pople, Gaussian, Inc.

20　S. Yamanaka, R. Takeda, T. Kawakami, S. Nakano, D. Yamaki, S. Yamada, K. Nakata, T. Sakuma, T. Takada and K. Yamaguchi, *Int. J. Quant. Chem.*, 2003, **95**, 512.

21　M. Takahata, M. Shoji, H. Nitta, R. Takeda, S. Yamanaka, M. Okumura, M. Nakano, K. Yamaguchi, *Int. J. Quant. Chem.*, 2005, **105**, 615.

22　T. Kawakami, T. Taniguchi, M. Shoji, Y. Kitagawa, S. Yamanaka, M. Okumura and K. Yamaguchi, Mol. Cryst. Liq. Cryst., in press.

Design and Synthesis of Functional Molecular Materials

RECENT PROGRESS OF $(EDO\text{-}TTF)_2PF_6$ AND RELATED COMPLEXES

H. Yamochi[1] and G. Saito[2]

[1]Research Center for Low Temperature and Materials Sciences, Kyoto University, Sakyo-ku, Kyoto 606-8502, Japan and JST, Exploratory Research for Advanced Technology (ERATO), 1-1 O-ho, Tsukuba 305-0801, Japan. E-mail: yamochi@kuchem.kyoto-u.ac.jp
[2]Division of Chemistry, Graduate Scool of Science, Kyoto University, Sakyo-ku, Kyoto 606-8502, Japan

1 INTRODUCTION

The discovery of the superconductivity in organic charge-transfer (CT) complexes has stimulated the exploit of conducting component molecules more than before. The first organic superconductors contained the heavy atom substituted tetrathiafulvalene (TTF) derivative, TMTSF, which had been followed by an all sulfur type heterocyclic compound of BEDT-TTF to expand the variety of complexes.[1] In the course of spreading the classes of superconducting component molecules, BEDO-TTF was synthesized, which is a partially oxygen substituted analogue of BEDT-TTF.[2] The thorough investigation of this donor molecule proved the self-assembling nature to form a particular packing pattern of the layered structure in the conducting CT complexes. The donor layer structure denoted as 'I_3-type' provided two-dimensional (2D) electronic structures and hence the stable metallic nature of the complexes. The origin of the self-assembling nature was assigned to the molecular shape in which the molecular width at the inner five and outer six membered rings were close to each other in addition to the existence of terminal ethylenedioxy groups which enabled the formation of the intermolecular C-H···O type weak hydrogen bonds.

The stability of the metallic states of BEDO-TTF complexes prohibited phase transitions in general. This is the reason why only a few kinds of BEDO-TTF complexes shows superconductivity. The phase transitions are regarded as a functionality of conducting CT complexes since the emergence can be utilized as the basis of the device actions. Furthermore, phase transition itself is a subject of scientific studies. In these viewpoints, BEDO-TTF is a poor conducting component molecule for the functional conducting molecular materials, although this donor had provided interesting materials such as metallic Langmuir-Blodgett films without any secondary treatment besides the compilation from the air-water interface[3] and the transparent metallic polymer composite

TTF: X = S, $R^1 = R^2 = H$
TMTSF: X = Se, $R^1 = R^2 = CH_3$
BEDT-TTF: X = S, $R^1R^1 = R^2R^2 = -SCH_2CH_2S-$
BEDO-TTF: X = S, $R^1R^1 = R^2R^2 = -OCH_2CH_2O-$
BEDODB-TTF: X = S, $R^1R^1 = R^2R^2$ = 4,5-ethylenedioxybenzo
EOET: X = S, $R^1R^1 = -OCH_2CH_2O-$, $R^2R^2 = -SCH_2CH_2S-$
EDO-TTF: X = S, $R^1R^1 = -OCH_2CH_2O-$, $R^2 = H$

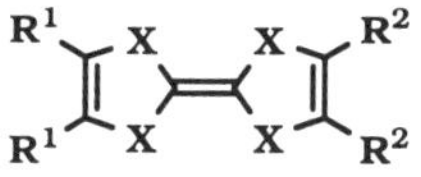

Scheme 1

films.[4] The chemical modification of BEDO-TTF was carried out to suppress the self-assembling nature partly. The molecular designs to introduce a bulky substituent and heteroatoms provided BEDODB-TTF[5] and EOET,[6] respectively. The CT complexes of the former donor were classified to insulators or semiconductors except an example which showed a faint metallic behavior, while the latter conducting component molecule exhibited the weaker self-assembling property than that of BEDO-TTF.

As another chemical modification of BEDO-TTF, the removal of one of the ethylenedioxy groups was examined, which is the key substituent to form the self-assembled 2D donor layers in the CT complexes. Although the target molecule of EDO-TTF had been known when the authors planned to study this donor molecule, only a few numbers of semiconducting or insulating CT complexes had been reported.[7] This report describes the potency of the self-assembling nature of EDO-TTF and how the authors had found out the thermally induced peculiar metal-insulator (MI) transition in the title complex along with the summary of the physical properties the compilation of which assisted the finding of the ultra-fast and highly efficient photo-induced insulator-to-metal phase transition reported in a separate paper of this book.

2 CT COMPLEXES OF EDO-TTF

2.1 Complexes with TCNQ derivatives

As our first examination of EDO-TTF for a conducting component molecule, the complex formation with TCNQ derivatives had been carried out.[8] Figure 1 shows the plot of the lowest CT transition energy vs. the difference between donor oxidation and acceptor reduction potentials of the component molecules for a complex. This type of examination had been applied to seek neutral-ionic transition systems[9] and later utilized also for the designing of conducting CT complexes. As is described in reference 9, CT complexes having ionic and neutral ground states appear near the V-shaped line in the left and right

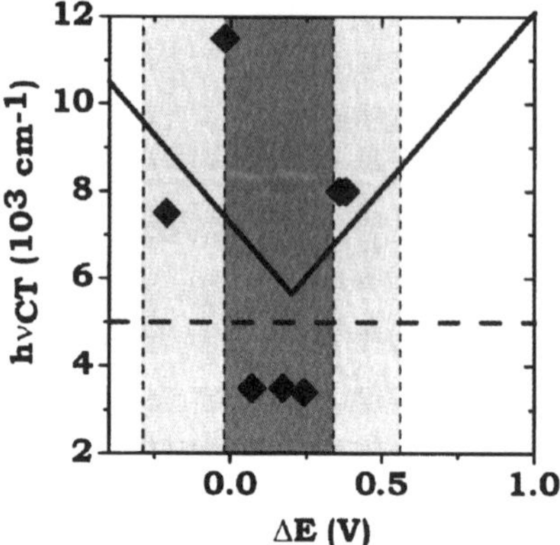

Figure 1 *Plot of the transition energy of the lowest energy charge-transfer band measured on KBr pellet (hv_{CT}) vs. the difference between the donor oxidation and acceptor reduction potentials (ΔE) for EDO-TTF complexes with TCNQ derivatives. The lightly and thickly shaded parts indicate the ΔE regions in which one-dimensional TTF-TCNQ complexes and two-dimensional BEDO-TTF complexes were obtained, respectively. The V-shaped line indicates the theoretically derived relation between hv_{CT} and ΔE for mixed stacked CT complexes.*

regions, respectively. When the CT complex having a segregated stack structure is a partially charge-transferred one, the plot is located below the horizontal line at around 5000 cm^{-1} in the moderate ΔE region. The one-dimensional (1D) TTF-TCNQ system provides metallic complexes in the region of the ΔE from -0.02 to +0.34 V (heavily shaded region in Figure 1).[10] Due to the self assembling nature to form stable 2D electronic states, BEDO-TTF has afforded metallic complexes in the ΔE region of almost twice of that for 1D system as shown by lightly shaded area in Figure 1.

The plots of EDO-TTF complexes indicated by closed squares demonstrated that this donor does not succeed to the self-assembling nature of BEDO-TTF, though one of the etylenedioxy groups remains.

2.2 Radical Cation Salts of EDO-TTF

As was described in the preceding section, the removal of an etylenedioxy group drastically modified the cohesive property of BEDO-TTF. To investigate the potential of EDO-TTF as a conducting component molecule and how the self-assembling nature is lost, the authors examined the radical cation salts, in which only the donor can provide the conducting path and itinerant charge carriers.

The preparation of the salts was carried out by employing the conventional electrocrystallization method, in which the donor was electrolyzed in the solution containing tetrabutylammonium salt. A trick of the usage of ethanol as solvent was applied. Since EDO-TTF is labile to acids, the commonly used halogenated solvents such as 1,1,2-trichloroethane could not be used. Other common solvents of CH_3CN, PhCN, and THF have not provided the solid salts maybe due to the high solubility of the salts.

The crystal structures of EDO-TTF radical cation salts were rich in variety reflecting the poor self-assembling nature. Along with the head-to-tail type columnar stack formed in the 2:1 PF_6 salt described later, twisted head-to-head one was observed in the ReO_4 and $GaCl_4$ salts.[11] The novel feature of the latter two isostructural salts is the co-existence of two types of the columnar stacking in a crystal. Interestingly, the tetrahedral anion seemed to stick out the apex into the cavity formed in the head-to-tail stacking donor layer ($z = 1/4$) and the flat basal plane fits to the flat surface of the head-to-head stacking donor layer ($z =$

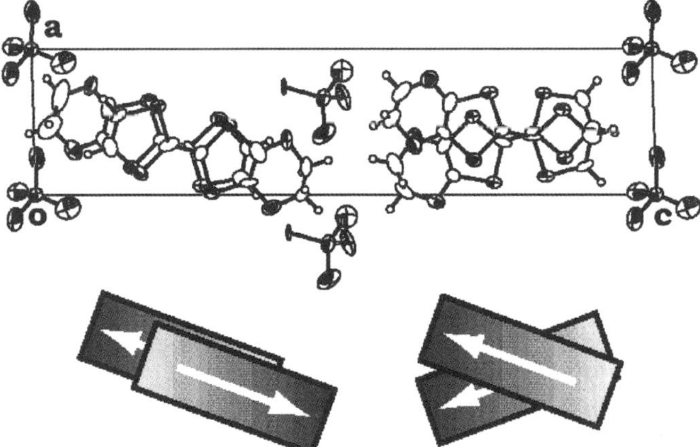

Figure 2 *The b-axis projection of the crystal structure of (EDO-TTF)$_2$ReO$_4$ (top) and the schematic representation of the stacking pattern of the donor molecules (bottom).*

3/4) as shown in Figure 2. Furthermore, 5:3 and 4:0.85:4(H_2O) stoichiometries were observed in the BF_4 and Sb_2F_{11} salts, respectively, the latter of which had obtained by the elecrtocrystallization in the presence of $Bu_4N\cdot SbF_6$. In the donor layer of the former salt, monomer and dimer of donor molecules were arranged orthogonally in the donor layer, while the face-to-face stacked tetramer also coexisted in the latter salt.[12] These results strongly suggest that EDO-TTF adjusts the packing pattern in the complexes to fit the shape of the counter components. This situation reminds the discussion what determines the donor packing pattern in the BEDT-TTF ambient pressure superconductors; the shape of the anion layer was assigned to the dominant factor in this case.[13]

2.3 Metal-Insulator Transition of (EDO-TTF)$_2$PF$_6$

As reported, (EDO-TTF)$_2$PF$_6$ exhibited first-order metal-insulator (MI) transition at approximately 280 K with the jump of five orders of resistivity.[14] The comparison of the crystal structures at above and below the MI transition temperature (T_{MI}) showed the cooperative feature of the mechanisms to cause this transition, while an MI transition is, in general, interpreted by one of the mechanisms such as Peierls transition, spin density wave transition and so on. Also, the molecular deformation of the conducting component molecule was of a peculiar characteristic. In the room temperature (RT) phase, only one EDO-TTF was crystallographically unique, which showed almost flat shape. Two donor molecules were, however, independent to each other in the low temperature (LT) phase, one of which exhibited distinct bending of the molecular plane while the flatness of another was enhanced from that in the RT phase as shown in Figure 3.

As for the transition mechanism, the emergence of a kind of density waves had been expected from the quasi-1D Fermi surface based on the RT crystal structure. In fact, the weakly dimerized donor stacking at RT was modulated to a tetramerized one in the LT phase, the direction and periodicity of the dimerization of dimer corresponded to those of the nesting vector. This modulation of the crystal structure associated with the MI transition assigned the mechanism to a Peierls transition. On the other hand, the comparison of the intramolecular bond lengths suggested that the flat and bent molecules in the LT phase were richly and poorly charged, respectively, while all the donor

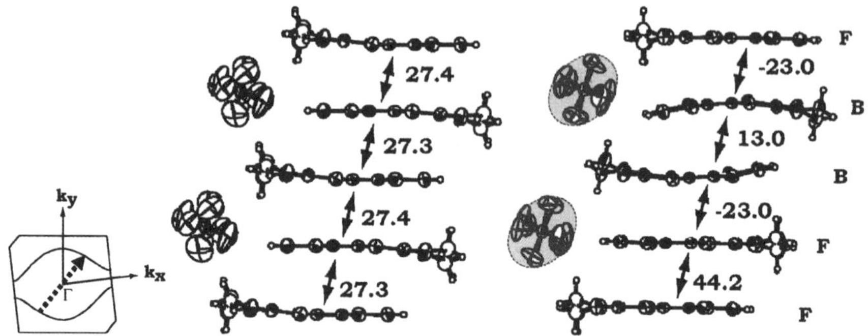

Figure 3   The calculated Fermi surface (left), the crystal structure of (EDO-TTF)$_2$PF$_6$ at room temperature (center) and at 260 K (right) viewed along the molecular short axis of the donor. The expected nesting vector is indicated with dotted arrow in the most left panel. The numbers between the donor molecules indicate the magnitudes of intermolecular overlap integrals in multiples of 10^{-3}. In the most right panel, the bent and flat donor molecules are designated with B and F, respectively. The plane of the uniaxial rotation of anion is shown by shaded ellipses.

molecules were charged +0.5 in the RT phase as summarized in Figure 4. This meant that this MI transition also showed the feature of charge-ordering (CO) transition. Furthermore, the almost freely rotating disorder of the anion in the RT phase was suppressed to an uniaxial one in the LT phase. The periodicity of the anion arrangement was doubled in the LT phase comparing to that in the RT phase corresponding to the nesting vector. The mechanism of this MI transition, hence, also contained that of the order-disorder (OD) transition of the counter component.

It was a difficult task to conclude that the MI transition was caused by the cooperation of the mechanisms since these three types of independent transitions could accidentally take place at the same temperature. Fortunately, AsF_6 anion provided the isostructural salt to $(EDO\text{-}TTF)_2PF_6$, which showed the same type of MI transition at a specified temperature of 268 K. These results strongly suggested the cooperative feature of the transition mechanisms in both of PF_6 and AsF_6 salts.

The occurrence of the CO was confirmed by successive two examinations. The comparison of the Raman spectra of the neutral donor molecule, 1:1 salt with IBr_2, and those of PF_6 salt at RT and at LT concluded that the charge on EDO-TTF in the PF_6 salt was disproportionated in 0.1 and 0.9 at 4.2 K.[15] The [0110] type of CO pattern corresponding to the arrangement of bent and flat donor molecules in the LT phase was directly proved by the accurate crystal structure analysis, in which the mapping of the electron density in the unit cell resulted in the charges on the bent and flat molecules as 0.2(1) and 0.8(1), respectively.[16] The CO pattern depicted schematically in Figure 5 is categorized into the diagonal stripe type according to the classifications for BEDT-TTF complexes,[17] however, further experimental and theoretical investigations should be needed to understand the CO pattern in $(EDO\text{-}TTF)_2PF_6$. The deformation of conducting component molecule and the Coulombic interaction between the cationic and anionic species were expected to play an important role to stabilize the LT phase,[15, 16] while these factors have not been examined thoroughly.

Thermodynamic measurements were carried out to examine the MI transition. The contribution of the OD transition to the latent heat associated with the MI transition of $(EDO\text{-}TTF)_2PF_6$ was confirmed, while the magnitude of 5.0 kJ/mol suggested that the disorder of the anion in the RT phase is not a free rotation but a restricted one.[18]

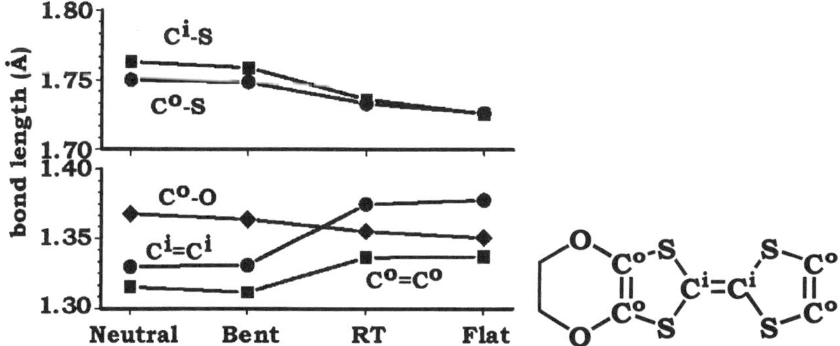

Figure 4 *Comparison of the averaged bond lengths of EDO-TTF. The molecular structure was taken from the crystal structures of neutral molecule (Neutral), (EDO-TTF)₂PF₆ at room temperature (RT), bent (Bent) and flat (Flat) molecules at 129 K. The right panel shows the definitions of the bonds.*

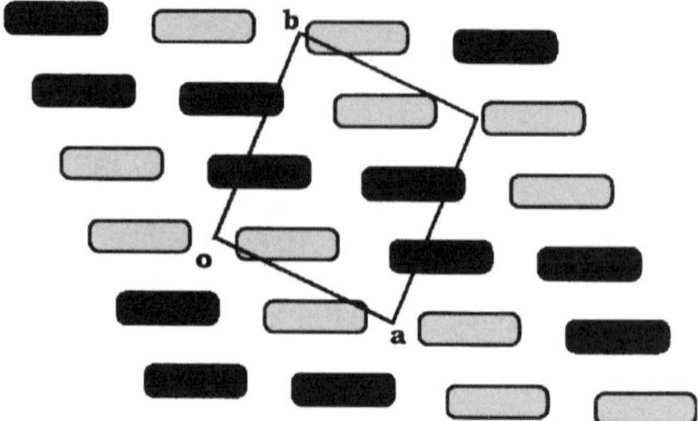

Figure 5 *Schematic representation of the charge-ordering pattern in the low temperature phase of (EDO-TTF)₂PF₆ projected along the donor longitudinal axis. The closed and shaded rectangulars correspond to the charge poor and rich donor molecules, respectively. The open parallelogram shows the c-axis projection of a unit cell of the low temperature phase.*

To find out the effect of the peculiarity of the MI transition, the anisotropic feature of the modulation of the unit cell associated with the transition was noted. The area and the thickness of the layered structure were increased and decreased, respectively, when the metal was transformed to insulator. Concerning the relative motions of donor and anion, the penetration of the latter into the donor layer brought about the transition from metal to insulator. This feature was proved by the observation of the uniaxial strain dependence of T_{MI}. The application of a small strain of ca. 4 kbar perpendicular to the layered structure of (EDO-TTF)₂PF₆ pushed up T_{MI} more than 60 K, while the same magnitude of strains applied parallel to the layers affected on T_{MI} only by ca. 10K.[19] These results seemed to correspond to the deformation of the donor molecule associated with the transition.

The CO feature of the first-order MI transition of (EDO-TTF)₂PF₆ facilitated the examination of the photo-induced phase transition (PIPT). As a result, this salt showed unexpectedly ultra-fast and highly efficient insulator-to-metal transition.[20] It is noteworthy that the temperature dependences of the reflectivity[15] and heat capacity[18] measured independently to this examination supported the emergence of the PIPT. Especially, the big variation of the reflectivity associated with the MI transition at around the *CT2* peak in Figure 6 had drawn attention in the PIPT examinations.

3 CONCLUDING REMARKS

The chemical modification to reduce the size of π-electron system of a self-assembling conducting component of BEDO-TTF resulted in the findings of the thermally induced peculiar MI transition associated with the distinct molecular deformation and ultra-fast and highly efficient PIPT of (EDO-TTF)₂PF₆. The molecular deformation is regarded to be the origin of the mixing of the mechanisms in the former transition. Also, this flexibility of the conducting component molecule is expected to serve the strong electron-phonon coupling which realized the ultra-fast and highly efficient PIPT.

A perturbation can efficiently modulate the electronic structure of a small sized conjugated system more than that applied on a big π-electron system just like as the substitution effects. The deformation of a π-electron system ought to modulate the

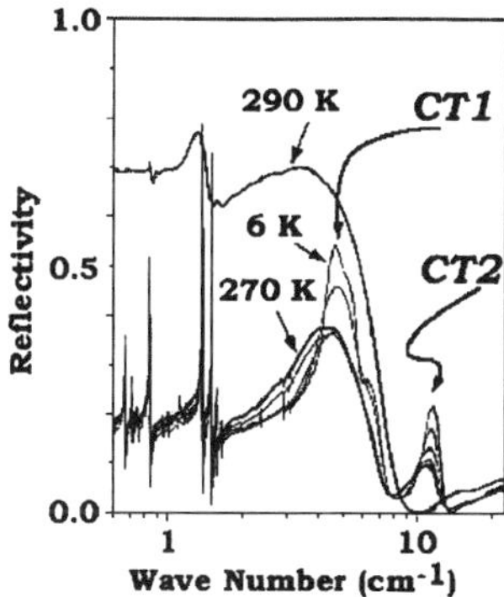

Figure 6   *Temperature dependence of the reflectance spectra of (EDO-TTF)$_2$PF$_6$. Only the spectrum at 290 K shows the metallic feature, while all the others correspond to the insulating phase. The peaks denoted as CT1 and CT2 observed in the low temperature phase were assigned to the intermolecular electronic transitions of $D^0D^+D^+D^0 \rightarrow D^+D^+D^0D^0$ and $D^0D^+D^+D^0 \rightarrow D^0D^0D^{2+}D^0$, respectively.*

electronic structure, and hence, the intramolecular vibrations corresponding to the deformation should show strong electron-molecular vibration couplings. When the molecular vibration contributes to the phonon, a flexible π-electron system is expected to provide the solids with strong electron-lattice couplings. (EDO-TTF)$_2$PF$_6$ is regarded as a typical example in which the small sized flexible π-electron system provided the huge response to a small stimulation of pulsed light. Although distinct molecular deformation associated with the thermally induced phase transition had been observed even in the BEDT-TTF complex with PtCl$_6^{2-}$,[21] the smaller size of the conjugated system should provide bigger possibilities of the molecular deformation in a solid concerning the void volume to allow the motions. As for the generation of the itinerant charge carriers in the CT complexes, a moderately small sized $7\pi+7\pi$ multi stage redox system with a few numbers of additional conjugated atoms such as EDO-TTF is regarded to be suitable to provide molecular conductors showing the huge response to a small stimulation.

Acknowledgments

The authors would express their sincere thanks to the colleagues worked with them. The works described in this article were partly supported by Grants-in-Aid for Scientific Research on Priority Areas of Molecular Conductors (No. 15073215), 21st Century COE on Kyoto University Alliance for Chemistry from the Ministry of Education, Culture, Sports, Science and Technology, Japan, and for Basic Research (No. 15205019) by JSPS.

References

1 T. Ishiguro, K. Yamaji and G. Saito, in Organic Superconductors 2nd. edition, Springer, Berlin, 1998.

2 S. Horiuchi, H. Yamochi, G. Saito, K. Sakaguchi and M. Kusunoki, *J. Am. Chem. Soc.*, 1996, **118**, 8604, H. Yamochi, in *TTF Chemistry Fundamentals and Applications of*

Tetrathiafulvalene, ed. J. Yamada and T. Sugimoto, Kodansha Springer, Tokyo, 2004, ch. 4.

3   T. Nakamura, G. Yunome, R. Azumi, M. Tanaka, H. Tachibana, M. Matsumoto, S. Horiuchi, H. Yamochi and G. Saito, *J. Phys. Chem.*, 1994, **98**, 1882,   J. K. Jeszka, A. Tracz, A. Sroczynska, M. Kryszewski, H. Yamochi, S. Horiuchi and G. Saito, *Synthetic Metals*, 1999, **106**, 75.

4   J. K. Jeszka, A. Tracz, A. Sroczynska, M. Kryszewski, H. Yamochi, S. Horiuchi and G. Saito, *Synthetic Metals*, 1999, **106**, 75.

5   T. Senga, K. Kamoshida, L. A. Kushch, G. Saito, T. Inayoshi and I. Ono, *Mol. Cryst. Liq. Cryst. Sci. Technol., Sect. A*, 1997, **296**, 97.

6   G. Saito, H. Sasaki, T. Aoki, Y. Yoshida, A. Otsuka, H. Yamochi, O. O. Drozdova, K. Yakushi, H. Kitagawa and T. Mitani, *J. Mater. Chem.*, 2002, **12**, 1640.

7   T. Mori, H. Inokuchi, A. M. Kini and J. M. Williams, *Chem. Lett.*, 1990, 1279. G. C. Papavassilliou, V. C. Kakoussis, D. J. Lagouvardos and G. Mousdis, *Mol. Cryst. Liq. Cryst.*, 1990, **181**, 171. See, also; C. Meziere, M. Fourmigue and J. P. Fabre, *C. R. Acad. Sci., Ser. IIc: Chim.*, 2000, **3**, 387. M. Iyoda, Y. Kuwatani, E. Ogura, K. Hara, H. Suzuki, T. Takano, K. Takeda, J. Takano, K. Ugawa, M. Yoshida, H. Matsuyama, H. Nishikawa, I. KIkemoto, T. Kato, N. Yoneyama, J. Nishijo, A. Miyazaki and T. Enoki, *Heterocycles*, 2001, **54**, 833.

8   A. Ota, H. Yamochi and G. Saito, *Mol. Cryst. Liq. Cryst.*, 2002, **376**, 177.

9   J. B. Torrance, J. E. Vanzquetz, J. J. Mayerle and V. Y. Lee, *Phys. Rev. Lett.*, 1981, **46**, 253, J. B. Torrance, A. Girlando, J. J. Mayerle, J. I. Crowley, V. Y. Lee, P. Batail and S. J. LaPlaca, *Phys. Rev. Lett.*, 1981, **47**, 1747.

10 G. Saito and J. P. Ferraris, *Bull. Chem. Soc. Jpn.*, 1980, **53**, 2141.

11 A. Ota, H. Yamochi and G. Saito, *J. Low Temp. Phys.*, in press.

12 A. Ota, H. Yamochi and G. Saito, *this book.*

13 H. Yamochi, T. Komatsu, N. Matsukawa, G. Saito, T. Mori, M. Kusunoki and K. Sakaguchi, *J. Am. Chem. Soc.*, 1993, **115**, 11319.

14 A. Ota, H. Yamochi and G. Saito, *J. Mater. Chem.*, 2002, **12**, 2600.

15 O. Drozdova, K. Yakushi, A. Ota, H. Yamochi and G. Saito, *Synthetic Metals*, 2003, **133-134**, 277, O. Drozdova, K. Yakushi, K. Yamamoto, A. Ota, H. Yamochi, G. Saito, H. Hashiro and D. B. Tanner, *Phys. Rev.*, 2004, **B 70**, 075107.

16 S. Aoyagi, K. Kato, A. Ota, H. Yamochi, G. Saito, H. Suematsu, M. Sakata and M. Takata, *Angew. Chem. Int. Ed.*, 2004, **43**, 3670.

17 H. Seo, *J. Phys. Soc. Jpn.*, 2000, **69**, 805.

18 K. Saito, S. Ikeuchi, A. Ota, H. Yamochi and G. Saito, *Chem. Phys. Lett.*, 2005, **401**, 76.

19 M. Sakata, M. Maesato, A. Ota, H. Yamochi and G. Saito, *Synthetic Metals*, 2005, **153**, 393.

20 N. Uchida, S. Koshihara, T. Ishikawa, A. Ota, S. Fukaya, C. Matthieu, H. Yamochi and G. Saito, *J. Phys. IV France*, 2004, **114**, 143, M. Chollet, L. Guerin, N. Uchida, S. Fukaya, H. Shimoda, T. Ishikawa, K. Matsuda, T. Hasegawa, A. Ota, H. Yamochi, G. Saito, R. Tazaki, S. Adachi and S. Koshihara, *Science*, 2005, **307**, 86.

21 M.-L. Doublet, E. Canadell and R. P. Shibaeva, *J. Phys. I*, 1994, **4**, 1479.

CRYSTAL STRUCTURES AND PHYSICAL PROPERTIES OF $(EDO\text{-}TTF)_5(BF_4)_3$
AND $(EDO\text{-}TTF)_4(Sb_2F_{11})_{0.85}(H_2O)_4$

A. Ota[1,2], H. Yamochi[3], G. Saito[1]

[1]Division of Chemistry, Graduate School of Science, Kyoto University, Sakyo-ku, Kyoto 606-8502, Japan
[2]Department of Chemistry, Graduate School of Science and Engineering, Tokyo Institute of Technology, 2-12-1-W4-1, O-okayama, Meguro-ku, Tokyo 152-8551, Japan. E-mail: aota@chem.titech.ac.jp
[3]Research Center for Low Temperature and Materials Sciences, Kyoto University, Sakyo-ku, Kyoto 606-8502, Japan

1   INTRODUCTION

$(EDO\text{-}TTF)_2PF_6$ (EDO; ethylenedioxytetrathiafulvalene) shows a drastic metal-insulator transition at around 280 K the mechanism of which consists of the cooperative nature of Peierls, charge ordering, and anion ordering transitions associated with the distinct molecular deformation. Moreover, photo-induced phase transitions (PIPT) from the insulating state into the metallic state was observed in this complex with the high speed and efficiency.[2] The search of the new series of EDO complexes is needed for the understanding of EDO as a component molecule of functional materials, and the tuning of the structures and properties has been examined by changing the counter component.[3] This report describes the structures and properties of $(EDO)_5(BF_4)_3$ (**1**) and $(EDO)_4(Sb_2F_{11})_{0.85}(H_2O)_4$ (**2**) as in the course of the search.

2   EXPERIMENTAL

1 and **2** were prepared by the electrochemical method with $[(C_4H_9)_4N]X$ ($X = BF_4$ for **1** and SbF_6 for **2**) in the ethanol solutions.[1] X-ray diffractions were measured with an imaging plate type diffractometer. UV-vis-IR spectra were measured with KBr disc method. DC conductivities were measured with a four-probe method on the single crystals for **1** and on the compressed pellets for **2**. Magnetic susceptibility (χ) was measured by a SQUID magnetometer on the polycrystalline sample of **2**.

EDO-TTF (EDO)

3   RESULTS AND DISCUSSION

3.1 Crystal and band structures

*3.1.1 $(EDO)_5(BF_4)_3$ (**1**).*   The crystallographic and refinement data of **1** and **2** are summarized in Reference 4. The single crystals of **1** were obtained as rhombic plates. One unit cell includes the five EDO molecules and three BF_4 anions, all of which are crystallographycally independent (Figure 1a, b). The crystal consists of the alternate stacks of the donor and anion layers. All the BF_4 anions show the disorder in all F atoms which have the two sites for one atom, and are accommodated in the channels between the donor layers. EDO forms the two kinds of the head-to-tail dimers (**D1** and **D2**). **D1** and **D2** forms the tetramer of EDO as **D1-D2** along the $a + b$ direction. The molecular plane of the rest donor molecule is perpendicular to those in the tetramer. The quite warped open Fermi surfaces are calculated (Figure 1c). The splits and deep dips of the density of states (DOS) are derived from the dimerizations of the donor molecules. As a result, the conduction band is effectively narrow, and this suggests that the electronic correlation is substantial in this system.

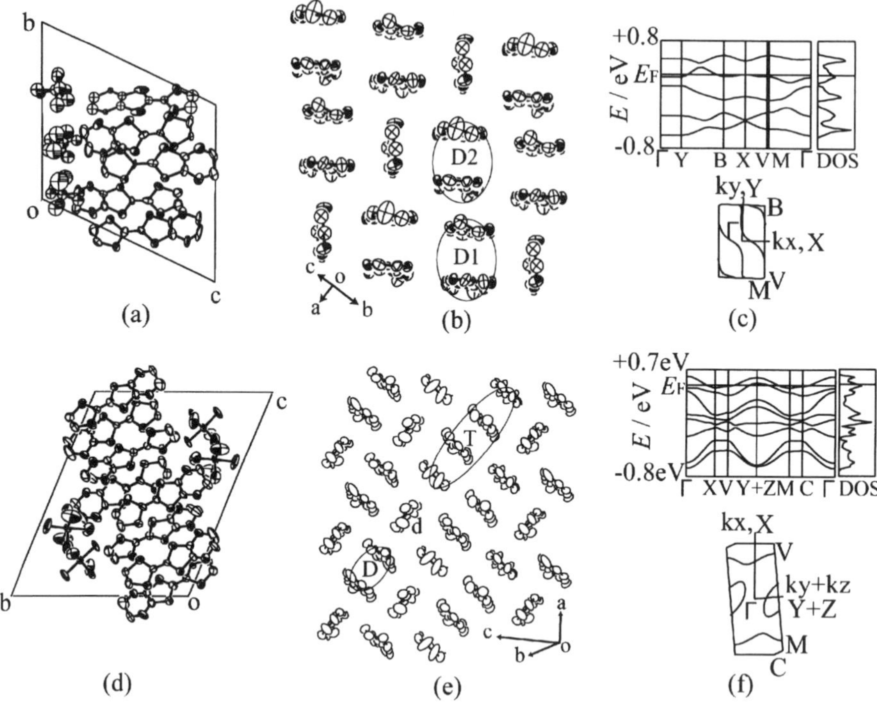

Figure 1 *Crystal structure of 1 viewed along (a) the a axis, and (b) long axis of EDO. (c) Calculated band structure and Fermi surfaces of 1. (d), (e), (f) Those corresponding in 2. E_F is Fermi energy. The definitions of D1, D2, T, D, d , and DOS are shown in the text.*

3.1.2 $(EDO)_4(Sb_2F_{11})_{0.85}(H_2O)_4$ (2).   **2** was obtained as thin and small (below 0.5 mm in the length) rohmbic crystals. Although SbF_6^- was employed for the electrocrystallization, $Sb_2F_{11}^-$ was found to be the counter anion (Figure 1d).[5] The stoichiometry was decided both by the elemental analysis and measured crystal density. The water molecules were not found in the crystal structure analysis.

The space group is $P\bar{1}$ and one unit sell contains eight donor molecules and two Sb_2F_{11}. The crystal consists of the alternate stacking of the donor and anion layers. Sb_2F_{11} is located in the channels between the donor stacking columns. The tetramer (**T**) and dimer

(**D**) of the donor molecules are formed in the conducting layer (Figure 1e). **T** and **D** form the stacking column as -**D-d-T-d-D-**, where **d** represents the donor of which molecular plane is perpendicular to those in **T** and **D**. The calculated band structure affords the deep dips in DOS with the quasi-one- and two-dimensional Fermi surfaces (Figure 1f).

3.2 UV-vis-IR spectra

3.2.1 (EDO)$_5$(BF$_4$)$_3$ (**1**).   Three electronic absorption bands were observed below 20 $\times10^3$cm^{-1} (Figure 2a). Comparing to the spectra of the related complexes,[6] the lowest energy absorption band at 3.6 $\times10^3$cm^{-1} is assigned to the transition, EDO$^+$ + EDO^0 → EDO^0 + EDO$^+$ (A-band). The absorption band at 10.1 $\times10^3$cm^{-1} (B-band) and at 15.0 $\times10^3$cm^{-1} (C-band) corresponds to the transitions of EDO$^+$ + EDO$^+$ → EDO^0 + EDO^{2+} and the intra-molecular transition of EDO$^+$, respectively.

3.2.2 (EDO)$_4$(Sb$_2$F$_{11}$)$_{0.85}$(H$_2$O)$_4$ (**2**).   The lowest energy absorption band was A-band (Figure 2a). Weak B-band was observed at 10.1 $\times10^3$cm^{-1}. The uniform distribution of the charge on the donor molecules can not account the appearance of B-band, since the averaged formal charge of +0.21 is less than +0.5. Thus the existence of the charge ordering with the neighboring charge rich molecules is suggested.

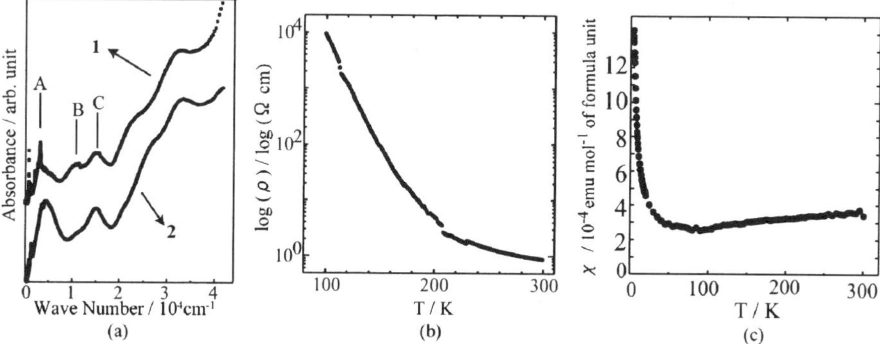

Figure 2 (a) UV-vis-IR spectra of **1** (up) and **2** (down). A, B, and C are corresponded to A-, B- and C-band described in the text, respectively. (b) Temperature dependence of resistivity (ρ) of **1**. (c) Temperature dependence of χ of **2**.

3.3 Conductivity

3.3.1 (EDO)$_5$(BF$_4$)$_3$ (**1**).   The conductivity measurement was performed along the *a* + *b* direction. The conductivity at room temperature (σ_{RT}) was 1 Scm^{-1}. The semiconducting behavior was observed with the activation energy (ε_a) of 62 meV from room temperature to 235 K (Figure 2b). The enhanced electronic correlation suggested from the band calculation is regarded to be the origin of the carrier localization. At around 210 K, the resistivity shows the small jump and the activation energy changes to 150 meV in the range of 200 K and 120 K.

3.3.2 (EDO)$_4$(Sb$_2$F$_{11}$)$_{0.85}$(H$_2$O)$_4$ (2). σ_{RT} is 2 Scm^{-1}, and the semiconducting behavior was observed with ε_a of 55 meV. The charge disproportionation suggested from the electronic spectrum should be related to the charge localization.

3.4 Static susceptibility of (EDO)$_4$(Sb$_2$F$_{11}$)$_{0.85}$(H$_2$O)$_4$ (2)

χ at RT is 3.6 $\times 10^{-4}$ emu mol^{-1}of formula unit (Figure 2c). The value is enhanced compared to the metallic EDO complexes at RT (2.5 $\times 10^{-4}$ emu mol^{-1}of formula unit for (EDO)$_2$PF$_6$ at room temperature)[1], corresponding to the calculated band structure with the effectively narrow bands. χ gradually decreases down to 90 K (2.5 $\times 10^{-4}$ emu mol^{-1}of formula unit at 90 K), and then increases, showing that the charge localization is not derived from a simple gap formation at the Fermi level.

4   Conclusions

1 and **2** showed the deviation from the 2 to 1 stoichiometry of (EDO)$_2$PF$_6$. The molecular planes of the donor molecule arrange perpendicularly in the donor layers in both salts. This shows that the donor molecules create the channels for the smaller (**1**) or larger (**2**) anions by the modifications of the packing pattern, showing that the flexibility of EDO in the formation of the conducting layer.

Non-metallic conducting behaviors were observed in **1** and **2**, though the calculations showed the Fermi surfaces. In **1**, this inconsistency is derived mainly from the following two factors. 1) The dimerizations of EDO results in the effectively narrow electronic bands. 2) The relatively small size of EDO is related to the enhanced on-site Coulomb energy in the complex. Thus the electronic correlation plays the important role for the localization of the carriers in **1**. The most plausible mechanism of the carrier localization in **2** is the emergence of the charge ordering in which the charge rich molecules are located closely, indicating that the sensitivity of EDO to the environment both on the molecular shape and the charge as shown in (EDO)$_2$PF$_6$.

References

1   A. Ota, H. Yamochi, G. Saito, *J. Mater. Chem.*, 2002, **12**, 2600.
2   M. Chollet, L. Guerin, N. Uchida, S. Fukaya, H. Shimoda, T. Ishikawa, K. Matsuda, T. Hasegawa, A. Ota, H. Yamochi, G. Saito, R. Tazaki, S. Adachi, S. Koshihara, *Science*, 2005, **307**, 86.
3   A. Ota, H. Yamochi, G. Saito, Proceedings of ISCOM 2005, Florida, *J. Low Temp. Phys.*, in press.
4   **1**: Triclinic P1, a = 7.9580(3) Å, b = 13.402(1) Å, c = 14.954(1.1) Å, α = 114.641(3)°, β = 90.291(4)°, γ = 90.787(0)°, V = 1448.7(1.6) Å^3, Z = 1, 5483 independent, 4655 observed ($> 2\sigma(I)$), 691 parameters, D_{calcd} = 1.802 g cm^{-3}, R_1 = 0.0689, and wR_2 = 0.1976 for all data. **2**: Triclinic P$\bar{1}$, a = 8.604(0.4) Å, b = 16.068(0.7) Å, c = 19.397(0.6) Å, α = 112.683(2)°, β = 93.496(2)°, γ = 92.300(2)°, V = 2463.7(2) Å^3, Z = 2, 9043 independent, 7757 observed ($F > 4\sigma(F)$), 622 parameters, D_{calcd} = 2.030 g cm^{-3}, D_{obs} = 2.001 g cm^{-3}, R_1 = 0.120, and wR_2 = 0.151 for $F > 4\sigma(F)$.
5   A. Otsuka, H. Yamochi, G. Saito, T. Sugano, M. Kinoshita, S. Sato, K. Honda, K. Ohfuchi, M. Konno, *Synth. Met.*, 1991, **41-43**, 1699.
6   T. Senga, K. Kamoshida, L. A. Kushch, G. Saito, T. Inayoshi, I. Ono, *Mol. Cryst. Liq. Cryst.*, 1997, **296**, 97.

IONICITY OF INTRAMOLECULAR CHARGE TRANSFER COMPOUNDS IN SOLUTION

G. Honda, K. Nishimura, Y. Enomoto and G. Saito

Division of Chemistry, Graduate School of Science, Kyoto University, Sakyo-ku, Kyoto, Japan

1 INTRODUCTION

Intramolecular charge transfer (CT) compounds, $D^{\delta+}$-π-$A^{\delta-}$, in which donor (D) and acceptor (A) moieties are linked by π-bond (δ, ionicity), have been studied in a wide range of functional materials such as nonlinear optics[1,2] and molecular rectifiers[3,4] from the view point of their large intramolecular dipole moment. In these compounds, the ionicity, δ, affects their electronic and optical properties, and the estimation of δ gives important information about their functionality. In addition, controlling δ is expected to open new opportunity for the development of new functional materials.

We have synthesized a series of intramolecular CT compounds, in which the 1,3,3-trimethyl-2-methyleneindoline moiety (D) and the TCNQ skeleton (A) are linked by π-bond (Figure 1(a), C_n-R3CNQ; n = 1; R = F_4, CF_3, F_2, F, H, Me, $(MeO)_2$ and $(EtO)_2$), and estimated their ionicities in solution by solvatochromic effect using various polar solvents.[5,6] We have disclosed that the ionicities of these intramolecular CT compounds increased according to the acceptor strengths of R3CNQ skeletons. We have also studied the C_{16}-H3CNQ compound, which showed higher energy shift of intramolecular CT transition with decreasing the polarity of the solvent below $E_T = 35$. This behavior was explained by the contribution of neutral ground state.[7]

In order to investigate the behaviour shown in C_{16}-H3CNQ in non-polar solvents, we have studied the solvatochromism of C_n-H3CNQs (n = 3, 10 and 20). Investigation of the C_n-R3CNQ derivatives having various length of alkyl chain enables us to control molecular arrangement and physical properties in the solid state. In this paper, we report on the synthesis of new C_n-H3CNQ derivatives (n = 3, 10 and 20) and the estimation of their ionicities by solvatochromism in various polar and non-polar solvents.

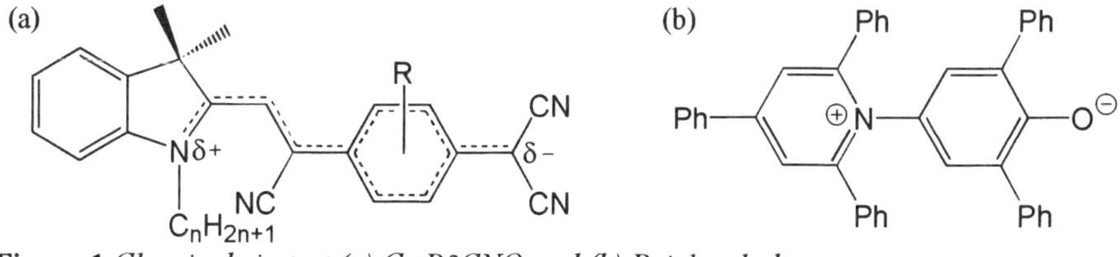

Figure 1 *Chemicals in text (a) C_n-R3CNQ and (b) Reichardt dye*

2 EXPERIMENTAL

C_n-H3CNQs were synthesized by a Stork enamine-type reaction between 1-alkyl-2,3,3-trimethylindolenium iodide and TCNQ according to the procedures reported previously.[7]

3 RESULTS AND DISCUSSION

3.1 UV-visible Spectra

UV-visible spectra of C_n-H3CNQs in various solvents are shown in Figure 2(a). The longest wavelength absorption bands of C_n-H3CNQs were assigned to be intramolecular CT transition from H3CNQ moiety to indoline moiety.[7] The C_n-H3CNQs exhibited two peaks in the range of 700-850 nm and two shoulders in the range of 550-700 nm in polar solvents. These peaks showed blue shift in non-polar solvents. This peak splitting is also observed in other intramolecular CT compounds.[8] The origin of appearance of several peaks might be the coupling of the electronic excitation with intramolecular vibration. The positions of these peaks were determined by line deconvolution into four Gaussian bands (Figure 2(b)). Figure 3 represents the plots of transition energies of these four peaks after Gaussian fitting vs Reichardt E_T values[9] of each solvent. Here, the E_T values are the intramolecular CT energies of the zwitterionic Reichardt dye (Figure 1(b)) in various solvents, and represent the polarity of each solvent.

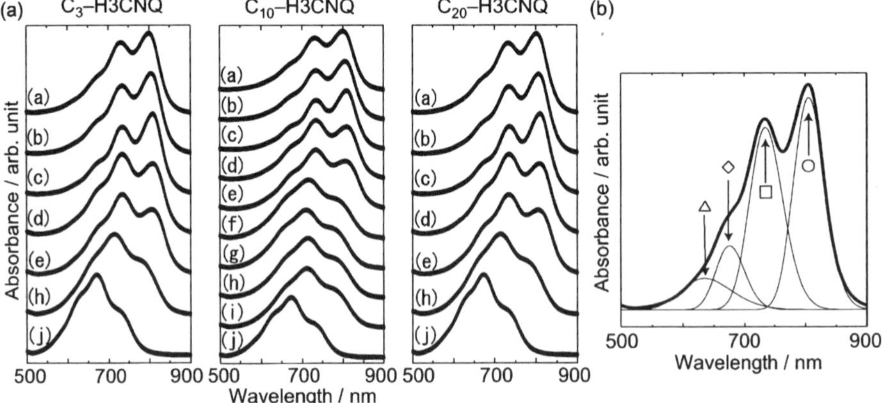

Figure 2 *(a) UV-visible spectra of C_n-H3CNQs (n = 3, 10 and 20) in various solvents (a, methanol; b, acetonitrile; c, acetone; d, chloroform; e, chlorobenzene; f, 1,4-dioxane; g, diethyl ether; h, benzene; i, toluene; j, hexane). (b) Gaussian fitting of absorption bands of C_3-H3CNQ in acetonitrile. Thick line is observed spectrum, and thin lines are Gaussian bands.*

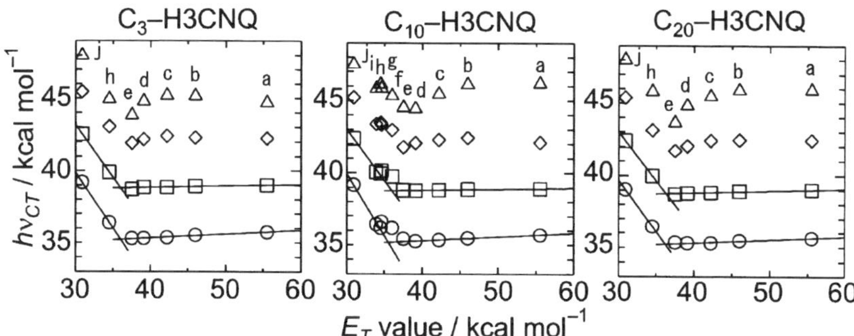

Figure 3 *Plots of transition energies of C$_n$-H3CNQs vs E_T values. Labels (a-j) are the same as the caption of Figure 2(a). The marks of plots are shown in Figure 2(b).*

3.2 Estimation of Charge Transfer Degree

In the UV-visible spectra of C$_n$-H3CNQs, a linear relation between $h\upsilon_{CT}$ and E_T is found in both small and large E_T values (Figure 3).

$$h\upsilon_{CT} = aE_T + b \tag{1}$$

In our previous report, the ionicities, δ, of C$_n$-R3CNQ derivatives in solution were estimated by equation (2).[5]

$$a/a_1 = 2\delta - 1 \tag{2}$$

a_1 is the slope of the $h\upsilon_{CT}$ vs E_T plot for $\delta = 1$. Since both δ and the slope a of the Reichardt dye are 1, equation (2) is modified to equation (3) on the assumption that the slope a of C$_n$-R3CNQ system for $\delta = 1$ is also 1.

$$a = 2\delta - 1 \tag{3}$$

Equation (3) gave the relative ionicity δ values for C$_n$-H3CNQs, and the estimated δ values are summarized in Table 1. The peaks used for estimation of δ were the ones of the lowest transition energy, because they showed the best linear relation between the transition energies and E_T values. We excluded some solvents such as chlorobenzene, 1,4-dioxane and diethyl ether for estimation, because the dependencies of transition energies on E_T values did not fit the linear relation.

The dependence of δ on solvent polarity showed no noticeable difference in these C$_n$-H3CNQ derivatives. In polar solvents such as methanol, acetonitrile, acetone and chloroform, the energies of intramolecular CT transition of C$_n$-H3CNQs were almost independent on the solvent polarities ($a \approx 0$), thus the ionicities were estimated to be around 0.5. This is the same behavior as our previous report about C$_1$-H3CNQ.[5] In non-polar solvents such as benzene, toluene and hexane, on the contrary, the energies of intramolecular CT transition and the solvent polarities had a linear relation with negative slope. The ionicities estimated from a values were around 0.1. These results indicate that the two slopes of the $h\upsilon_{CT}$ vs E_T plot represent the change of ionicities from ionic to neutral state with decreasing solvent polarities.

Table 1 *Estimated δ values of C_n-H3CNQs*

compounds	slope a of $h\nu_{CT}$ vs E_T plot (Figure 3)		estimated δ	
	non-polar[a]	polar[b]	non-polar[a]	polar[b]
C_3-H3CNQ	−0.78	0.026	0.11	0.51
C_{10}-H3CNQ	−0.87	0.026	0.07	0.51
C_{20}-H3CNQ	−0.72	0.021	0.14	0.51

[a] In less polar solvents than diethyl ether. [b] In more polar solvents than chlorobenzene.

4 CONCLUSION

We prepared new intramolecular CT compounds, C_n-H3CNQs (n = 3, 10 and 20), and estimated their ionicities in both polar and non-polar solvents. C_n-H3CNQs showed change of ionicities depending on the solvent polarity; more ionic in polar solvents and neutral in non-polar solvents. Here, the ionicities of C_n-R3CNQ system in non-polar solvents were discussed for the first time. C_n-R3CNQ system exhibits a neutral-ionic transition depending on polarity of solvent. The boundary of the transition is found between diethyl ether and chlorobenzene. The solid state properties will be reported in another paper.[10]

Acknowledgements

This work was supported by a Grant-in-Aid (21st Century COE program on Kyoto University Alliance for Chemistry) from the Ministry of Education, Culture, Sports, Science and Technology, Japan.

References

1   M. Szablewski, P.R. Thomas, A. Thornton, D. Bloor, G.H. Cross, J.M. Cole, J.A.K. Howard, M. Malagoli, F. Meyers, J-L. Brédas, W. Wenseleers and E. Goovaerts, *J. Am. Chem. Soc.*, 1997, **119**, 3144.

2   H.E. Katz, K.D. Singer, J.E. Sohm, C.W. Dirk, L.A. King and H.M. Gordon, *J. Am. Chem. Soc.*, 1987, **109**, 6561.

3   R.M. Metzger, B. Chen, U. Höpfner, M.V. Lakshmikantham, D. Vuillaume, T. Kawai, X. Wu, H. Tachibana, T.V. Hughes, H. Sakurai, J.W. Baldwin, C. Hosch, M.P. Cava, L. Brehmer and G.J. Ashwell, *J. Am. Chem. Soc.*, 1997, **119**, 10455.

4   G.J. Ashwell, R. Hamilton and L.R.H. High, *J. Mater. Chem.*, 2003, **13**, 1501.

5   G. Saito, C.-H. Chong, M. Makihara, A. Otsuka and H. Yamochi, *J. Am. Chem. Soc.*, 2003, **125**, 1134.

6   Ionicity and solvatochromism of $D^{\delta+}$-π-$A^{\delta-}$, hexadecylquinolinium tricyanoquinodimethanide, were discussed, see: (a) J. W. Baldwin, B. Chen, S. C. Street, V. V. Konovalov, H. Sakurai, T. V. Hughes, C. S. Simpson, M. V. Lakshmikantham, M. P. Cava, L. D. Kispert and R. M. Metzger, *J. Phys. Chem. B*, 1999, **103**, 4269. (b) F. Terenziani, A. Painelli, A. Girlando and R. M. Metzger, *J. Phys. Chem. B*, 2004, **108**, 10743.

7   C.-H. Chong, M. Makihara, G. Saito, *Mol. Cryst. Liq. Cryst.*, 2002, **376**, 183.

8   N.A. Bell, D.J. Crouch, D.J. Simmonds, A.E. Goeta, T. Gelbrich and M.B. Hursthouse, *J. Mater. Chem.*, 2002, **12**, 1274

9   C. Reichardt, *Chem. Rev.*, 1994, **94**, 2319.

10 Y. Enomoto, K. Nishimura, G. Honda, G. Saito, "Structures and thermal behavior of C_n-R3CNQ: Intramolecular charge transfer compounds with a long alkyl chain", in this proceeding.

STRUCTURES AND THERMAL BEHAVIOR OF INTRAMOLECULAR CHARGE TRANSFER COMPOUNDS Cₙ-H3CNQ DERIVED FROM THE REACTION BETWEEN 1-ALKYL-3,3-DIMETHYL-2-METHYLENEINDOLINE AND 7,7,8,8-TETRACYANOQUINODIMETHANE (TCNQ)

Y. Enomoto, K. Nishimura, G. Honda, G. Saito

Division of Chemistry, Graduate School of Science, Kyoto University, Kyoto 606-8502, Japan

1 INTRODUCTION

We have studied on intramolecular charge transfer (CT) compounds D-π-A composed of an electron-donor (D) and an acceptor (A) groups connected with π-conjugated system. Owing to their large electronic polarization, such intramolecular zwitterionic compounds have been studied as the functional materials of non-linear optics,[1,2] rectification,[3,4] memory,[5] and optoelectronic switching,[6] etc. As a model of intramolecular CT compounds, we have synthesized Cₙ-R3CNQ (Figure 1) between 1-alkyl-2-methylene-3,3-dimethylindoline and TCNQ derivatives.[7,8] In order to observe the effects of the alkyl chain, we fixed the acceptor part as R=H (Cₙ-H3CNQ).

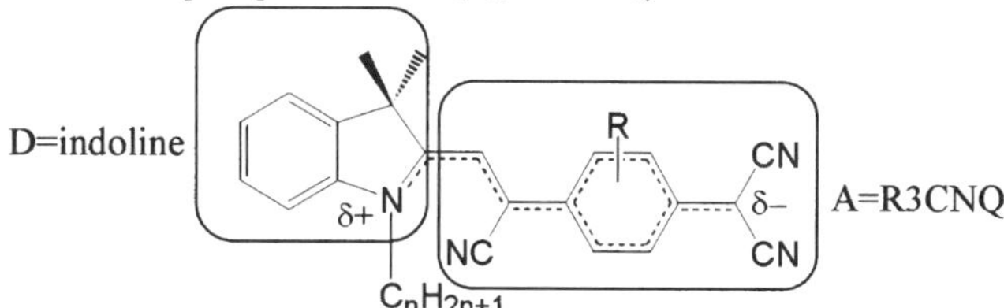

Figure 1 *Intramolecular charge transfer compound Cₙ-R3CNQ. The electron-donating indoline part and the electron-accepting R3CNQ part are encircled by rectangles.*

The ionicity of intramolecular CT compounds (δ) is a basic parameter closely related to the strength of molecular dipole moments, so the determination of molecular structure is essential to discuss microscopic electronic structure in these materials.[9,10] We obtained a series of Cₙ-H3CNQ single crystals and estimated the values of the ionicity of these compounds in solid state. We found that the crystal structures of them show two kinds of molecular conformation and a variety of intermolecular arrangements depending on the length of their alkyl chains.

Generally, intramolecular CT compounds have high melting points because of their electronic polarization. We have observed a gradual decrease of melting point of Cₙ-H3CNQ with increasing n up to 18.

2 SYNTHESIS

C_n-H3CNQs were synthesized by a Stork enamine-type reaction between 1-alkyl-2,3,3-trimethylindolenium iodide and TCNQ derivatives according to the procedures reported previosly.[7]

3 RESULTS AND DISCUSSION

3.1 Crystal structures of C_n-H3CNQ

Crystal structures of C_n-H3CNQ (n=3, 6, 7, 8, 10, 20) have been determined and are classified into three types according to the length of the alkyl chain. When the alkyl chain is short (n=3, 6, 7, 8), 3CNQ dimers are formed by cohesion of 3CNQ moieties. With increasing n, molecules make alternating stack of 3CNQ moieties and alkyl chains in the C_{10}-H3CNQ crystal, due to the effect of steric repulsion of alkyl chain. A two-dimensional sheet of alkyl chains is formed, due to the dominance of self-cohesion effect of alkyl chains in the crystal of C_{20}-H3CNQ. An example of crystal structures of C_n-H3CNQs is shown in Figure 2 for C_8-H3CNQ.

In C_3-H3CNQ and C_{10}-H3CNQ, the CN group close to the indoline moiety lies near the alkyl chain (Type-I conformation in Figure 3), whereas near the dimethyl group in the others (Type-II conformation in Figure 3). An estimation of the ionicity of C_n-H3CNQ based on the bond length[9,10] revealed that the molecular conformation of C_n-H3CNQ affects the ionicity of molecule, which is larger in Type-I conformation (δ~0.68) than that in Type-II conformation (δ~0.53).

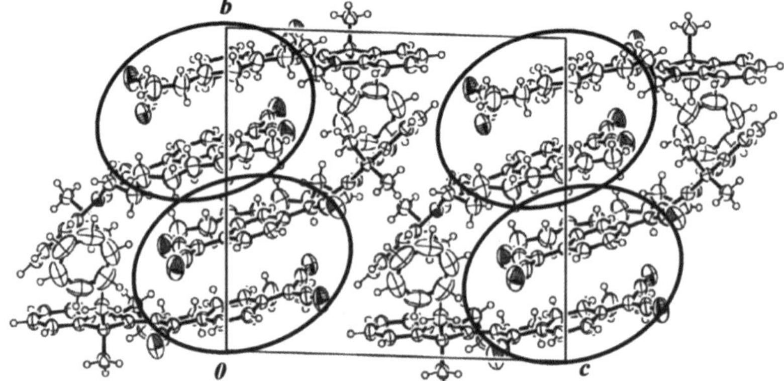

Figure 2 *Crystal structure of C_8-H3CNQ. This crystal contains benzene molecules.*

Figure 3 *Two different conformations of C_n-H3CNQ.*

3.2 Thermal properties of C_n-H3CNQ

The melting points of C_n-H3CNQ powder for n=1, 14, 16, 18 and 22, and crystals for n=3, 6, 7, 8, 10 and 20 are depicted in Figure 4. A monotonous decrease was observed from 234°C (n=3) to 95°C (n=18) followed by an increase, similar to the tetrakis(alkylthio)-tetrathiafulvalene (TTC$_n$-TTF).[11] These low melting points allow us to investigate intramolecular CT compounds in a liquid state. Thermal properties of C_n-H3CNQ compounds were investigated by DSC measurements. For example, C_{20}-H3CNQ (Figure 5a), which was obtained as golden plates melted at 105°C then turned to purple liquid. No exothermic peaks were observed in a cooling process of liquid sample, and its appearance did not change down to room temperature. When the sample was subjected under slow thermal change (Figure 5b), a broad exothermic peak was observed in the 2nd heating process around 70°C, and then golden luster was recovered. This peak was assigned to a phase transition from supercooled liquid to crystalline solid based on powder X-ray diffraction. The supercooled liquid was able to be stretched into a wire with the thickness of 70 to 120µm at 120°C. Then the wire was annealed at 70°C to have the solid wire with golden luster.

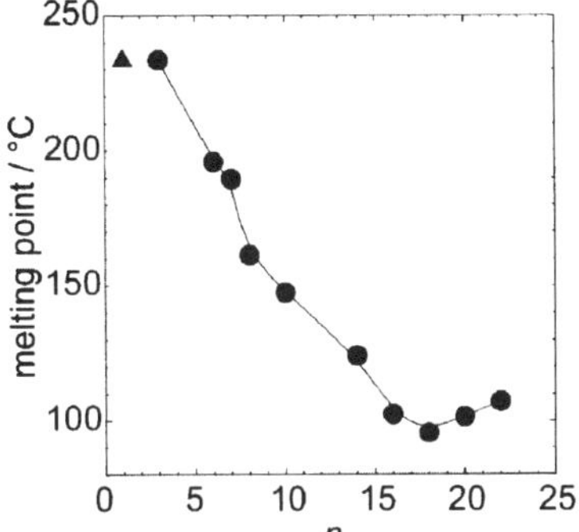

Figure 4 *Melting point (circle) of C_n-H3CNQ. Minimum is at 95°C for n=18. C_1-H3CNQ decomposed during the melting (~250°C, triangle).*

a)   b)

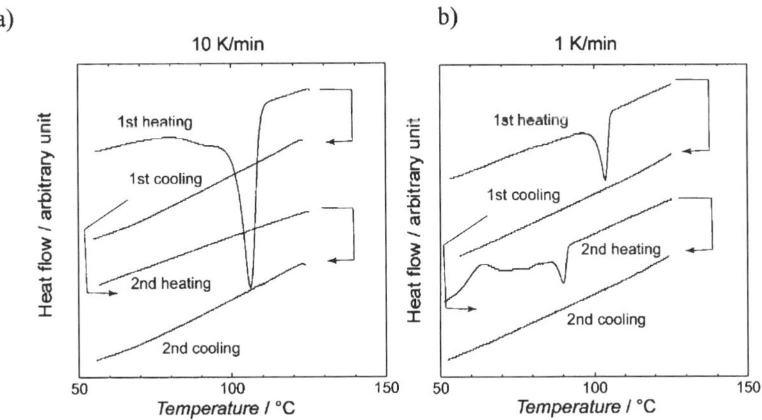

Figure 5 *Thermal properties of C_{20}-H3CNQ investigated by DSC measurements by rapid (a) and slow (b) thermal change.*

4 CONCLUSION

We prepared new intramolecular CT compounds C_n-H3CNQ (n=1, 3, 6~8, 10, 14, 16, 18, 20, 22). Some of them gave single crystals, which have a variety of molecular arrangements and two different molecular conformations concerning with the relative position of indoline moiety and CN group close to it. Crystal structure analysis show that a conformation having close separation between N atom in indoline part and the CN group (Type-I) possesses higher degree of intramolecular charge transfer than the other (Type-II). We measured the melting points of C_n-H3CNQ. The minimum value was found at 95°C for R=18. DSC measurements revealed the presence of the supercooled liquid phase. The supercooled liquid can be stretched with the thickness of 70 to 120μm and be annealed to have the solid wire with golden luster.

Acknowledgment

The work was supported by a Grant-in-Aid (21st Century COE program on Kyoto University Alliance for Chemistry) from the Ministry of Edition, Culture, Sports, Science and Technology, Japan.

References

1 M. Szablewski, P. R. Thomas, A. Thornton, D. Bloor, G. H. Cross, J. M. Cole, J. A. K. Howard, M. Malagoli, F. Meyers, J.-L. Bredas, W. Wenseleers and E. Goovaerts, *J. Am. Chem. Soc.*, 1997, **119**, 3114.

2 H. E. Katz, K. D. Singer, J. E. Sohm, C. W. Dirk, L. A. King and H. M. Gordon, *J. Am. Chem. Soc.*, 1987, **109**, 6561.

3 R. M. Metzger, B. Chen, U. Höpfner, M. V. Lakshmikantham, D. Vuillaume, T. Kawai, X. Wu, H. Tachibana, T. V. Hughes, H. Sakurai, J. W. Baldwin, C. Hosch, M. P. Cava, L. Brehmer and G. J. Ashwell, *J. Am. Chem. Soc.*, 1997, **119**, 10455.

4 G. J. Ashwell, R. Hamilton and L. R. H. High, *J. Mater. Chem.*, 2003, **13**, 1501.

5 D. S. Noyce and S. K. Brauman, *J. Am. Chem. Soc.*, 1968, **90**, 5218.

6 A. E. Riley, G. W. Mitchell, P. A. Koutentis, M. Bendikov, P. Kaszynki, F. Wudl and S. H. Tolbert, *Adv. Funct. Mater.*, 2003, **13**, 531.

7 C.-H. Chong, M. Makihara and G. Saito, *Mol. Cryst. Liq. Cryst.*, 2002, **376**, 183.

8 G. Saito, C.-H. Chong, M. Makihara, A. Otsuka and H. Yamochi, *J. Am. Chem. Soc.*, 2003, **125**, 1134.

9 P. S. Flandoris and D. Chasseau, *Acta Cryst. B.*, 1977, **33**, 2744.

10 P. Guionneau, C. J. Kepert, G. Bravic, D. Chasseau, M. R. Truter, M. Kurmoo and P. Day, *Synth. Met.*, 1997, **86**, 1973.

11 Z. Shi, T. Enoki, K. Imaeda, K. Seki, P. Wu, H. Inokuchi and G. Saito, *J. Phys. Chem.*, 1998, **92**, 5044.

SYNTHESIS AND PROPERTIES OF DT-TTF DERIVATIVES POSSESSING DICYANOMETHYLENE GROUP AS THE ACCEPTOR UNIT

S. Matsumoto[1,3], M. Hayashi[1,3], H. Fueno[1,3], K. Tanaka[1,3], H. Miyamoto[2] and Y. Misaki[2,3]

[1]Department of Molecular Engineering, Kyoto University, Kyoto 615-8510, Japan.
[2]Department of Applied Chemistry, Ehime University, Matsuyama 790-8577, Japan.
[3]JST-CREST

1 INTRODUCTION

Tetrathiafulvalene (TTF) derivatives with acceptor moiety are of current interest as single-component molecular conductors as well as optoelectronic materials.[1-3] We have noted oligomeric TTFs possessing acceptor moieties as molecular nanowires by use of a conduction path through the π-π stack of TTF units,[4] if they adopt cyclophane-type conformations (Figure 1).[5] In these Proceedings, we report the synthesis and properties of the 1,3-dithiol-2-ylidene fused (DT-TTF) derivatives possessing dicyanomethylene group (**1a**, **b**) and dimeric ones (**2A**, **B**) as the prototypes of molecular wires.[6]

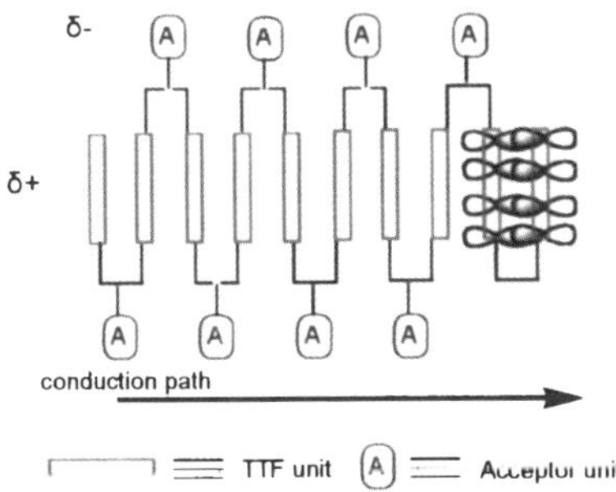

Figure1. Schematic structure of molecular nanowire based on π-π stack in TTF oligomer

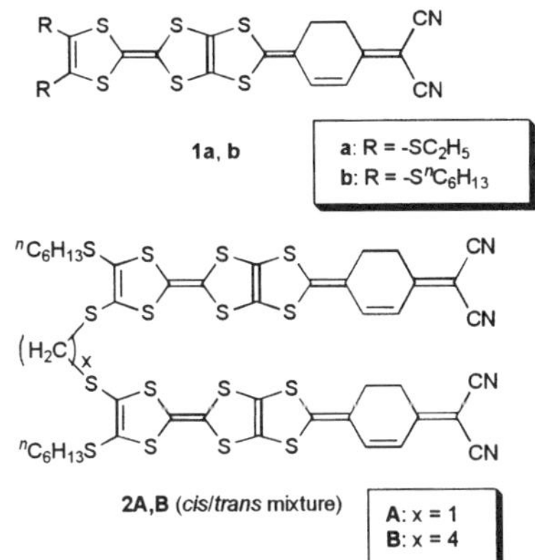

2 RESULTS AND DISCUSSIONS

2.1 Synthesis

Synthesis of TTF derivatives **1a**, **b** was achieved according to Scheme 1. The TTF derivatives with phosphonates **3a**, **b** were reacted with diketone **4** in the presence of *n*-BuLi to provide monoadducts **5a**, **b** in 35 and 37% yields, respectively. Introduction of dicyanomethylidene moiety was achieved by the treatment of **5a, b** with malononitrile in the presence of acetic acid in refluxing toluene. In this case, the retro-Diels-Alder reaction also proceeded at the same time to give **1a, b** in 51 and 88% yields, respectively.

The synthesis of dimeric-1 (**2A, B**) was achieved according to Scheme 2. A derivative of **3** with cyanoethylthio group (**3c**) and **4** were reacted with *n*-BuLi to provide the monoadduct **5c** in 46% yield. The compound **5c** was reacted with $CsOH \cdot H_2O$, followed by treatment with an excess of diiodomethane to afford dimeric TTF **6A** in 62%. On the other hand, a monoadduct **5d** was obtained in 66% yield by use of a large amount of 1,4-bromobutane instead of diiodomethane. Treatment of **5c** with $CsOH \cdot H_2O$, followed by the reaction with **5d** gave **6B** in 94% yield. The target molecules **2A, B** were obtained in 36 and 24% yields by the treatment of **6A, B** with malononitrile and acetic acid.

Scheme 1

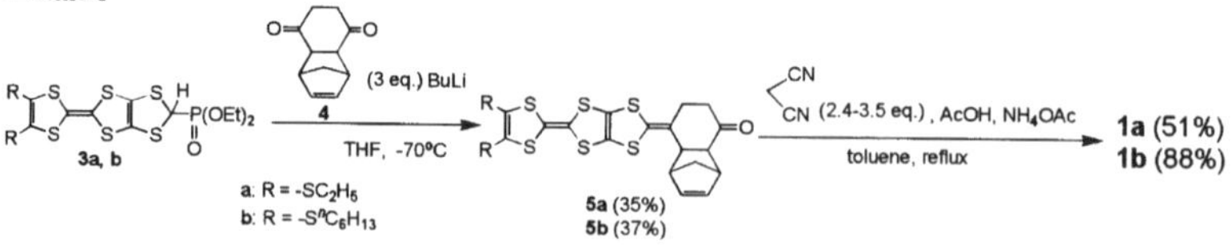

Scheme 2

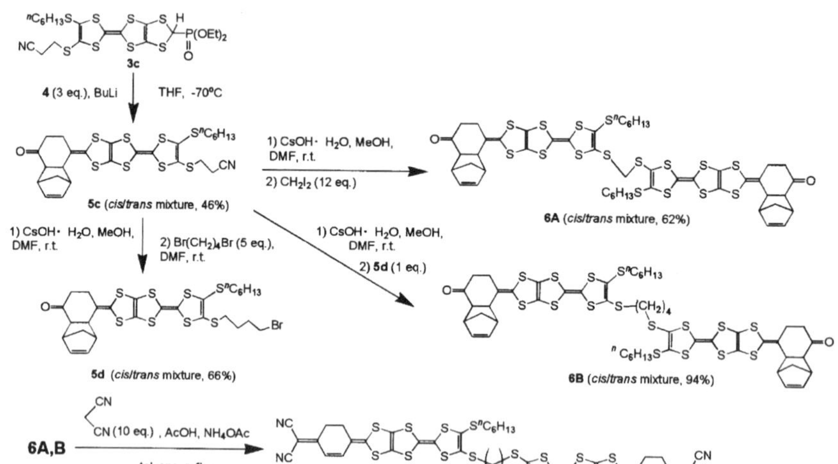

2.2 UV-visible spectra

Figure 2 shows UV-visible absorption spectra of **1b** and **2A, B**. The compound **1b** exhibits the absorption maxima at 319, 440 and 560 nm, respectively. In comparison of **1b** with **8**, a new absorption band was observed around 560 nm. This indicates that the the lowest unoccupied molecular orbital (LUMO) is lowered by electron-withdrawing effect of the dicyanomethylene group. The compound **2A** shows the absorption bands at 317, 438 and 556 nm, respectively. No significant shift of absorption maxima to longer wavelength region has been observed compared with those of monomeric analogue **1b**. In a similar manner, **2B** which has more flexible spacer between TTF units, also exhibits no distinct shift of absorption maxima compared with **1b**. This fact indicates there is little intramolecular interaction between TTF units.

2.3 Electrochemical Properties

Electrochemical properties of **1** and **2** have been investigated by cyclic voltammetry. Their redox potentials are summarized in Table 2. The compound **1** has exhibited three-pairs of redox waves corresponding to the oxidation of DT-TTF unit. In a similar manner, **2** has shown three-pairs of redox waves, each of which corresponds to two-electron transfer process. The first redox potential of **1** is higher by 0.05 V than that of **9** because **1a** is more polarized due to conjugation between donor moiety and dicyanomethylidene group. The redox potentials of **1** and **2B** are different from each other. This indicates there is little intramolecular interaction between cationic TTF units of **2B** owing to a long alkyl spacer. In contrast, the second redox potential of **2A** is higher by 0.04 V than that of **1b**. Furthermore the first redox wave of **2A** is broader than that of **1**. These results strongly indicate that two TTF units in **2A** interact with each other in the one- and two-electron oxidized states.

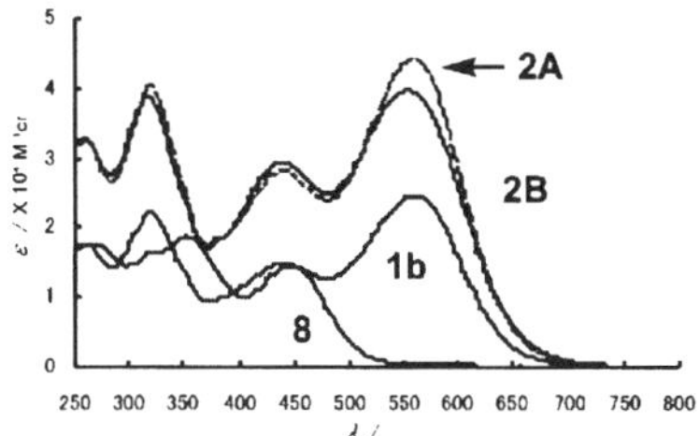

Figure 2. UV-vis absorption spectra of **1b** and **2A, B** in CH_2Cl_2

Table 1. Redox Potentials of **1** and **2** in CH_2Cl_2 containing 0.1 M Bu_4PF_6 (V vs. Fc/Fc^+)

Compound	E_1	E_2	E_3
1a	0.16	0.47	1.05[a]
1b	0.18	0.48	1.12[a]
2A	0.18 (2e)	0.52 (2e)	1.09[a] (2e)
2B	0.18 (2e)	0.47 (2e)	1.01[a] (2e)
8	0.15	0.48	1.11[a]
1a[b]	0.14	0.40	1.02[a]
9[b]	0.09	0.37	1.05[a]

[a] Irreversible wave. Anodic peak potential.
[b] Measured in PhCN containing 0.1 M Bu_4PF_6

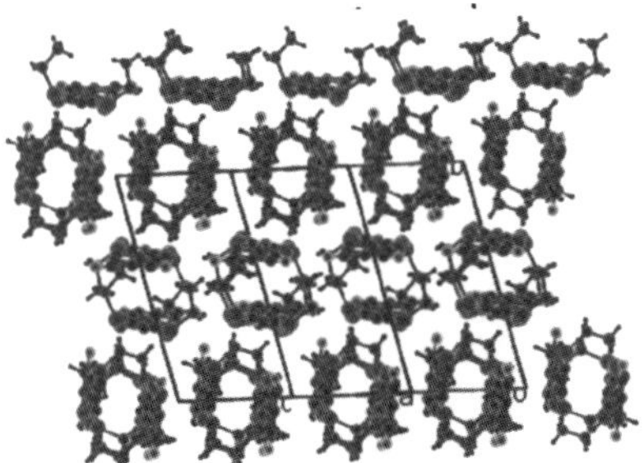

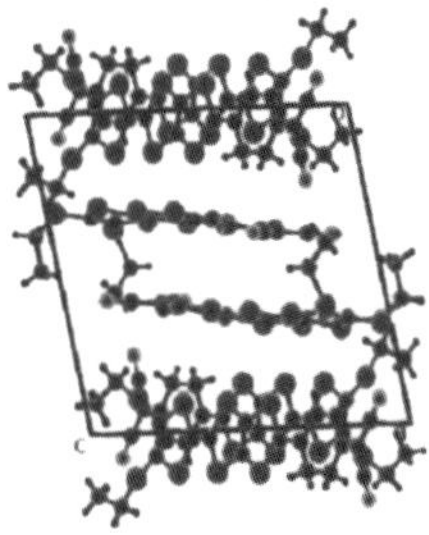

2.4 Crystal structure of 1a

Figure 3 shows crystal structure of **1a** determined by X-ray structure analysis.[7] The crystal is composed of two different layers of the molecules. The unsymmetrical molecules are stacked along the *a*-axis in a head-to-tail-manner. One the other hand, the dimerized molecules form a sheet along the *ac* plane. The molecules in the column and sheet are orthogonally arranged with each other. The donor moieties face the acceptor ones in all cases (intracolumn, intradimer, between column and sheet). Thus, there is unfortunately little conduction path in the present crystal.

Figure 3. Molecular arrangement of **1a** viewed from molecular long axis (left), and viewed onto the *bc* plane (right).

2.5 Calculation of the electronic structure

The electronic structure of unsubstituted-**1** has been calculated by means of density functional theory (DFT) at the B3LYP/6-31G* level.[8] The highest occupied molecular orbital (HOMO) and the lowest unoccupied molecular orbital (LUMO) are shown in Figure 4. The HOMO mainly distributes on the TTF unit, while LUMO is localized on the 1-dicyanomethylidene-4-(1,3-dithiol-2-ylidene)-*p*-quinodimethane moiety. The HOMO-LUMO gap (2.41 eV) is narrower by 0.3 eV than that of unsubstituted-**9** thanks to direct π-conjugation between donor and acceptor moieties through cyclohexene spacer.

Figure 4. The highest occupied molecular orbital of unsubstituted-**1a**(left), and the lowest unoccupied molecular orbital (right).

2.6 Conductivity

Electrical conductivity measurements have been carried out for **1a, b** using two-probe technique. Both of them are insulators (**1a**: $<10^{-8}$, **1b**: $<10^{-9}$ S cm^{-1}) as is anticipated by X-ray structure analysis for **1a**.

3 REFERENCE

1   H. Tanaka, Y. Okano, H. Kobayashi, W. Suzuki and A. Kobayashi, *Science*, 2001, **291**, 285.

2   S. Sheib, M.P. Cava, J. W. Baldwin and R. M. Metzger, *J. Org. Chem.*, 1998, **63**, 1198; E. Tsiperman, L. Regev, Y. Becker, J. Bernstein, A. Ellern, V. Khodorkovsky, A. Shames and A. L. Shapiro, *Chem. Commun.*, 1999, 1125.

3   D. F. Perepichka, M. R. Bryce, C. Pearson, M. C. Petty, E. J. L. McInners and J. P. Zhao, *Angew. Chem. Int. Ed. Engl.*, 2003, **42**, 4635.

4   S. Matumoto, W. Matsuda, H. Fueno, Y. Misaki and K. Tanaka, *Synth. Met.*, 2005, **153**, 429.

5   K. Takimiya,Y. Aso, F. Ogura and T. Otsubo, *Chem. Lett.*, 1995 735; S. Yunoki, K. Takimiya, Y. Aso and T. Otsubo, *Tetrahedron Lett.*, 1997, **38**, 3017; Takimiya, A. Oharuda, Y. Aso, F. Ogura and T. Otsubo, *Chem. Mater.*, 2000, **12**, 2196.

6   R. E. Martin and F. Diederich: Angew. Chem. Int. ed. Engl. 38, 1350 (1999); J. M. Tour: Acc. Chem. Res. 33, 791 (2000); N. Sumi, H. Nakanishi, S. Ueno, K. Takimiya, Y. Aso and T. Otsubo: Bull. Chem. Soc. Jpn. 74, 979 (2001).

7   Crystallograohic data for **1a**: triclinic, space group P^-1, $a = 11.475(3)$, $b = 14.024(4)$, $c = 15.443(4)$ Å, $\alpha = 77.569(9)$, $\beta = 74.961(9)$, $\gamma = 81.230(10)°$, $V = 2331.3(12)$ Å^3, $Z = 2$, $\rho_{calc} = 1.541$ g/cm^3, $R (R_w) = 0.076 (0.093)$.

8   The molecular orbital calculation was carried out using the Gaussian 98 program. M.J. Frish et al., Gaussian 98, Gaussian, Inc., Pittsburgh, PA (1998).

NMP TCNQ BASED MOLECULAR WIRE

K. Ueda [1,2], T. Suzuki [2] and R. Kita [1,2]

[1] Faculty of Engineering, Shizuoka University, Johoku 3-5-1, Hamamatsu, Shizuoka 432-8561, Japan
[2] Innovative Joint Research Center, Shizuoka University, Johoku 3-5-1, Hamamatsu, Shizuoka 432-8561, Japan

1   INTRODUCTION

Fabrication of molecular materials on several substrates is an attractive field not only from the fundamental scientific interests but also their potential applications for molecular electronic devices.   For the electronic devices applications, the preparation of molecule-based conductors, superconductors and magnets in a practical usable form, such as thin films, molecular wires and so on, is a challenging subject.   Among molecular conductors many techniques have been proposed for fabrication of molecular wires.   In these techniques preparation of surface of substrates or modification of molecules for introducing self-assembling were employed before the fabrication. [1-3]   Our goal for fabrication of molecular wires is focused on the construction of molecular wires without any structural change during fabrication.   This means that we can easily estimate the properties of wires referring huge physical properties libraries so far known in the molecular materials.

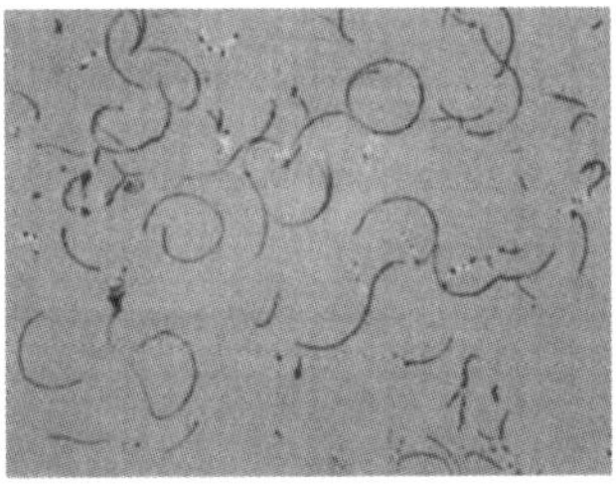

Figure 1 *A microscopic image of NMP TCNQ molecular wires*

Here we present results on the fabrication of molecular wires of TCNQ radical anion salts, NMP TCNQ (NMP = *N*-methylphenazinium, TCNQ = 7,7,8,8-tetracyano-quinodimethanide), known as the high electrical conductor, [4,5] just only controlling their recrystallization conditions (Figure 1).

2   EXPERIMENTAL

2.1   Preparation and Fabrication of NMP TCNQ

The 1:1 salt of NMP TCNQ was prepared as follows.   First, TCNQ was reacted with 3 equiv. of LiI in hot acetonitrile, and the precipitated Li^+ salt of $TCNQ^-$ was treated with 1.1 equiv of NMP^+ chloride in water to afford the corresponding NMP^+ salt of $TCNQ^-$.   After recrystallization of the crude salt from hot acetonitrile, NMP TCNQ was obtained as dark blue crystals.   Molecular wires of NMP TCNQ were obtained by rapid recrystallization (from 100 h to 30 min for solvent evaporation) from the acetone solution (3.4 µmol/l – 0.8 µmol/l) of the corresponding single crystals on glass substrates in the solution.

2.2   Spectroscopic Measurements of NMP TCNQ

The morphology of molecular wires was examined with an analytic scanning electron microscope (JSM-6360LA, JEOL Ltd.) and a scanning probe microscope (SPI 3700, SII Nanotechnology Inc.).   The electronic structures of TCNQ radical anions were confirmed by infrared spectra of the samples dispersed in KBr matrix using an IR spectrophotometer (FT/IR-550, JASCO Corp.).   Crystallinity of the prepared wires was identified by X-ray diffraction (XRD).   XRD was performed on a Rigaku X-ray diffractometer (RINT 2200) with Cu-$K\alpha$ radiation (40 kV, 40 mA).   The scanning rate was $0.02°$ s^{-1} and the 2θ ranged from $5°$ to $50°$.

3   RESULTS AND DISCUSSION

3.1   Relationship between Width and Recrystallization Condition

Figure 2a shows a relationship between width and concentration.   As the concentration was lower, the width was gradually minimized from 3.8 to 1.3 µm.   When the concentration was below 0.8 µmol/mL, only platelet crystals were obtained.

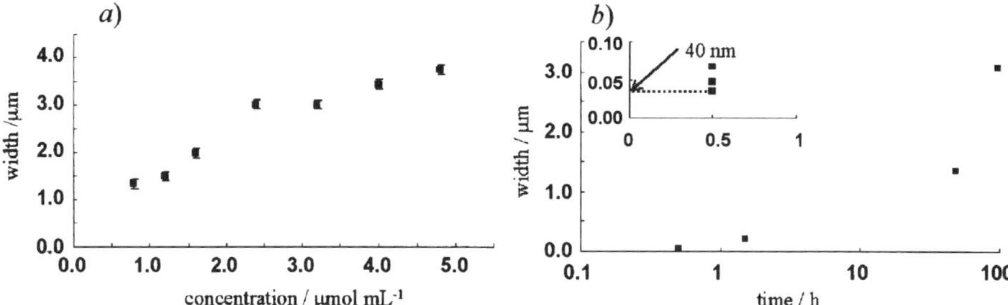

Figure 2 *Relationship between a) width and concentration, and b) width and solvent evaporating time (Inset: relationship under 1 h)*

Furthermore as the evaporating time became shorter, the width was minimized. The initial concentration was fixed to 0.8 µmol/ml during these experiments. The width of the wires largely depended on its solvent evaporating time (Figure 2b). When the evaporating time was 100 h (1.7 µl/min), the width was about 3 µm. When the evaporating time was 30 min (167 µl/min), the width was minimized to 40 nm (Inset of Figure 2b). This result indicates that evaporating time can control the width of the wires.

3.2 Morphology and electronic structure comparison between the wires and the corresponding single crystals

NMP TCNQ showed their nitrile stretching mode at 2195 cm^{-1} for the single crystals and at 2195 cm^{-1} for the wires, respectively. This result revealed any electronic structural change did not occur during the fabrication of the wires. Figures 3 shows the XRD patterns of the obtained molecular wires on the substrate and the corresponding single crystals.

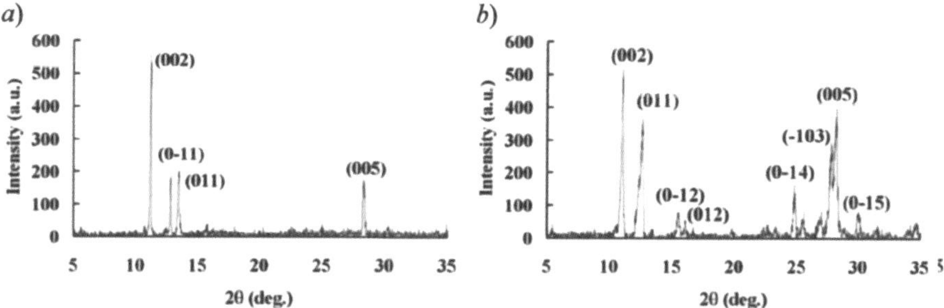

Figure 3 *XRD patterns of (a) the NMP TCNQ wires and (b) the corresponding single crystals. Indexing was made by the CMPR program* [6]

All diffraction peaks match well with the characteristic peaks of the single crystal ones.

The main reflection peaks in Figure 3 were labeled and could be indexed to the corresponding triclinic phase of NMP TCNQ with cell parameters $a = 3.87$ Å, $b = 7.80$ Å, $c = 15.78$ Å, $\alpha = 91.67°$, $\beta = 92.67°$ and $\gamma = 95.38°$, which were strong intensities of (00l) and (0kl) peaks of the triclinic phase of NMP TCNQ single crystals. Other relatively strong reflection peaks were absent in these patterns in comparison with the reported data. The above results strongly indicate that the NMP TCNQ wires are preferentially grown along the (100) direction, which is parallel to the stacking direction of TCNQ molecules.

4 CONCLUSION

NMP TCNQ molecular wires have been successfully synthesized by controlling the solvent evaporating time from acetone solution. Microscopic images showed that changing the evaporating time from 100 h (1.7 µl/min) to 30 min (167 µl/min) reduced the width of the wires from about 1 µm to about 40 nm. XRD patterns of NMP TCNQ wires revealed that the preferentially grown direction was parallel to the stacking direction of TCNQ molecules. This result also means that conducting path of NMP TCNQ would align to the well-grown direction.

Acknowledgement

This work was partly supported by a grant-in-aid for scientific research (No. 16750114) from JSPS and Foundation for Engineering Promotion from Faculty of Engineering, Shizuoka University.

References

1 D. de Caro, J. Sakah, M. Basso-Bert, C. Faulmann, J.-P. Legros, T. Ondarcuhu, C. Joachim, L. Aries, L. Valade, and P. Cassoux, *C. R. Acad. Sci. Paris, Série IIc, Chimie*, 2000, **3**, 674.

2 G. Cao, C. Ye, F. Fang, X. Xing, H. Xu, D. Sun, and G. Chen, *Mat. Sci. Eng. B*, 2005, **119**, 41.

3 T. Akutagawa, T. Ohta, T. Hasegawa, T. Nakamura, C. A. Christensen, and J. Becher, *PNAS*, 2002, **99**, 5028.

4 C.J. Fritchie Jr., *Acta Cryst.*, 1966, **20**, 892.

5 H. Kobayashi, *Bull. Chem. Soc. Jpn.*, 1975, **48**, 1373.

6 B. Toby, in *CMPR, A Graphical Manual Powder Indexing and Powder Diffraction Toolkit*, National Institute of Standards and Technology, USA, 2004.

ETHYLENEDITHIO-TTF-IMIDAZOLE: CONSTRUCTION OF PLURI-
DIMENSIONAL NETWORK BY HYDROGEN-BONDING AND S···S
INTERACTIONS

T. Murata,[1,2] Y. Morita,[2] Y. Nishimura,[2] Y. Yakiyama,[2] and K. Nakasuji[2]

[1] Division of Chemistry, Graduate School of Science, Kyoto University, Sakyo-ku, Kyoto,
Japan
[2] Department of Chemistry, Graduate School of Science, Osaka University, Toyonaka,
Osaka, Japan

1 INTRODUCTION

Among the research topics in the development of organic conductors, introduction of
hydrogen-bonding (H-bonding) interaction into charge-transfer (CT) complexes and salts
has been recognized as a useful strategy to control molecular arrangement and to increase
the dimensionality of network structure.[1~4] The studies for the design and synthesis of
tetrathiafulvalene (TTF) derivatives with H-bonding functionalities afforded a variety of
well-defined assembled structures.[1~3] In addition to this structural aspect, we have recently
disclosed the electronic effects of H-bonding interaction to control ionicity of each
component in CT complexes: 1) regulation of donor-acceptor ratio by forming D-A-D H-
bonded triad and 2) enhancement of electron-affinity of acceptor molecule[5] in TTF-
imidazole (TTF-Im)–p-chloranil complex,[5] and 3) modulation of the electron-donating
ability in diamino-dibenzo-TTF system.[6] In the TTF-Im–p-chloranil complex, these new
roles of H-bonding interaction resulted in the first demonstration of purely organic
molecular metal based on the CT complexes composed of H-bonded TTF derivatives.[5] In
order to construct further assembled structures in CT complexes by cooperation of H-
bonding, π-stacking and S···S interaction and to increase the dimensionality of conduction
paths, we have newly designed an ethylenedithio (EDT)-substituted TTF-Im derivative,
EDT-TTF-Im. Herein we report the synthesis, electrochemical properties and crystal
structure of EDT-TTF-Im.

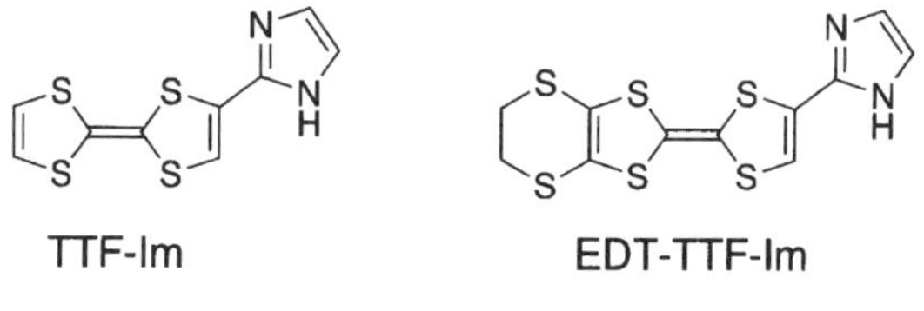

TTF-Im   EDT-TTF-Im

Chart 1

2 METHOD AND RESULTS

2.1 Synthesis

Synthesis of EDT-TTF-Im was achieved by the Stille-type cross coupling reaction of tributylstannylated-EDT-TTF with *N*-tosyl protected 2-iodoimidazole in the presence of Pd(PPh$_3$)$_4$ and CuI catalysts followed by removal of protection group using KOH (Scheme 1).[7] The single crystals of EDT-TTF-Im suitable for X-ray analysis were obtained as orange-colored platelet crystals by the vapor diffusion method using THF-hexane.[8]

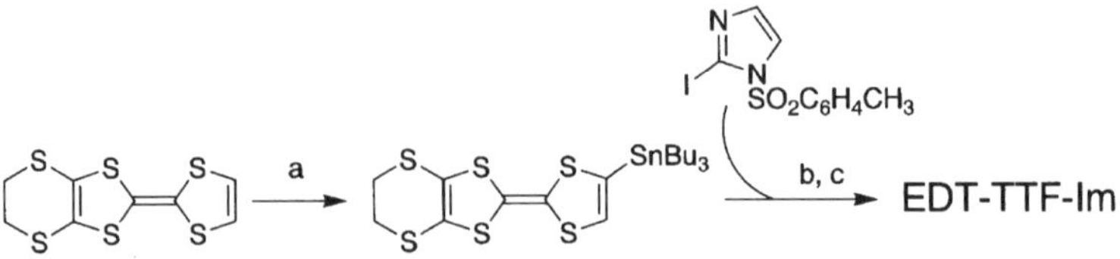

Scheme 1 *Synthetic method of EDT-TTF-Im. Reagent and conditions: a) n-BuLi, THF, –78 °C then Bu$_3$SnCl, –78 °C; b) 50 mol % Pd(PPh$_3$)$_4$, 5 mol % CuI, toluene, 100 °C; c) KOH, MeOH, rt.*

2.2 Electrochemical Measurement

Cyclic voltammogram of EDT-TTF-Im was measured in a DMF solution with Et$_4$NClO$_4$ as supporting electrolyte, showing two-stage one-electron reversible oxidation peaks. The half-wave oxidation potentials of EDT-TTF-Im and EDT-TTF calibrated with ferrocene/ferrocenium coupling were summarized in Table 1. The first oxidation potential of EDT-TTF-Im was slightly higher than that of EDT-TTF, and showed that EDT-TTF-Im was somewhat weaker electron-donor molecule than EDT-TTF (Table 1). This behavior was similar to that shown in TTF-Im.[5] The difference between first and second oxidation potentials of EDT-TTF-Im was slightly smaller than that of EDT-TTF.

2.3 Crystal Structure

EDT-TTF-Im crystallizes in the space group *Pbca* with eight molecules in a unit cell. The imidazole moiety was twisted by 28.0° from TTF skeleton (Figure 1a). The folding angles of TTF skeleton were 19.6° and 12.7° for α and β, respectively (Figure 1b). N1–H···N2 H-bonding interaction (2.86 Å, sum of van der Waals radius of N atom: 3.10 Å[9]) on the imidazole-ring formed a one-dimensional chain along the *a*-axis in a zigzag fashion similar

Table 1 *Half-wave oxidation potentials of EDT-TTF-Im and EDT-TTF*

	$E_{1/2}^{ox1}$ (V)	$E_{1/2}^{ox2}$ (V)	ΔE (V)
EDT-TTF-Im	+0.05	+0.25	0.20
EDT-TTF	+0.01	+0.23	0.22

Experimental conditions: solvent, DMF; concentration 5 mM; [Et$_4$NClO$_4$] = 0.1 M; scan rate, 100 mV/s; reference electrode, Ag/AgNO$_3$ (0.01 M); counter electrode, Pt wire; working electrode, glassy carbon; the results were calibrated with ferrocene/ferrocenium couple.

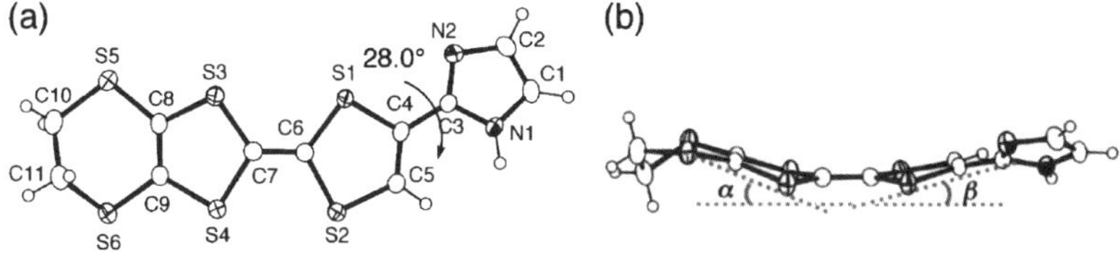

Figure 1 *Molecular structure and atomic numbering scheme of EDT-TTF-Im (a), and sideview of the molecular structure (b). α and β show the folding angles of TTF skeleton.*

to that of TTF-Im[5] (Figure 2). EDT-TTF-Im molecules in the H-bonded chain were inclined by 61.9° to neighboring molecule. The side-by-side S···S contacts of S2, S4 atoms on TTF skeleton and S6 atom belonging to the ethylenedithio group connected the H-bonded chains to construct a two-dimensional structure parallel to the *ac*-plane (Figure 2). The distances of S···S contacts in EDT-TTF-Im were 3.47 and 3.40 for S2···S4 and S2···S6, respectively (sum of van der Waals radius of S atoms: 3.60 Å[9]). In this two-dimensional structure, the close π-π interaction having small overlap with 3.44 Å of face-to-face distance was observed.

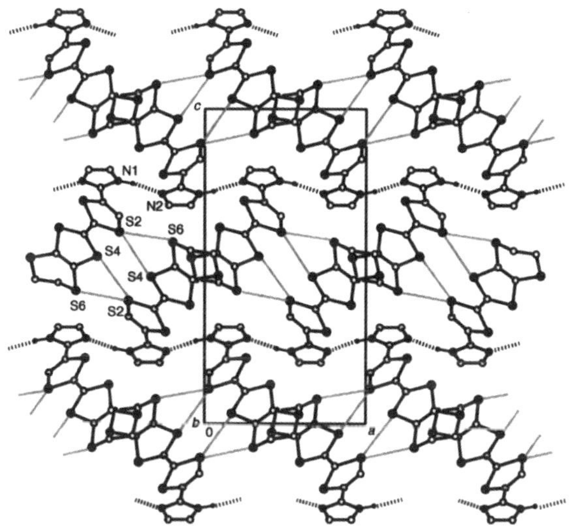

Figure 2 *Two-dimensional network constructed by N–H···N H-bonding and multiple S···S interactions viewed along the b-axis. The dashed lines and gray solid lines show N–H···N H-bonds and S···S contacts, respectively.*

3 CONCLUSION

We have newly synthesized the ethylenedithio-substituted derivative of TTF-Im. Similarly to TTF-Im, EDT-TTF-Im constructed the one-dimensional H-bonded chain in a zigzag fashion. This result suggests that EDT-TTF-Im has a high potential for forming a well-defined H-bonded structures in CT complexes as shown in TTF-Im–p-chloranil complex, and can exhibit electronic effects of H-bonding interaction. Furthermore, the introduction of ethylenedithio group was found to serve to form multiple S···S contacts and to increase the dimensionality of network structure. Our studies are currently focusing on the preparation of CT complexes of this novel electron-donor molecule and investigation of the structural and electronic effects of S···S contacts and H-bonding interactions in the CT complexes.

References

1   P. Batail, K. Boubekeur, M. Fourmigué and J.-C.P. Gabriel, *Chem. Mater.*, 1998, **10**, 3005.
2   M.R. Bryce, *J. Mater. Chem.*, 1995, **5**, 1481.
3   M. Fourmigué and P. Batail, *Chem. Rev.*, 2004, **104**, 5379.
4   Y. Morita, T. Murata, K. Fukui, S. Yamada, K. Sato, D. Shiomi, T. Takui, H. Kitagawa, H. Yamochi, G. Saito and K. Nakasuji. *J. Org. Chem.*, 2005, **70**, 2739.
5   T. Murata, Y. Morita, K. Fukui, K. Sato, D. Shiomi, T. Takui, M. Maesato, H. Yamochi, G. Saito and K. Nakasuji, *Angew. Chem. Int. Ed.*, 2004, **43**, 6343.
6   Y. Morita, E. Miyazaki, K. Fukui, S. Maki and K. Nakasuji, *Bull. Chem. Soc. Jpn.*, 2005, **78**, 2014.
7   Selected physical data of EDT-TTF-Im: mp 154–155 °C; TLC R_f 0.53 (1:3 hexane/ethyl acetate); ^{1}H NMR (DMSO-d_6, room temperature) δ 3.38 (s, 4), 6.94 (s, 1), 7.14 (s, 1), 7.26 (s, 1), 12.7 (s, 1); IR (KBr) 3400–2200 cm^{-1}; UV (KBr) 310 nm; EI-MS m/z 124 (M$^+$–C$_6$H$_6$S$_5$, 100%), 360 (M$^+$, 15%); Anal. Calcd for C$_{11}$H$_8$N$_2$S$_6$: C, 36.64; H, 2.28; N, 7.77. Found: C, 36.89; H, 2.28; N, 7.64.
8   Crystal Data for EDT-TTF-Im: C$_{11}$H$_8$N$_2$S$_6$, M_w = 360.56; orthorhombic, space group *Pbca*; a = 9.933(9), b = 15.47(1), c = 18.90(1) Å, V = 2904(4) Å^3, Z = 8, D_{calcd} = 1.649 g cm^{-1}, μ(Mo Kα) = 9.26 cm^{-1}, T = 200 K, 20660 reflections were measured of which 3224 independent, R_1 = 0.023 [2475 data with $I > 3.00\sigma(I)$], R_w = 0.060, and GOF = 1.36.
9   A. Bondi, *J. Phys. Chem.*, 1964, **68**, 441.

NEW STRATEGIES FOR DESIGNING TTP-BASED DONOR MOLECULES

Y. Bando,[1] M. Ashizawa,[1] T. Matsuzawa,[1] N. Takashita,[1] T. Kawamoto,[1] T. Mori,[1] T. Kakiuchi,[2] H. Sawa,[2,3] K. Takimiya,[4] T. Otsubo,[4] Y. Misaki[5] and J. Yamada[6]

[1]Tokyo Institute of Technology, Meguro-ku, Tokyo 152-8552, Japan
[2]The Graduate University for Advanced Studies,Tsukuba, Ibaraki 305-0801, Japan
[3]High-Energy Accelerator Research Organization, Tsukuba, Ibaraki 305-0801, Japan
[4]Hiroshima University, Higashi-Hiroshima, Hiroshima 739-8527, Japan
[5]Ehime University, Matsuyama , Ehime, 790-8557, Japan
[6]University of Hyogo, Ako-gun, Hyogo 678-1297, Japan

1   INTRODUCTION

Radical-cation salts of TTF (tetrathiafulvalene) derivatives have been investigated for a long time owing to their high conductivity and superconductivity. Among them, bis-fused TTF, called TTP (tetrathiapentalene), has been prepared aiming at designing a donor molecule with a more extended π-skeleton.[1] As an attempt to control the intermolecular interaction, we have investigated a variety of TTP donors such as selenium substituted ones and those with different terminal units. Although selenium substitution to the outer 1,3-dithiole rings has been carried out,[2] selenium substitution to the inner TTP-skeleton is comparatively difficult. We have recently succeeded in selenium substitution of two of the four sulfur atoms for the inner TTP skeleton in TMET-TTP (Scheme 1). We have also attempted to control the steric hindrance effect on the donor arrangement in bis(ethylthio)-TTP salts by introducing an MDT (methylenedithio) or 1,3-dithiane unit (C_nTMT-TTP; n = 1, 2 and C_2TDA-TTP in Scheme 1).[3] In this paper, we present the structural and electrical properties of these rasical-cation salts.

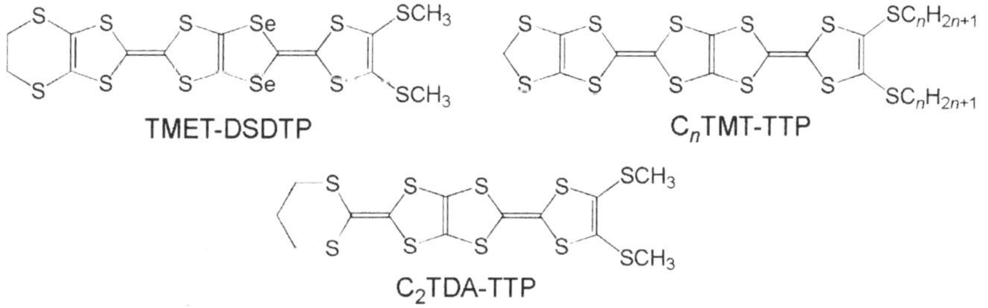

Scheme 1

2   METHOD AND RESULTS

2.1 TMET-TTP derivative

Charge-transfer complexes of selenium substituted TMET-TTP derivative were prepared by the electrochemical oxidation of the donors in the presence of tetrabutylammonium salts of various anions (PF_6, AsF_6 and AuI_2). Crystal structure analyses were carried out for these salts, which forms θ-type arrangement similarly to the other TMET-TTP derivatives (Figure 1(a)).[2]

For (TMET-DSDTP)$_x$AuI$_2$, the modulated structure was investigated by using the synchrotron radiation x-ray at BL-1B of the Photon Factory, KEK, Tsukuba. A two-fold modulation was observed along the c^* axis (Figure 1(b)).

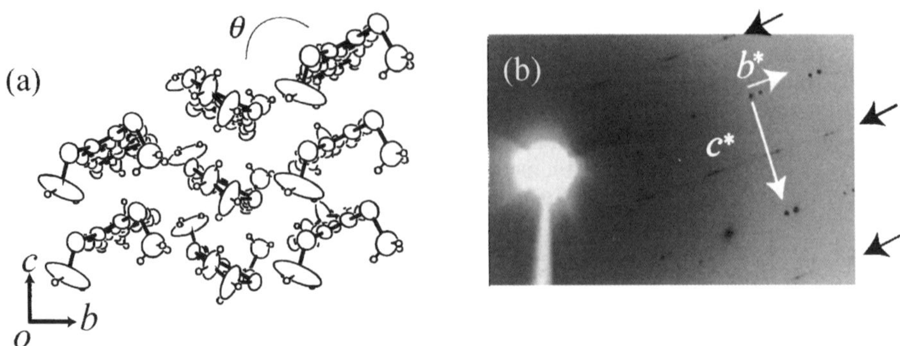

Figure 1 *(a) Crystal structure of the TMET-TTP derivatives projected along the donor long axis. (b) Oscillation photograph of (TMET-DSDTP)$_x$AuI$_2$ at room temperature. White arrows in this figure show the reciprocal lattice vectors of the average structure without two-fold superstructure; monoclinic, P2$_1$/a, a = 39.956(9), b = 11.146(2), c = 5.000(1) Å, β = 93.71(2)° and V = 2222.1(7) Å^3.*

The resistivity of all salts increases below room temperature, but the resistivity behavior of (TMET-DSDTP)$_y$PF$_6$ is different from those of the other salts. Although this salt has a large dihedral angle ($\theta = 125°$), this salt shows flat resistivity behavior between 200 K and 50 K. This disagrees with the universal phase diagram of the θ-phase. The extended Hückel molecular orbital calculation indicates that the selenium substitution changes the distribution of HOMO. The HOMO is more largely populated on the TTF unit which does not contain the Se atoms. This result is in agreement with the small change of the bond lengths associated with the change of the charge distribution. In the TMET-DSDTP complexes, the HOMO is more largely located on the ET part. Qualitatively speaking, the ET part forms more two-dimensional electronic structure, and stabilizes the metallic state.

2.2 Bis(ethylthio)-TTP molecules with the steric hindrance unit

We have prepared (C$_n$TMT-TTP)$_2$PF$_6$ ($n = 1, 2$), (C$_2$TDA-TTP)$_2$PF$_6$ (monoclinic, C2/m, a = 32.769(8), b = 12.445(6), c = 5.711(6) Å, β = 97.67(4)° and V = 2308(2) Å^3) and (C$_2$TDA-TTP)GaCl$_4$ (orthorhombic, Pnnm, a = 9.253(4), b = 18.342(9), c = 18.046(7) Å, and V = 3062(2) Å^3) in the same way as the salts of the TMET-TTP derivatives.[3] Crystal structures of these salts were analyzed.[3] These salts construct uniform stacking structures (Figures 2(a), 2(b) and 2(c)). The Fermi surface of (C$_2$TDA-TTP)$_2$PF$_6$ is open as shown in Figure 2(d). (C$_2$TDA-TTP)GaCl$_4$ forms strongly dimerized one-dimensional structure and has an additional strong two-fold modulation along the *a* axis (Figure 3). We could not analyze this situation, so the average structure is shown in Figure 3.

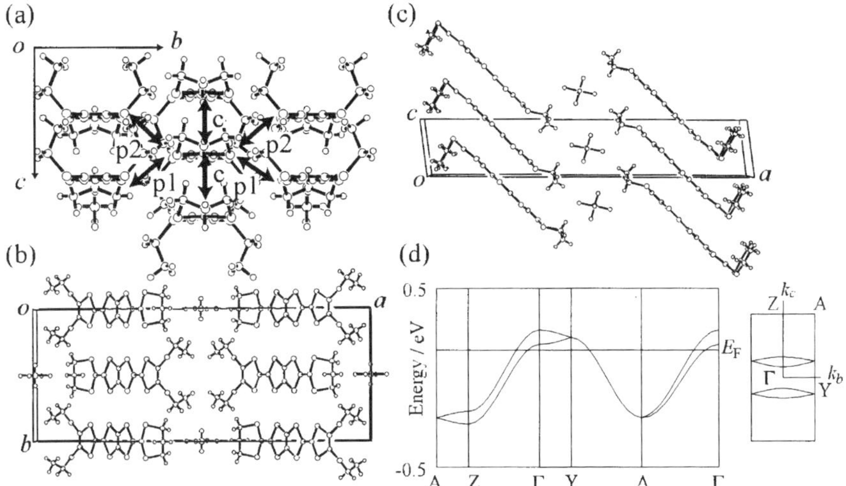

Figure 2 *Crystal structure of $(C_2TDA\text{-}TTP)_2PF_6$, (a) view along the donor long axis, (b) projection along the c axis, and (c) projection along the molecular short axis. (d) Energy band structure and the Fermi surface of $(C_2TDA\text{-}TTP)_2PF_6$ without sulfur 3d atomic orbitals. The intermolecular overlap integrals S_i between the HOMOs are $S_c = -11.1$, $S_{p1} = 0.134$, and $S_{p2} = 1.86 \times 10^{-3}$.*

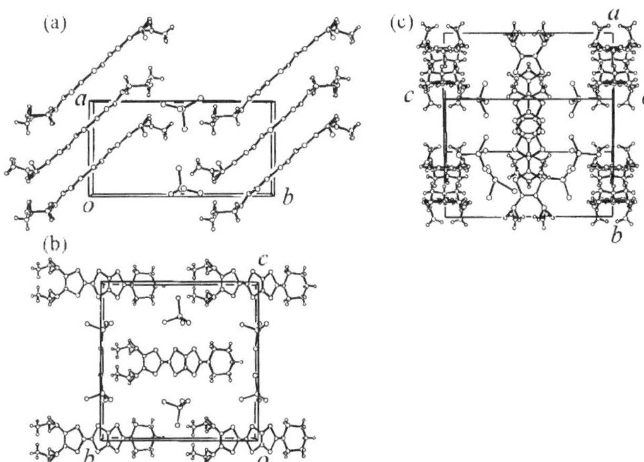

Figure 3 *Crystal structure of $(C_2TDA\text{-}TTP)GaCl_4$, (a) projection along the c axis, (b) projection along the a axis, and (c) view along the donor long axis.*

$(C_nTMT\text{-}TTP)_2PF_6$ shows metallic conductivity down to 1.4 K. This salt shows linear temperature dependence of the thermoelectric power and the Dysonian-like lineshape of an ESR spectrum.[3] The resistivity of $(C_2TDA\text{-}TTP)_2PF_6$ decreases with decreasing temperature, but increases slightly below 17 K (Figure 4(a)). In order to clarify the low temperature state of $(C_2TDA\text{-}TTP)_2PF_6$, we have carried out the ESR measurement below 105 K. In all region, upon cooling, the linwidth decreases linearly and the normalized spin susceptibility is almost constant (Figure 4(b)). The magnetic transition was not observed at low temperature. $(C_2TDA\text{-}TTP)GaCl_4$ is an insulator because this salt is a dimerized 1 : 1 salt ($\sigma_{rt} < 10^{-6}$ S cm^{-1}).

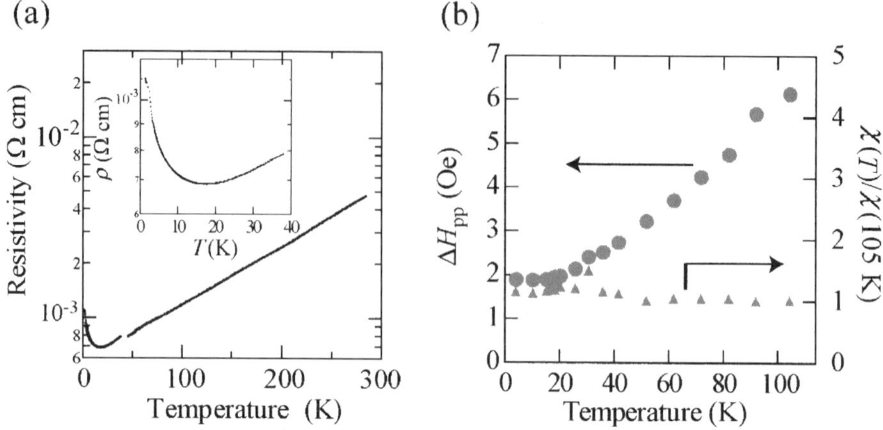

Figure 4 *(a) Electrical resistivity of $(C_2TDA\text{-}TTP)_2PF_6$ between room temperature and 1.3 K. The inset shows the electrical resistivity at low temperatures. (b) Temperature dependence of the peak-to-peak linewidth ΔH_{pp} (square points) and the normalized spin susceptibility (circle points) of $(C_2TDA\text{-}TTP)_2PF_6$ (H // c).*

3 CONCLUSION

We have prepared new TTP molecules with selenium substitutions to the TMET-TTP molecule and with a 1,3-dithiane or an MDT unit. In the TMET-TTP derivatives, the conducting properties are influenced by the change of the HOMO distribution induced by the selenium substitution. All C_nTMT-TTP complexes indicate that the TTP molecules with $-SC_2H_5$ (ethylthio) or MDT units have a strong tendency to form the uniform stacking structure regardless of the counter part. C_2TDA-TTP, however, gives both a dimerized 1 : 1 insulating salt and metallic uniformly stacked salts isostructural to other ethylthio-containing salts.

References

1 Y. Misaki, H. Nishikawa, K. Kawakami, S. Koyanagi, T. Yamabe, and M. Shiro, *Chem. Lett.*, **1992**, 2321.

2 M. Aragaki, H. Hoshino, T. Mori, Y. Misaki, K. Tanaka, H. Mori, and S. Tanaka, *Adv. Mater.*, 2000, **12**, 983; M. Ashizawa, H. Nii, T. Mori, Y. Misaki, K. Tanaka, K. Takimiya, and T. Otsubo, *Bull. Chem. Soc. Jpn.*, 2003, **76**, 2091.

3 Y. Bando, T. Matsuzawa, N. Takashita, T. Kawamoto, T. Mori, and J. Yamada, *Bull. Chem. Soc. Jpn.*, 2005. **78**. 1442.

SYNTHESIS, PHYSICAL PROPERTIES AND CRYSTAL STRUCTURES OF A NEW ASYMMETRICAL MDOB-EDT-TTF AND ITS CHARGE-TRANSFER COMPLEXES

T. Inayoshi and E. Harima

Department of Chemistry, College of Science and Engineering, Aoyama Gakuin University, 5-10-1 Fuchinobe Sagamihara, Kanagawa, 229-8558, Japan. E-mail: inayoshi@chem.aoyama.ac.jp

1 INTRODUCTION

The presence of intermolecular interactions through short contacts due to chemical modifications has been shown to be important for conductive charge-transfer (CT) complexes. BMDO-DBTTF donor does not afford metallic compounds in the same as BEDO-DBTTF and no short S⋯S contacts were observed in its CT complexes.[1,2] As an attempt to form the intermolecular interactive effects, we have combined the features of BEDO-DBTTF and BEDT-TTF (ET) into a new donor, EDOB-EDT-TTF, which was formed by conjugating half of a BEDO-DBTTF and half of an ET. S⋯S short contacts were found to exist in PF_6 radical salt composed of this hybrid donor.[3] Herein, a new asymmetrical (methylendioxybenzo)(ethylenedithio)tetrathiafulvalene (MDOB-EDT-TTF) donor formed by conjugating half of a BMDO-DBTTF and half of an ET was synthesized. We report synthesis, properties and crystal structure of MDOB-EDT-TTF and its CT complexes.

2 RESULTS AND DISCUSSION

The donor was synthesized by cross-coupling reaction of 5, 6-methylenedioxy-1,3-benzodithiol-2-thione (5) and 4,5-ethylenedithio-1,3-dithiol-2-one (6) as shown Scheme 1. The mixture of products was purified by column chromatography with CS_2 as the eluent. The crystals of neutral MDOB-EDT-TTF were recrystallized from benzene and were obtained as orange needles.

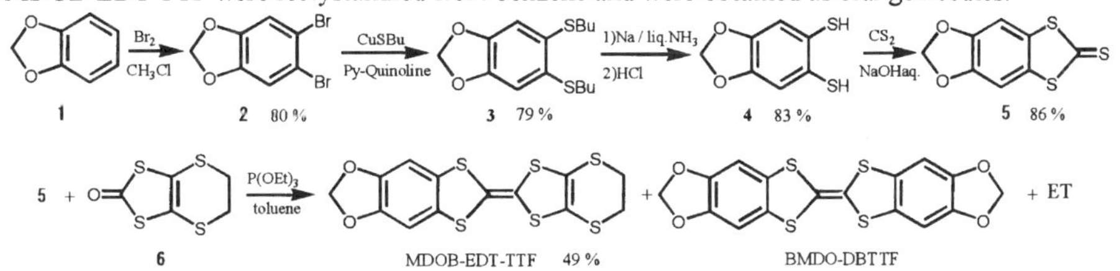

Scheme 1 *Synthetic route to MDOB-EDT-TTF*

The redox potentials measured by cyclic voltammetry and absorption spectrum of the donor are shown in Table 1. The redox process exhibited two reversible one-electron waves. As compared with BMDO-DBTTF, the present electron donor is weaker than BMDO-DBTTF as a result of conjugating half of an ET. The peak positions of the absorption spectrum of the present donor were red-shifted by the introduction of half of an ET.

Table 1 *Redox potentials and absorption spectra of MDOB-EDT-TTF*

Donor	Redox potentials [a]			UV-Vis (logε) [c]	Reference
	$E_{1/2(1)}$/V	$E_{1/2(2)}$/V	ΔE /V [b]	λ/ nm	
MDOB-EDT-TTF	0.52	0.86	0.34	315(4.19), 339(4.12), 452(2.55), 482sh	This study
BMDO-DBTTF	0.48	0.83	0.35	304(4.02), 327(3.93), 444(2.30), 470(2.26)	1

[a] The half-wave redox potentials vs. SCE, Pt working electrode, 0.1M Bu_4NClO_4 in CH_2Cl_2, scan rate 50 and 200 mV/s, at $22\pm1°C$. [b] $\Delta E=E_{1/2(2)}-E_{1/2(1)}$. [c] In 1,1,2-trichloroethane.

The CT complexes were prepared by directly mixing hot benzene solutions of the donor and the acceptors. Radical salts were prepared by electrochemical oxidation in the presence of n-$Bu_4N^+X^-$ (X=ReO_4, PF_6, ClO_4, I_3, Br) as supporting electrolyte.

Single crystal structure analyses have been carried out for neutral MDOB-EDT-TTF, MDOB-EDT-TTF(Me_2TCNQ)$_{0.5}$ complex and (MDOB-EDT-TTF)ReO_4 radical salt. The crystal data are listed in Table 2. A computer program, maXus,[4] was used for the determination and refinement of the crystal structures. The structures were solved by a direct method using SIR-97,[5] and were refined using a full-matrix least-squares technique.

Table 2 *Crystallographic data of the crystals*

	MDOB-EDT-TTF(C_6H_6)$_{0.5}$	MDOB-EDT-TTF(Me_2TCNQ)$_{0.5}$	(MDOB-EDT-TTF)ReO_4
Chemical formula	$C_{16}H_{11}O_2S_6$	$C_{20}H_{12}O_2N_2S_6$	$C_{13}H_8O_6ReS_6$
Formula weight	427.66	504.73	638.81
Shape	Orange needle	Black needle	Black block
Crystal system	Triclinic	Monoclinic	Triclinic
Space group	P -1	$P\,2_1/c$	P -1
a /Å	6.365(2)	11.887(2)	8.082(1)
b /Å	16.559(3)	11.970(1)	9.801(2)
c /Å	17.138(3)	14.558(2)	12.044(3)
α /°	100.11(1)	90	97.03(2)
β /°	98.77(2)	92.61(1)	106.36(2)
γ /°	94.86(2)	90	93.76(2)
V /Å^3	1745.8(6)	2069.1(5)	903.5(3)
Z	4	4	2
$D_{calc}(D_{obs})$/gcm^{-3}	1.63(1.65)	1.61(1.63)	2.35(2.36)
Diffractometer	Mac Science MXC18	Mac Science MXC18	Mac Science MXC18
Temperature	r.t.	r.t.	r.t.
Radiation	Mo Kα (λ=0.71073Å)	Mo Kα (λ=0.71073Å)	Mo Kα (λ=0.71073Å)
Reflection used	4227 ($I>3.0\sigma(I)$)	3411 ($I>3.0\sigma(I)$)	3678 ($I>3.0\sigma(I)$)
$R(R$w$)$	0.043(0.086)	0.043(0.056)	0.034(0.076)
Weighting function[a]	FA=1.00, FB=0.001	FA=3.00, FB=0.01	FA=2.00, FB=0.01

[a] Counting statistic weights of the form: $w=exp[FA \cdot sin^2\theta/\lambda^2]/[\sigma^2(F_0)+FB(F_0)^2]$

Table 3 *Properties of CT complexes and radical cation salts of MDOB-EDT-TTF*

Acceptor or anion	Solvent	Appearance	m.p.(°C)	Found (calcd.) (%)			Stoichiometry[a] D:A:Solvent	CT band [b] ×10^3 cm^{-1}	Conductivity[c] σ_{rt}(Scm^{-1})	E_a(meV)
				C	H	N				
TCNQ	Benzene	Black powder	>300	50.81(50.65)	2.07(2.04)	9.32(9.45)	1:1	<3.13, 9.76	2.1	68
FTCNQ[d]	Benzene	Black powder	>300	49.06(49.16)	2.07(1.82)	9.01(9.17)	1:1	<3.13, 9.56	6.0×10^{-1}	80
Me_2TCNQ[e]	Benzene	Black needle	>247~8				2:1 [f]	6.67	2.1×10^{-6} [g]	
PF_6	PhCl	Black crystal	>300	29.43(29.26)	1.79(1.51)		1:1	5.90, 9.98	1.1×10^{-5} [h]	220
ReO_4	PhCl	Black crystal	>300				1:1 [f]	6.76	5.5×10^{-5} [h]	348
ClO_4	PhCl	Black crystal	>300	38.45(38.00)	2.18(2.18)		1:1:1	5.88sh, 10.01	1.9×10^{-4} [h]	139
I_3	PhCl	Black crystal	>300	23.76(23.38)	1.26(1.28)		1:1:0.5	5.98sh, 9.03	4.3×10^{-4} [h]	129
Br	PhCl	Black crystal	>300	36.70(36.61)	2.10(2.02)		1:1:0.5	6.94, 10.00sh	Insulator	

[a] Ratio determined by elemental analysis. [b] Measured with a KBr disk method. [c] Measured on compressed pellet. [d] 2-Fluoro-TCNQ [e] 2,5-Dimethyl-TCNQ. [f] Determined by X-ray analysis. [g] Measured on crystal with a four-probe method. [h] Measured on crystal with a two-probe method.

Properties, the electrical conductivity (σ_{rt}) and activation energy (E_a) of the CT complexes and radical salts are summarized in Table 3. 1:1 powder TCNQ complex exhibited a conductivity of 2.1 Scm^{-1}, and E_a of 68 meV and a semiconductive behavior.

The crystal structure of neutral MDOB-EDT-TTF is shown in Figure 1. There are two crystallographically independent donor molecule A and B and one benzene molecule. Molecule A is bent at the sulfur atoms (167.4°) of one side of the TTF skeleton, however molecule B is plane. Benzene used for recrystallization solvent is contained in the crystal structure. S···S short contacts[6] (3.46, 3.50 and 3.56 Å) exist in the crystal structure of neutral MDOB-EDT-TTF.

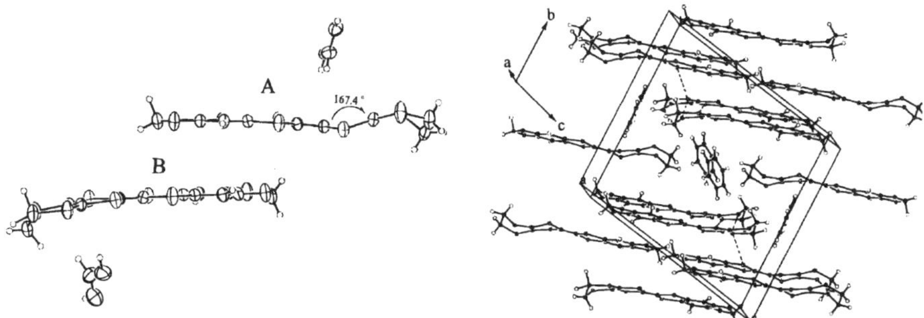

Figure 1   *ORTEP drawing of side view and the crystal structure of neutral MDOB-EDT-TTF, the intermolecular short S···S (< 3.60 Å) contacts are indicated by dotted lines.*

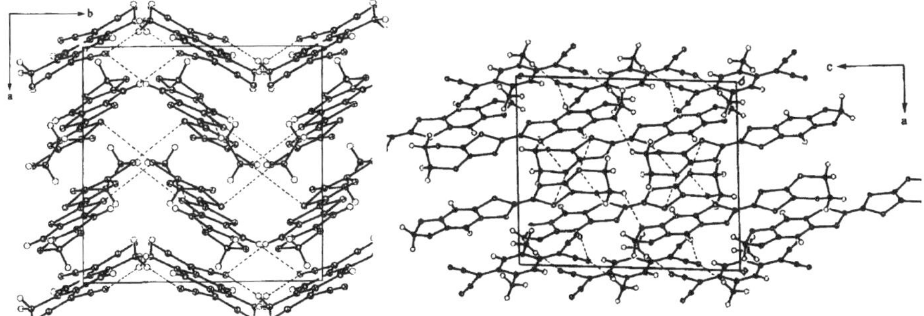

Figure 2   *Two different projections of MDOB-EDT-TTF(Me₂TCNQ)₀.₅, dotted lines indicate S···S, N···H (< 2.75 Å) and O···H (< 2.72 Å) short contacts.*

Figure 2 shows the crystal structures of MDOB-EDT-TTF(Me$_2$TCNQ)$_{0.5}$. There are not only S···S contacts (3.40 and 3.57 Å) and N···H contacts (2.62 and 2.62 Å), but also O···H contact (2.67 Å). The complex is composed of columns in which the donor and acceptor molecules stack alternately to form AD$_1$D$_2$AD$_1$D$_2$A mixed column along a-axis, and this explains the non-conductivity. The D$_1$ and D$_2$ molecules are stacked in alternating orientations and the mean separation of them is 3.63 Å. The face-to-face distances of the A and the D$_1$ molecules, and the D$_2$ and the A molecules are 3.45 and 3.49 Å, respectively. The donor-acceptor interaction are stronger than those of the donor-donor. Figure 3 shows the crystal structure of (MDOB-EDT-TTF)ReO$_4$ radical salt. Two donor molecules are connected by two S···S contacts (3.52Å) between two S atoms of TTF frames and form a head-to-tail dimer, the face-to-face distance of which is 3.51 Å. The short S···S contact between two S(5) atoms of ethylenedithio groups is present at 3.36 Å, however, the mode of intermolecular overlap is unfavorable (The calculated overlap integral t$_5$ is 0.29×10^{-3}). The existence of hydrogen bonds between the donors and each O atoms of ReO$_4^-$ anions is suggested, since there exist seven short H···O contacts with 2.305~2.652 Å distances. Additionally, there are five short O···S contacts (3.046~3.195 Å) between donors and anions.

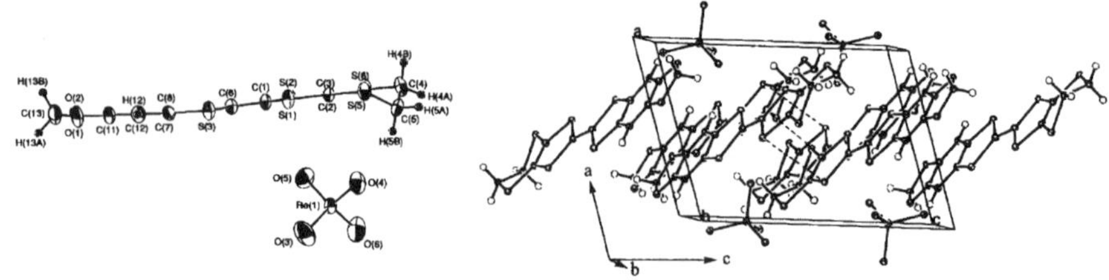

Figure 3   *ORTEP drawing of side view and crystal structure of (MDOB-EDT-TTF)ReO₄, dotted lines indicate S···S short contacts*

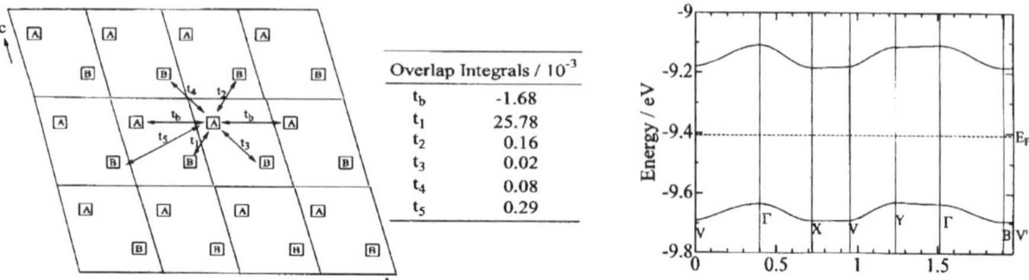

Figure 4   *Donor arrangement, overlap integrals and band structure of (MDOB-EDT-TTF)ReO₄*

Intermolecular overlaps of the HOMO are calculated by extended Hückel method (Figure 4). The band structure is calculated on the basis of the tight-binding approximation using the intermolecular overlap integrals of HOMO.[7] As a result of calculated overlap integrals, the t_1 interaction is large. This is suggested that the donor molecules form dimers. The calculated band structure is in accordance with the semiconductive property.

Acknowledgements

Fruitful collaboration with emeritus professor S. Matsumoto and Mr. T. Takahashi are greatly appreciated.

References

1   T. Inayoshi and I. Ono, *Synth. Met.*, 2000, **110**, 153.
2   T. Senga, K. Kamoshida, L. A. Kushch, G. Saito, T. Inayoshi and I. Ono, *Mol. Cryst. Liq. Cryst.*, 1997, **296**, 97.
3   T. Inayoshi, S. Matsumoto and I. Ono, *Synth. Met.*, 2003, **133-134**, 345.
4   *Programs of Structure Determination Package*, MAC Science, Yokohama, 1998.
5   A. Altomare, G. Cascarano, C. Giacovazzo, A. Gaugliardi, M. C. Burla, G. Polidori and M. Camalli, *J. Appl. Crystallogr.*, 1994, **27**, 435.
6   The van der Waals radii employed in this paper are as follows; H 1.20, N 1.55, O 1.52, S 1.80 Å. A. Bondi, *J. Phys. Chem.*, 1964, **68**, 441.
7   T. Mori, A. Kobayashi, Y. Sasaki, H. Kobayashi and G. Saito, *Bull. Chem. Soc. Jpn.*, 1984, **57**, 627.

THE FIRST METALLIC SALT CONTAINING FERROCENE, β''-(BEDT-TTF)$_4$(FE-(CPCONHCH$_2$SO$_3$)$_2$)·2H$_2$O, AND NOVEL FERROCENE CONTAINING ANIONS, CPFECP-CONH-(m-, p-)C$_6$H$_4$SO$_3^-$

K. Furuta,[1] H. Akutsu,[1] J. Yamada,[1] S. Nakatsuji,[1] and S.S. Turner[2]

[1]Graduate School of Material Science, University of Hyogo, 3-2-1 Kouto, Kamigori-cho, Ako-gun, Hyogo 678-1297, Japan. E-mail: akutsu@sci.u-hyogo.ac.jp
[2]Department of Chemistry, Warwick University, Gibbet Hill Road, Coventry CV4 7AL, U.K.

1 INTRODUCTION

For more than five years, we have focussed on creating new anions that are combinations of an organic functional molecule and the sulfonate (-SO$_3^-$) group. They were prepared for use as a counterion in organic conductors and provided multifunctional systems combining electrical conductivity with another property.[1,2,8,9] Initially we used derivatives of an organic free radical, TEMPO, as the anion in purely organic magnetic conductors. We have prepared several TEMPO-based anions, one of which TEMPO-NHCOCH$_2$SO$_3$ (atsa) provided the first genuine BEDT-TTF organic magnetic conductor, α-(BEDT-TTF)$_3$(atsa)$_2$·6H$_2$O.[1] The salt is a semiconductor with relatively low electrical resistivity.

Secondly we considered ferrocene, which is a fascinating functional molecule due to its redox properties and its use as a cation in ferromagnets.[3] For these reasons, many researchers have tried to introduce the ferrocenyl group into conducting materials over the last couple of decades, but the resultant charge transfer salts showed only poor conductivities.[4-7] We have been able to synthesize a sulfo derivative of the ferrocenyl moiety. One of the novel dianions, Fe(Cp-CONHCH$_2$SO$_3$)$_2$ (fdnm), provided a BEDT-TTF salt, α-(BEDT-TTF)$_4$(fdnm)·4H$_2$O.[8] It is semiconductive and shows a much higher conductivity than in salts synthesized by other research groups. Recently, we obtained another polymorph, β''-(BEDT-TTF)$_4$(fdnm)·2H$_2$O, which is the first metallic salt containing the ferrocenyl group, the structure and properties of which are reported briefly in this paper. We also report the structures and properties of new anions, CpFeCp-CONH-(m-, p-)C$_6$H$_4$SO$_3$ (fcma and fcpa, respectively), and the TTF salt of the fcma anion.

atsa   fdnm   fcma (m-) and fcpa (p-)

2 METHOD AND RESULTS

β''-(BEDT-TTF)$_4$(Fe-(CpCONHCH$_2$SO$_3$)$_2$)·2H$_2$O

Single crystals were obtained by the electrocrystallization method previously reported in Ref. 8. Using strictly dry solvents seems to provide the β''-type salt. The crystal structure is shown in Figure 1.

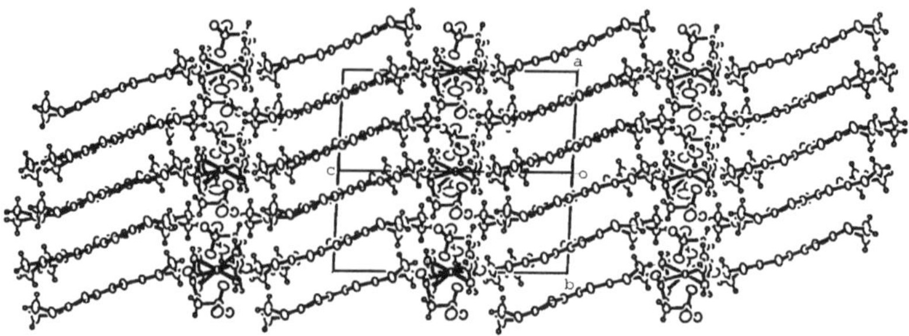

Figure 1 *Crystal structure of β''-(BEDT-TTF)$_4$(fdnm)·2H$_2$O.*

The structure consists of alternating layers of BEDT-TTF donors with a β''-type packing motif, and the dianions, with water molecules in the anionic layers. In the anion layer, the Fe atom of the ferrocenyl moiety is located on an inversion centre. The distance between Fe and the cyclopentadienyl ring is 1.652(3) Å which is similar to that in unsubstituted neutral ferrocene (1.664 Å),[10] suggesting the ferrocenyl moiety is in the neutral state. The electrical resistivity was measured by a standard four probe method (Figure 2). The salt shows metallic behaviour from room temperature to 62 K, at which point there is a sharp metal-insulator transition. The details will be published elsewhere.

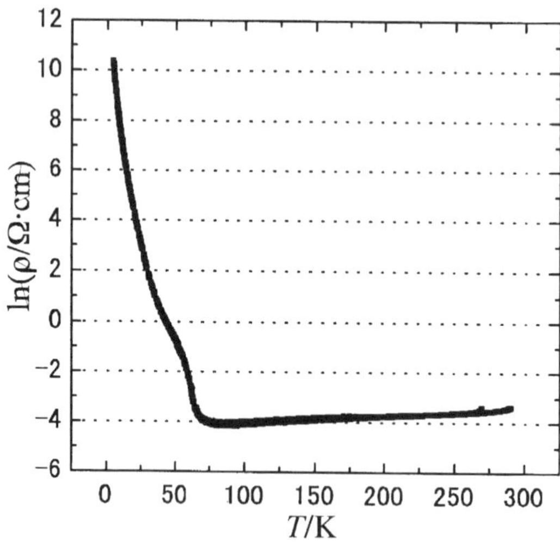

Figure 2 *Temperature dependence of electrical resistivity of β''-(BEDT-TTF)$_4$(fdnm)·2H$_2$O.*

2.2 New anions CpFeCp-CONH-(*m*-, *p*-)C$_6$H$_4$SO$_3^-$, fcma and fcpa

The acidic anions fcma and fcpa are prepared by the condensation reactions of *m*- or *p*-anilinesulfonate and ferrocenecarboxyllic acid in the presence of N,N'-dicyclohexylcarbodiimide (DCC) and 4-dimethylaminopyridine (DMAP) in CH$_2$Cl$_2$. Exchanges of the counterions with N(PPh$_3$)$_2$Cl for fcma and PPh$_4$Br for fcpa yielded orange crystals of N(PPh$_3$)$_2$fcma and PPh$_4$fcpa, respectively. The synthesis of the *ortho*-derivative was attempted but the reaction did not proceed. The structural characterization was performed by an single crystal X-ray analysis. The crystal data are as follows; N(PPh$_3$)$_2$fcma·2CH$_3$CN·2H$_2$O: triclinic $P\bar{1}$, a = 12.300(1), b = 14.333(2), c = 16.041(2) Å, α = 104.161(7), β = 111.155(1), γ = 92.282(3)°, V = 2531.4(6) Å^3, Z = 2, D_c = 1.342 g/cm^3, R = 0.110, R_W = 0.107, PPh$_4$fcpa·2H$_2$O, triclinic $P\bar{1}$, a = 13.192(4), b = 16.253(7), c = 9.470(2) Å, α = 106.83(3), β = 108.06(1), γ = 97.90(3)°, V = 1789(1) Å^3, Z = 2, D_c = 1.393 g/cm^3, R = 0.057, R_W = 0.049. Figure 3 shows the molecular structures of the anions in the salts. The anion fcma forms a dimer around the centre of symmetry, with hydrogen bonds between -NH and –SO$_3$. The fcpa molecules are also interlinked by hydrogen bonds from –SO$_3$ to –SO$_3$ and from –SO$_3$ to –NH via incorporated water molecules.

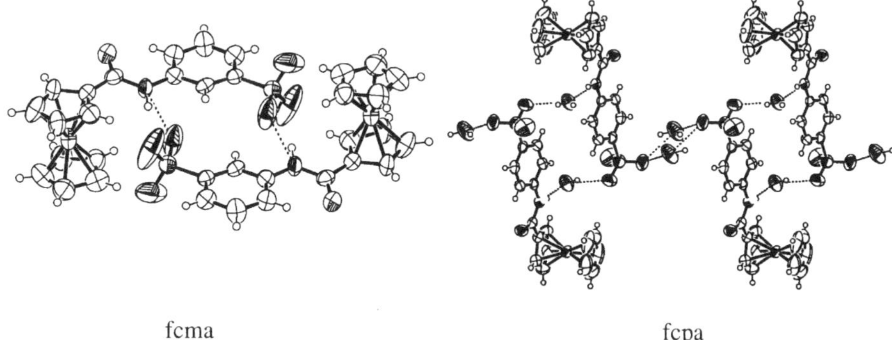

Figure 3 *Molecular structures of the fcma and fcpa anions in the N(PPh$_3$)$_2$fcma and PPh$_4$fcpa salts. Dashed lines indicate hydrogen bonds (< 3.05 Å).*

Metathesis of (TTF)$_3$(BF$_4$)$_2$ and N(PPh$_3$)$_2$fcma in a mixed solvent of CH$_3$CN and CH$_3$OH gave black crystals with a needle morphology. The crystal structure was determined by an X-ray analysis. The crystal data are as follows; (TTF)(fcma), triclinic $P\bar{1}$, a = 9.921(1), b = 11.403(1), c = 12.099(1) Å, α = 102.646(3), β = 99.743(2), γ = 114.467(1)°, V = 1162.5(2) Å^3, Z = 2, D_c = 1.681 g/cm^3, R = 0.076, R_W = 0.034. The crystal structure is shown in Figure 4 in which one TTF cation and one fcma anion are independent. The anion forms a hydrogen bonded dimer similar to that in N(PPh$_3$)$_2$fcma. The TTF monocation also forms a face-to-face dimer, which is surrounded by the relatively large anions (see Figure 4). There is no TTF interdimer short contact, suggesting poor conductivity and the measured electrical resistivity is more than 10^6 Ω·cm. In addition, the distances between the least square plane of each Cp ring and the Fe atom are 1.644(3) and 1.641(1) Å, which are similar to that in ferrocene (1.664 Å),[10] indicating that the ferrocenyl moiety is in the neutral state. We also prepared a Zwitter-ion, CpFe$^+$Cp-CONH-*m*-C$_6$H$_4$SO$_3^-$, by the oxidation of N(PPh$_3$)$_2$fcma with I$_2$ as a dark green powder. Preliminary magnetic measurements show that this obeys Curie-Weiss law with C = 0.69 emu K mol^{-1}

and $\theta = -3.0$ K. The Curie constant corresponds well to that of the ferrocene cation (0.72 emu K mol^{-1}), indicating the ferrocenyl part has a charge of +1.

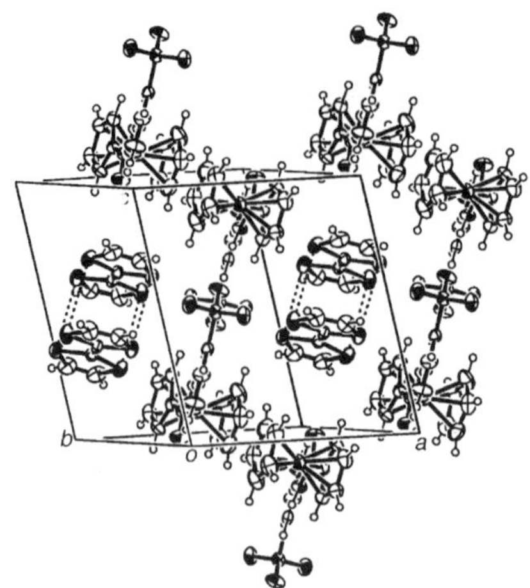

Figure 4 *Crystal structure of (TTF)(fcma). Dashed lines indicate short S...S contacts below 3.70 Å.*

3 CONCLUSIONS

We have briefly described the structure and properties of the first metallic BEDT-TTF salt containing ferrocene, β''-(BEDT-TTF)$_4$(Fe-(CpCONHCH$_2$SO$_3$)$_2$)·2H$_2$O, which shows a sharp metal-insulator transition at 62 K. The details will be published elsewhere. We also report novel ferrocene-based anions, CpFeCp-CONH-(m-, p-)C$_6$H$_4$SO$_3^-$. The *meta*-derivative provides a TTF salt, in which TTF forms a dimer that is surrounded by the anions. The preparation of BEDT-TTF salts by electrocrystallization is in progress.

References

1   H.Akutsu, J. Yamada and S. Nakatsuji, *Chem. Lett.*, 2003, **32**, 1118.
2   H.Akutsu, J. Yamada and S. Nakatsuji, *Synth. Met.*, 2005, **152**, 377.
3   J.S. Miller, A.J. Esptein and W.M. Reiff, *Chem. Rev.*, 1988, **88**, 201.
4   S. B. Wilkes, I. R. Butler, A. E. Underhill, M. B. Hursthouse, D. E. Hibbs and K. M. A. Malik, *J. Chem. Soc. Dalton Trans.*, 1995, 897.
5   A. J. Moore, M. R. Bryce, P. J. Skabara, A. S. Batsanov, M. L. Goldenberg, and J. A. K. Howard, *J. Chem. Soc. Perkin Trans. 1*, 1997, 3443.
6   H. J. Lee, D. -Y. Noh, A. E. Underhill, and C. -S. Lee, *J. Mater. Chem.*, 1999, **9**, 2359.
7   M. Iyoda, T. Takano, N. Otani, K. Ugawa, M. Yoshida, H. Matsuyama, and Y. Kuwatani, *Chem. Lett.*, 2001, 1310.
8   K. Furuta, H.Akutsu, J. Yamada and S. Nakatsuji, *Chem. Lett.*, 2004, **33**, 1214.
9   K. Furuta, H.Akutsu, J. Yamada and S. Nakatsuji, *Synth. Met.*, 2005, **152**, 381.
10   B. M. Yamin, H.-K. Fun, K. Sivakumar, B.-C. Yip, and O. B. Shawkataly, *Acta Cryst. C*, 1996, **C52**, 600.

Organic/Inorganic Hybrids and Photoinduced Phenomena

UNCONVENTIONAL MAGNETOTRANSPORT IN 1D TTF-BASED π-d INTERACTION SYSTEMS

Toshiaki Enoki, Kazuki Okabe, and Akira Miyazaki

Department of Chemistry, Tokyo Institute of Technology, Ookayama, Meguro-ku, Tokyo, 152-8551 Japan

1 INTRODUCTION

TTF-based organic charge transfer complexes are featured with low dimensional (D) π-electron systems, in which low D electronic instability creates a variety of insulating electronic phases such as CDW, SDW, etc.[1] When the insulating low temperature phase is magnetic, the incorporation of magnetic anions having localized d-electron spins into the charge transfer complexes is expected to affect the low-D electronic instability, resulting in interesting interplay between the electron transport of the π-electrons and the magnetism of the d-electron spins.[2-6] Among the phenomena appearing in the interplay in these π-d interaction systems, a competition of magnetic insulating state and superconductivity has intrigued us in relation to the discovery of field-induced superconductivity in λ-$(BETS)_2FeCl_4$.[4,5] The coexistence of ferromagnetism of the d-electron in metallic π-electron phase $(BEDT\text{-}TTF)_3[MnCr(C_2O_4)_3]$ showed a clue for paving a way toward molecular ferromagnetic metal, which is not a mimic of traditional metal magnets such as iron. Under these perspectives, we have investigated the correlations of the electron transport and magnetism of various low-D molecular magnetic conductors such as $(DMET)_2FeBr_4$,[2,3,7] $(EDTDM)_2FeBr_4$,[8,9] $(EDO\text{-}TTFBr_2)_2FeBr_4$,[3] $(EDT\text{-}TTFI_2)M(mnt)_2$ $(M=Ni, Pt)$,[10] $(EDT\text{-}TTFBr_2)FeBr_4$,[11] etc, in which the changes in the magnetic ground state in the applied field strongly affect the transport properties of the π-electron systems. Actually, the analysis of field-induced singularities in the magnetoresistance of $(DMET)_2FeBr_4$ tells us how to tailor the low-D molecular magnetic conductors for intentionally controlling the magnetotransport. This gives a clue to make a roadmap toward molecular electronics/spintronics driven by magnetic stimulus.

In the present paper, we focus on two interesting examples of the π-d interacting system; $(EDTDM)_2FeBr_4$ and $(VT)_2FeBr_4$ (see Figure1 for the molecular structures of EDTDM and VT), where anomalous magnetotransport phenomena are observed in relation to the 1D electronic instability.

Figure 1 *The molecular structures of EDTDM (left) and VT (right).*

2   PRESSURE-INDUCED SINGULARITY OF THE NEGATIVE MAGNETORESISTANCE IN (EDTDM)$_2$FeBr$_4$

The crystal structure shown in Figure 2 is characterized with an alternating stack of donor and anion sheets along the c-axis, where the donor sheet consists of a 2D array of quasi-1D slightly dimerized EDTDM donor chains directed along the $(a \pm b)$ direction, and the FeBr$_4$ anions having Fe^{3+} 3d spins (S=5/2) form a 2D square lattice. The strength ratio of the intra-chain to inter-chain transfer integral is estimated as ca.10. The Fermi surface calculated in the tight-binding scheme indicates a lens-shaped quasi-1D Fermi surface, in which the presence of a nesting vector of $Q \sim (a*+b*)/2$ ($|Q| = (1/4)2\pi/d_\pi$) threatens the metallic π-electronic state to be subjected to the electronic instability, where d_π is the intra-chain donor-donor distance. It is worth noting the presence of short inter-donor-anion S-Br atomic contact (0.374 nm), which is responsible for the π-d interaction.

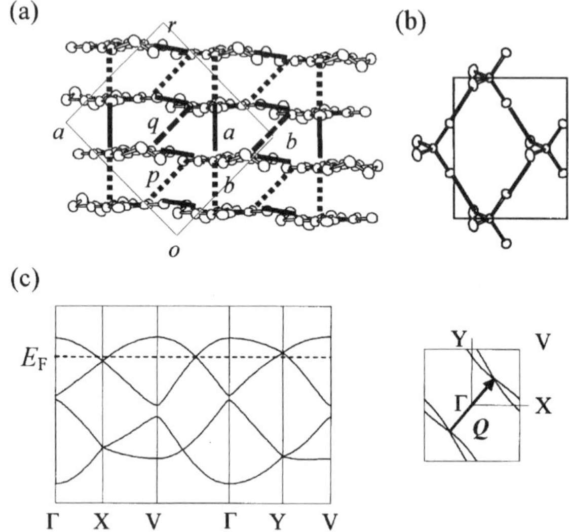

Figure 2   *Structure of (EDTDM)$_2$FeBr$_4$. (a) The transfer integral network between donor molecules on the ab-plane. The inter-molecular transfer integrals are estimated as a=24.92, b=22.19, p=1.502, q=2.877 and r=2.360×10^{-2} eV. (b) Fe^{3+} anion layer projected on the ab-plane. (c) Band structure and Fermi surfaces. The solid arrow indicates a nesting vector $Q \sim (a*+b*)/2$.*

Here, we extract information on the electronic structure of the π-electron system on the basis of isostructural compound (EDTDM)$_2$GaBr$_4$, in which no d-electron spin is present. According to the conductivity, susceptibility, and ESR measurements of the nonmagnetic (EDTDM)$_2$GaBr$_4$, the donor system is metallic down to $T_{MI}\sim$11 K in the ambient pressure, although the trend in the temperature dependence of the resistivity is switched from metallic-like to semiconducting-like with a resistivity minimum being around $T_{min}\sim$250 K. The susceptibility has a weak decreasing trend upon the lowering of the temperature from room temperature to ca. 20 K. It drops abruptly below 20 K. The weakly decreasing trend in the susceptibility suggests that the magnetism is not described simply in terms of Pauli paramagnetism, but is reminiscent of antiferromagnetic (AF) short range order effect. The ESR linewidth showing a steep divergence below T_{MI} in addition to the susceptibility drop at ca. 20 K proves that the low-temperature insulating state below T_{MI} is characterized as an SDW state associated with the 1D electronic instability. This is confirmed by the ^{13}C-NMR result, which suggests the presence of localized magnetic moments.[12] The apparent semiconducting behavior and the weakly decreasing susceptibility in the intermediate temperature range T_{min} to T_{MI} is considered to be due to the AF spin fluctuation effect with fractional localized spin feature. This is evident from the presence of the anisotropy in the susceptibility which is reminiscent of the susceptibility of 1D Heisenberg antiferromagnet. From the temperature dependence of the susceptibility, the strength of the AF exchange interaction and the value of magnetic moment in the donor chain are estimated as $|J_{\pi\pi}|\sim$250 K and 0.29 μ_B per donor molecule, respectively. In order to confirm the important role of the electron correlation effect, which is responsible for the SDW state with spin fluctuations, the tight-binding electronic structure calculation including the on-site Coulomb interaction U is carried out. The result of the calculation demonstrates the presence of spin polarization with a fractional magnetic moment of 0.29 μ_B per donor molecule and the four-fold periodicity along the chain direction when U exceeds 0.5 eV, in good agreement with the experimental finding.

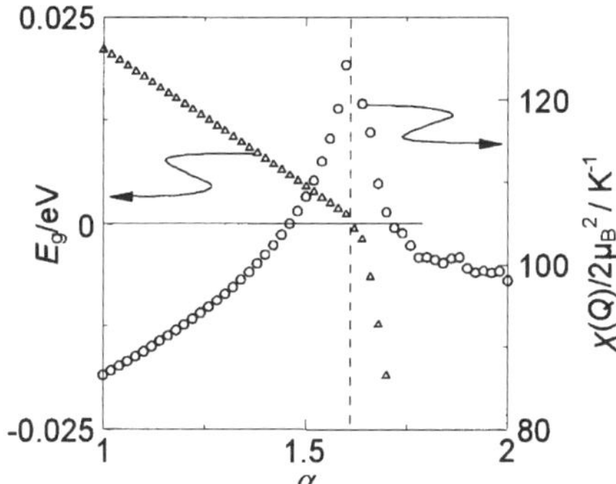

Figure 3 *The SDW gap E_g and susceptibility $\chi(Q)$ calculated at U=0.5 eV for various pressures represented by α. α is defined the ratio of the inter-chain transfer integrals between under pressure and at ambient pressure; that is, $\alpha = t_\perp(P)/t_\perp(0)$, where $t_\perp$ indicates the inter-chain transfer integras p, q, and r. The dashed line represents the MI boundary at $\alpha\sim$1.6, which is defined as the value of α, at which E_g becomes zero.*

The application of pressure modifies the electronic structure of the π-electron system as the resistivity indicates a drastic change upon the elevation of the pressure. The position of T_{min} is lowered accompanied with the decrease in the resistivity and the lowering of T_{MI} as the pressure increases. Above $P_c\sim$11.2 kbar and $\sim$9.2 kbar for the GaBr$_4$ and FeBr$_4$ complexes, respectively, the MI transition is suppressed although a weak resistivity increase still remains below ca. 10 K. It should be noted that, in the vicinity of the MI critical pressure P_c, the derivative of the resistivity is sharply peaked at 4 K, at which an AF long range ordering takes place in the Fe^{3+} spin system as will be discussed later. The MI transition behavior as a function of applied pressure is calculated in the tight-binding scheme with the involvement of U. Figure 3 presents the calculated SDW energy gap E_g

and the susceptibility vs pressure. The energy gap monotonically decreases toward zero at the critical pressure of the MI transition. The susceptibility becomes divergent at the critical pressure.

The susceptibility of the $FeBr_4$ complex, in which the magnetic moments of Fe^{3+} spins are the major contribution, shows a 3D AF transition of the Fe^{3+} spins at $T_N{\sim}3$ K, below which the appearance of anisotropy in the susceptibility indicates the spin easy axis oriented along the a-axis. The magnetization below T_N shows a spin-flop transition at $H_{sf}{\sim}1.8$ T and is saturated above ca. 5 T. Interestingly, the susceptibility has an anisotropy with the spin easy axis oriented along the b-axis in the whole temperature range above T_N, and the anisotropy is pronounced as the temperature is lowered. Eventually, the trend in the spin orientation is switched from the easy b-axis feature to the easy a-axis one at T_N. This proves the presence of interaction between the d-electron spins and the π-electron spins, the latter of which are subjected to the AF short range ordering.

Figure 4 shows the intra-layer magnetoresistace for the $GaBr_4$ and $FeBr_4$ complexes at 1.8 K. Due to the large resistivity values of these samples, the mangetoresistance can be detected only above 7 kbar. The $GaBr_4$ complex has a weak positive magnetoresistance. In contrast, the magnetoresistance in the $FeBr_4$ complex has large negative values. At $P{\sim}7.0$kbar, the magnetoresistance decreases slowly in the low field range up to ca. 2 T, and then it has a steep drop. Above the saturation field of the magnetization around 5 T, the magnetoresistance becomes saturated with no further change in higher fields. The derivative of the magnetoresistance reveals the presence of anomalies in the magnetoresistance. In the field applied along the a-axis, which is the spin easy axis, the derivative shows a steep dip at the spin-flop transition of 1.8 T. In addition, another

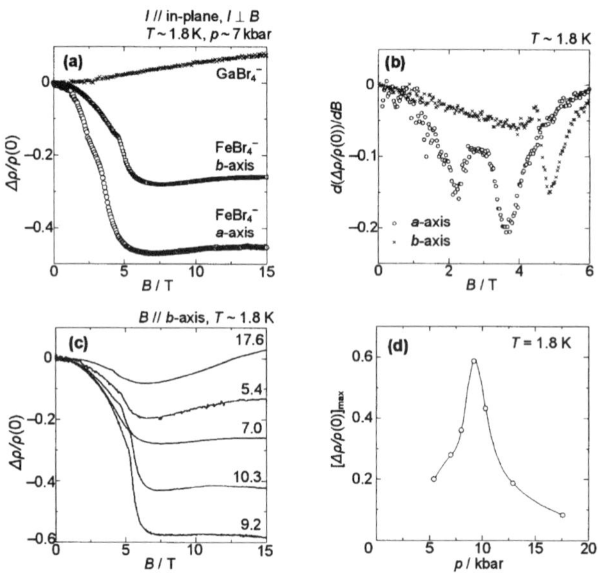

Figure 4 *(a) The field dependence of the intra-layer magnetoresistance at T~1.8 K for (EDTDM)₂MBr₄ (M=Fe, Ga) under a pressure of 7.0 kbar. (b) The numerical derivatives of the intra-layer magnetoresistance at P~7.0 kbar as a function of applied field, (c) the field dependence at T~1.8 K under pressures of 5.4, 7.0, 10.3 and 17.6 kbar, (d) the pressure dependence of the maximum of the absolute value of the negative magnetoresistance for (EDTDM)₂FeBr₄.*

anomalous dip appears around 4 T. In contrast, an anomaly is found around 5 T in the field applied along the spin-hard axis (*b*-axis). These experimental findings suggest that the internal field of the Fe^{3+} spin is responsible for the large negative magnetoresistance and its anomalies. As the pressure is elevated, the absolute value of the negative magnetoresistance increases, reaching a maximum value of about 60 % at the MI transition critical pressure $P_c{\sim}9.2$ kbar, and then it tends to be small in the metallic regime upon the further increase in the pressure. As summarized in Figure 4(c), the maximum of the absolute value of the negative magnetoresistance has a sharp peak at P_c. This is suggestive of the important role of the low-D electronic instability in the appearance of the peaked negative magnetoresistance.

We analyse the origin of the negative magnetoresistance in $(EDTDM)_2FeBr_4$ on the basis of the electronic structure calculated above. The internal field $H_{\pi d}$ of the d-electron spins, which affect the π-electron spins, is described as

$$H_{\pi d}(Q_d) = -\frac{2z_{\pi d}J_{\pi d}S_d(Q_d)}{g\mu_B},\tag{1}$$

where $S_d(Q_d)$ is the spin operator of the Fe^{3+} d-electron with wavenumber Q_d. π-electron spin $s_\pi(Q)$ is modified by the perturbation of $H_{\pi d}$ as given in the following equation;

$$g\mu_B\delta s_\pi(Q) = -\chi(Q)H_{\pi d}(Q),\tag{2}$$

where $\delta s_\pi(Q)$ is the fraction of the π-electron spin induced by the perturbation of the $H_{\pi d}(Q)$, and $\chi(Q)$ is the π-electron susceptibility. $\chi(Q)$ is sensitive to the electronic structure around the Fermi energy and becomes divergent at the MI transition point as shown in Figure 3. In general, the periodicity Q_d of the d-electron spin arrangement is not necessarily the same to that of the π-electron spins Q. However, if $Q=Q_d$, the spin polarization of the π-electron is enhanced due to the contribution of $H_{\pi d}(Q)$. This therefore contributes to increasing the energy gap E_g of the SDW insulating state, where the increment $\Delta E_{\pi d}$ is given in the following equation;

$$E_g = E_0 + \Delta E_{\pi d} = E_0 + \frac{1}{2}J_{\pi\pi}[\delta s_\pi(Q)]^2 = E_0 + \frac{1}{2}\chi(Q)^2\frac{J_{\pi\pi}J_{\pi d}^2 z_{\pi d}^2 S(Q)^2}{(g\mu_B)^2}.\tag{3}$$

Here E_0 and $z_{\pi d}$ are the SDW gap of the unperturbed system and the number of neighboring Fe^{3+} sites surrounding a donor, respectively. The resistivity is obtained from eq.(3) as expressed in the following equation;

$$\rho(P,H,T) = \rho_0 \exp\left(\frac{E_g(P,H)}{2k_BT}\right).\tag{4}$$

In $(EDTDM)_2FeBr_4$ with the 2:1 stoichiometry, the 3/4-filled state with a quasi-1D feature produces a lens-shaped Fermi surface with a nesting vector of $Q \sim (1/4)2\pi/d_\pi$, the four-fold periodicity of the donor arrangement being suggested in the low temperature insulating phase. In the meantime, the Fe^{3+} spins are aligned with the same periodicity in the AF ordered state; $Q_d=Q$. Therefore, the potential of the d-electron spins antiferromagnetically arranged on the $FeBr_4$ anion layers stabilizes the SDW state of the π-electrons. In contrast, the application of the field, which reduces the Q-component of S_d, tends to decrease $\Delta E_{\pi d}$, resulting in the appearance of negative magnetoresistance as shown in Figure 4.

Using eqs.(1)-(4), the magnetoresistance is calculated as functions of applied field and pressure as shown in Figure 5. The calculated magnetoresistance shows a large negative magnetoresistance that tends to be saturated above the saturation field of the magnetization of the d-electron spins as shown in Figure 5(a). This is in good agreement with the observed behavoir of the magnetoresistance shown in Figure 4(c). Figure 5(b) demonstrates that the maximum of the absolute value of the negative magnetoresistance grows steeply as the pressure is elevated and has a divergence at the critical pressure of the MI transition. This originates from the divergent feature of $\chi(Q_d)$, suggesting that the 1D electronic instability is responsible for the anomalous large negative mangetoresistance. In other words, the 1D electronic instability can strongly amplifies the effect of the d-electron spins on the electron transport of the π-electron of the donor system through the π-d interaction.

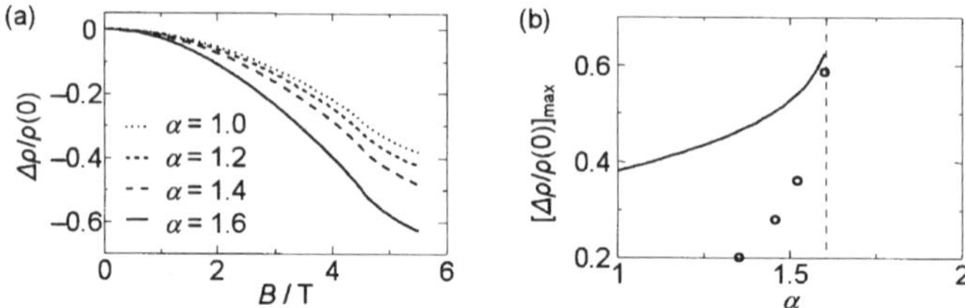

Figure 5 *(a) The calculated field dependence of the magnetoresistance for (EDTDM)$_2$FeBr$_4$ at various pressures. (b) The calculated (solid line) and observed (circle) pressure dependence of the maximum of the absolute value of the negative magnetoresistance in the SDW insulating region. The dashed line indicates the MI boundary at $\alpha\sim1.6$.*

3   FIELD-INDUCED SINGULARITY OF THE POSITIVE MAGNETORESISTANCE IN (VT)$_2$FeBr$_4$

(VT)$_2$FeBr$_4$ has an alternating sandwich layered structure of conducting donor and magnetic anion sheets. The donor sheet consists of a 2D array of 1D donor chains having $t_\perp/t_{//} \sim 0.1$. The band calculation suggests the presence of a quasi-1D open Fermi surface with a nesting vector which can create a long periodicity modulation in the insulating state. The FeBr$_4$ anion sheet forms a quasi-square lattice, where the inter-molecular Br-Br distance considerably larger than the corresponding van der Waals distance is suggestive of the absence of direct exchange interaction between the Fe^{3+} spins. In contrast, there is a short donor-anion inter-molecular S-Br atomic contact, which is responsible for π-d interaction. This means that the arrangement of the d-electron spins is subjected to the π-electron system.

The conductivity shows metallic behavior with a resistivity minimum around 150 K. It shows a MI transition at 20 K, below which a SDW insulating state is stabilized due to the 1D electronic instability, as evidenced by the divergence in the ESR linewidth. The susceptibility of nonmagnetic isostructural (VT)$_2$GaBr$_4$ increases monotonically upon the lowering of the temperature down to ca. 40K and has a broad peak around 40 K. The absolute value of the susceptibility, which is estimated as 7×10^{-4} emu/mol at room temperature, is higher than that of ordinary molecular conductors,[13,14] suggesting the important contribution of the on-site Coulomb interaction in the π-electronic structure. The deviation from the behavior of ordinary Pauli susceptibility indicates antiferromagnetic spin fluctuation feature in the donor chain. From the temperature dependence of the susceptibility, the fitting with the model of 1D Heisenberg AF magnet gives estimates of the intra-chain AF exchange interaction and the magnetic moment as J=-31 K and 0.55 μ_B per donor molecule. The apparently independent behavior of the magnetism and the electron transport is reminiscent of the system subjected to the spin-charge separation in Tomonaga-Luttinger liquid.[15,16]

The d-electron spins of the FeBr$_4$ anion layers take an AF long range order below T_N~4 K as seen in the behavior of the susceptibility exhibited in Figure 6 (a). The susceptibility has a broad maximum around 7 K, and then shows the emergence of well defined anisotropy below T_N. Interestingly, the anisotropy does not indicate any specific axis as the spin easy axis. This is confirmed by the magnetization curves in which no well-defined spin flop-transition of collinear AF magnet appears. Therefore, the magnetically ordered state has a feature of noncollinear arrangement such as helical structure. In this connection, it should be noted that the interaction between the d-electron spins is mainly mediated through the π-electrons with the direct interaction between the d-electron spins contributing as the minority. The long periodicity of the SDW insulating state expected from the π-electron band structure is favorable for the noncollinear spin arrangement in the ordered state.

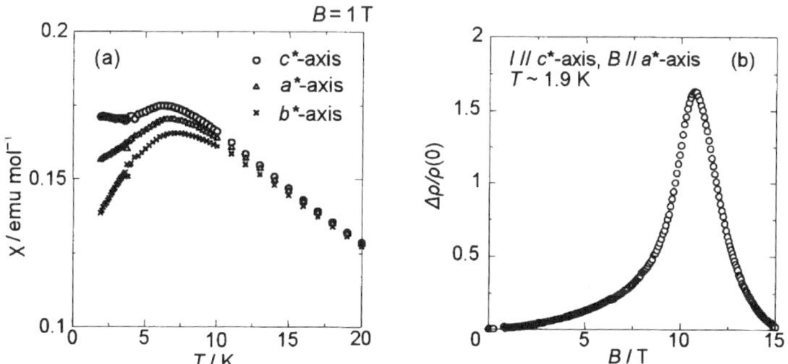

Figure 6 *(a)Temperature dependence of the magnetic susceptibility of (VT)$_2$FeBr$_4$. (b)Field dependence of the magnetoresistance of (VT)$_2$FeBr$_4$ at 1.9 K.*

This feature of the spin arrangement affects the magnetotransport in the π-electron system. Figure 6 (b) shows the field dependence of the magnetoresistance at 1.9 K well below T_N. It shows positive values in the whole field range investigated. This suggests that the application of the field destabilizes the SDW insulating state, in contrast to the behavior of (EDTDM)$_2$FeBr$_4$. Interestingly, the magnetoresistance has a sharp peak at 10.6 T, where the absolute value of the magnetoresistance is indeed 160 %. The divergent feature of the magnetoresistance at a specific field value is suggestive of the presence of a field-induced phase transition. The SDW insulating state having a long periodicity is stabilized less than that with a shorter periodicity. In general, therefore, the application of field can effectively perturb the 1D electronic instability responsible for the SDW state, resulting in a transition to another SDW state having a different periodicity. The divergent behavior of the magnetoresistance at a critical field of H_c~10.6 T is considered to originate from the spin fluctuations generated by the change in the periodicity in the vicinity of the transition.

4 SUMMARY

1D electronic instabilities play an important role in the appearance of a variety of electronic phases. SDW state that takes place as a result of the 1D electronic instability can be perturbed by the localized d-electron spins incorporated into molecular magnetic conductors. Therefore, the competition of the internal field of the d-electron spins and the applied magnetic field can modify the SDW state. Anomalous magnetotransport phenomena induced by this competition are presented for two interesting examples; $(EDTDM)_2FeBr_4$ and $(VT)_2FeBr_4$. The former shows anomalous negative magnetoresistance which has a peak at the pressure-induced MI transition. In contrast, the latter has a positive magnetoresistance peak at the critical field of the field-induced SDW-SDW transition. It is expected that the magnetic-field-induced function associated with the low-D electronic instability in molecular magnetic conductors provide a new aspect in developing molecular electronics/spintronics driven by magnetic stimulus.

Acknowledgement

This work was supported by the Grant-in-Aid (No.15073211) from the Ministry of Education Culture, Sport, Science and Technology.

References

1. T. Ishiguro, K. Yamaji, and G. Saito, *Organic Superconductors*, 1998, Springer, Berlin
2. L. Ouahab and T. Enoki, *Eur. J. Inorg. Chem.* 2004, 933.
3. T. Enoki and A. Miyazaki, *Chem. Rev.* 2004, **104**, 5449.
4. H. Kobayashi, A. Kobayashi, and P. Cassoux, *Chem. Soc. Rev.* 2000, **29**, 325.
5. S. Uji, H. Shinagawa, T. Terashima, T. Yakabe, Y. Terai, M. Tokumoto, A. Kobayashi, H. Tanaka, and H. Kobayashi, *Nature* 2001, **410**, 908.
6. E. Coronado, J. R. Galan-Mascaros, C. J. Gomez-Garcia, V. Laukhim, *Nature* 2000, **408**, 447.
7. K. Enomoto, A. Miyazaki, and T. Enoki, *Bull. Chem. Soc. Jpn.* 2001, **74**, 459.
8. K. Okabe, K. Enomoto, A. Miyazaki, and T. Enoki, *Mol. Cryst. Liq. Cryst.* 2002, **376**, 513.
9. K. Okabe, J.-I. Yamaura, A. Miyazaki, and T. Enoki, *J. Phys. Soc. Jpn.* 2005, **74**, 1508.
10. J. Nishijo, E. Ogura, J. Yamaura, A. Miyazaki, T. Enoki, T. Takano, Y. Kuwatani, and M. Iyoda, *Solid State Commun.* 2000, **116**, 661.
11. J. Nishijo, A. Miyazaki, T. Enoki, R. Watanabe, Y. Kuwatani, and M. Iyoda, *Inorg. Chem.* 2005, **44**, 2493.
12. K. Miyagawa and K. Kanoda, private communication.
13. K. Kanoda, T. Takahashi, K. Kikuchi, K. Saito, I. Ikemoto, and K. Kobayashi, *Phys. Rev.* 1988, **B39**, 3996.
14. H. J. Pedersen, J. C. Scott, and K. Bechgaard, *Phys. Rev.* 1981, B24, 5014.
15. J. M. Luttinger, *J. Math. Phys.* 1963, **15**, 1154.
16. F. D. M. Haldane, *J. Phys.* 1981, **C14**, 2585.

MAGNETIC CONDUCTORS BASED ON NEW ETHYLENEDIOXY-SUBSTITUTED DONORS WITH A BENT MOLECULAR STRUCTURE

H. Fujiwara,[1,2] K. Wada,[1] T. Hiraoka,[1] T. Hayashi,[1] T. Sugimoto,[1,2] H. Nakazumi,[3] M. Teramura,[4] K. Yokogawa,[4] S. Yasuzuka[4] and K. Murata[4]

[1] Department of Chemistry, Osaka Prefecture University, Osaka 599-8570, Japan
[2] CREST, Japan Science and Technology Agency, Saitama 332-0012, Japan
[3] Department of Applied Chemistry, Osaka Prefecture University, Osaka 599-8531, Japan
[4] Department of Molecular Materials Science, Osaka City University, Osaka 558-8585, Japan

1 INTRODUCTION

Recently, a considerable interest has been concentrated on the magnetic molecular conductors that possess both magnetic and conducting properties to understand the interactions between conducting π electrons and localized spins.[1] Among them, we have investigated using donor molecules with a bent molecular structure, and recently discovered a metal-semiconductor transition around 170 K and a ferromagnetic ordering of Fe(III) d spins in $(EDT-TTFVO)_2 \cdot FeBr_4$ (EDT-TTFVO = ethylenedithiotetrathiafulvalenoquinone-1,3-dithiolemethide).[2] To suppress such a metal-to-semiconductor transition and to realize the coexistence of metallic conductivities and magnetic orderings, we noticed the introduction of an ethylenedioxy group because of its strong tendency to form stable metallic states. Here we report the synthesis of ethylenedioxytetrathiafulvaleno-thioquinone-1,3-dithiolemethide, EDO-TTFVS (**1**) and its quinone derivative, EDO-TTFVO (**2**) (see Figure 1), and discuss the crystal structures and magnetic/conducting properties of an antiferromagnetic semiconductor, **1**$\cdot$FeBr$_4$, with a high Néel temperature of 15 K[3] and an antiferromagnetic metal, **2**$_2 \cdot$FeCl$_4$, which clears the coexistence of stable metallic behavior down to 0.3 K and antiferromagnetic ordering of the Fe(III) d spins at 3 K.[4]

2 RESULTS AND DISCUSSION

2.1 Synthesis of donor molecules 1 and 2

Donor molecules **1** and **2** were synthesized according to the previously reported method.[5] Thus, 4,5-ethylenedioxy-4',5'-bis(cyanoethylthio)TTF[6] was converted to a bis(tetra-n-butylammonium, n-Bu$_4$N$^+$) salt of a dianionic bis(4,5-ethylenedioxytetrathiafulvalene-4',5'-dithiolato)zinc complex by the successive treatment with sodium methoxide, zinc(II) chloride and n-Bu$_4$N$\cdot$Br in methanol. This salt was immediately reacted with an excess of 2-methylthio-1,3-dithiolanium fluoroborate in DMF at room temperature under argon for

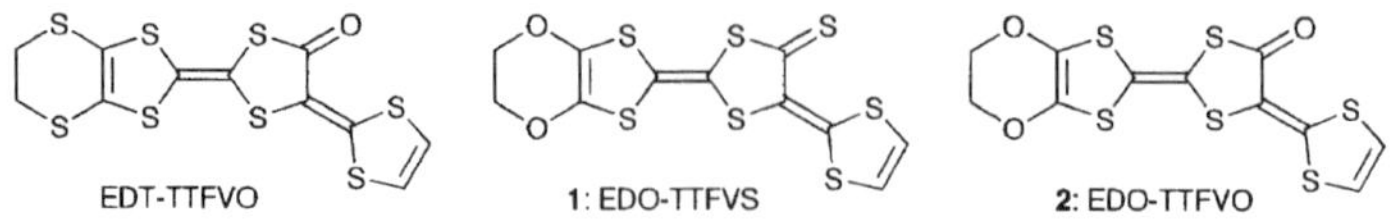

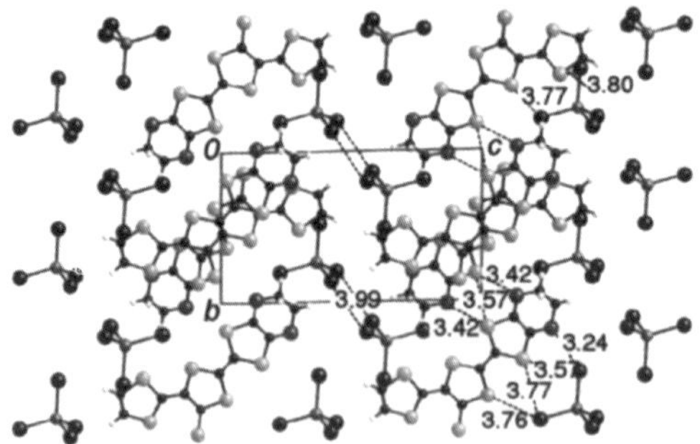

Figure 1 *Structures of EDT-TTFVO, EDO-TTFVS (1) and EDO-TTFVO(2)*

6 hours. After separation of the reaction mixture by column chromatography on silica gel, **1** was obtained as greenish microcrystals in 44% yield. Subsequently, **1** was converted into **2** in 90% yield as a reddish powder by the treatment with an excess of mercury(II) acetate in THF-acetic acid (3:1, v/v) at room temperature overnight.

2.2 Preparation, Crystal Structure and Physical Properties of 1•FeBr₄

Electrochemical oxidation of **1** was performed in chlorobenzene/ethanol (9:1, v/v) at 25 °C in the presence of n-Bu₄N•FeBr₄ under a constant current of 0.6 μA. Black block-like single crystals obtained have a 1:1 composition of donor molecule **1** and the FeBr₄⁻ ion. Figure 2 shows the crystal structure of **1**•FeBr₄.[7] In the unit cell, one donor molecule and one FeBr₄⁻ ion are crystallographically independent. The donor molecules form a strongly dimerized structure with a short interplanar distance of 3.36 Å. In the donor layer along the ab-plane, the dimers construct a columnar array along the a-axis and there are short S•••S (3.57 Å) and S•••O (3.42 Å) contacts along the side-by-side direction. The FeBr₄⁻ ions also construct a magnetic anion layer along the ab-plane with short Br•••Br contacts of 3.85 and 3.99 Å. Because the anions are located on the side position of **1**, there are several short Br•••S (3.57-3.77 Å) and Br•••O (3.24 Å) contacts between the donor molecule and the FeBr₄⁻ ion, suggesting a very strong π–d interaction.

The electrical conductivity measured on the single crystals exhibited a very low value of 10^{-4} S cm^{-1} at room temperature, as expected from the strongly dimerized structure with a 1:1 stoichiometry. The magnetic susceptibilities (χ) of **1**•FeBr₄ were measured using a SQUID magnetometer. The temperature dependence of χ measured on a polycrystalline sample at 10 kOe well obeyed a Curie–Weiss (C–W) law with a Curie constant (C) of 4.85 emu K mol^{-1} and a Weiss temperature (θ) of –38 K estimated for the data between 30 and 300 K. This large and negative θ value suggests that the magnetic interaction between the Fe(III) d spins are strongly antiferromagnetic. Since a sharp decrease of χ was observed below 20 K, an anisotropy of magnetic properties was

Figure 2 *The bc-projection of the crystal structure of 1•FeBr₄*

investigated using the single crystal. A rapid decrease of χ was observed only when the magnetic field was applied along the $(b+c)$-direction, indicating an antiferromagnetic ordering of the Fe(III) d spins at a high Néel temperature (T_N) of *ca.* 15 K. Judging from the fact that there is a strong π–d interaction along the magnetic easy-axis [the $(b+c)$-direction], an indirect magnetic interaction between the anions through the π–d interaction with the donor molecules plays an important role to cause the antiferromagnetic ordering.

2.3 Preparation, Crystal Structure and Physical Properties of 2_2•FeCl$_4$

Black needle-like crystals of 2_2•FeCl$_4$ were electrochemically grown up on the anode surface in chlorobenzene/ethanol (9:1, v/v) at 25 °C in the presence of Et$_4$N•FeCl$_4$ under a constant current of 0.1 μA. The crystal structure analysis of this salt was performed using Rigaku AFC-8 Mercury CCD diffractometer equipped with a confocal X-ray mirror system.[8] There are two crystallographically-independent donor molecules (**A** and **B**), and they form **A–B–A–B** type side-by-side arrays along the *c*-axis with several short S•••S(O) contacts, suggesting strong intermolecular interactions along the *c*-axis as shown in Figure 3. In the *bc*-layer, the donor molecules construct two identical diagonal stackings like **A–A'–B–B'** along [0 2 1] and [0 2 -1] directions. This donor array resembles a β'-type packing motif. The FeCl$_4^-$ ions form a two-dimensional rectangular arrangement in the *bc*-plane with a short Cl•••Cl contact (3.48 Å) along the *b*-axis. In addition, there is a short Cl•••S contact (3.60 Å) between the 1,3-dithiole ring of the donor molecule and the FeCl$_4^-$ ion, suggesting a possibility of a strong π–d interaction. Band structure calculation suggested that this salt has a two-dimensional Fermi surface due to relatively large overlap integrals along the side-by-side and diagonal directions.

The electrical conductivities of the single crystal measured along the conduction plane and perpendicular to the conduction plane are 8.4 and 0.02 S cm^{-1} at room temperature, respectively. This salt maintained a metallic behavior down to 0.3 K, giving the first stable metallic salt among the salts based on our donor molecules with a bent molecular structure, since all the salts had shown semiconducting behaviors or metal-to-semiconductor transitions at relatively high temperatures above 100 K. This stable metallic state is attributable to the achievement of the two-dimensional Fermi surface by the introduction of the ethylenedioxy group. Measurements of χ were carried out on the oriented several pieces of the single crystals at 1 kOe. The temperature dependence of χ above 15 K can well be fitted with the C–W law ($C = 4.45$ emu K mol^{-1} and $\theta = -9.8$ K), suggesting a relatively strong antiferromagnetic interaction between the Fe(III) d spins. Below *ca.* 3 K, a sudden decrease of χ was observed only when H was applied along the

(a)   *(b)*

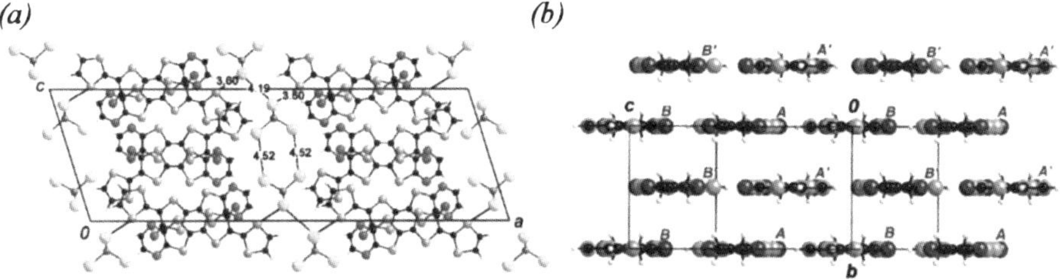

Figure 3 *(a) Crystal structure of 2_2•FeCl$_4$ projected down to the ac-plane. (b) Donor array in the bc-conduction plane*

a-axis, indicating an antiferromagnetic ordering at $T_N \approx 3$ K with an easy-spin axis parallel to the a-axis. Since the metallic state is still kept even when the antiferromagnetic ordering occurs, $2_2 \cdot FeCl_4$ surely is a new antiferromagnetic molecular metal at ambient pressure.

The field dependence of the magnetoresistances (MR) of the out-of-plane resistivities ($\rho_\perp$) was measured up to 14 T ($H \perp bc$-plane).[9] At 11.0 K, the MR increased quadratically with increasing H, which is simply explained in terms of the orbital effect by Lorentz force. Below T_N (≈ 3 K), a dip anomaly is observed near 8 T and becomes more prominent with lowering temperature, suggesting that the dip anomaly near 8 T is attributable to the transition from the antiferromagnetic to paramagnetic states. This observation of the dip structure suggests the presence of the finite π-d interaction in this system.

Acknowledgements

This work is in part financially supported by a Grant-in-Aid for Scientific Research on Priority Areas of Molecular Conductors (No. 16038222) from the Ministry of Education, Culture, Sport, and Technology, Japan.

References

1   (*a*) L. Ouahab and T. Enoki, *Eur. J. Inorg. Chem.*, 2004, 933; (*b*) H. Kobayashi, H.-B. Cui and A. Kobayashi, *Chem. Rev.*, 2004, **104**, 5265; (*c*) E. Coronado and P. Day, *Chem. Rev.*, 2004, **104**, 5419; (*d*) T. Enoki and A. Miyazaki, *Chem. Rev.*, 2004, **104**, 5449; (e) H. Fujiwara and H. Kobayashi, *Bull. Chem. Soc. Jpn.*, 2005, **78**, 1181.

2   (*a*) S. Noguchi, A. Matsumoto, T. Matsumoto, T. Sugimoto and T. Ishida, *Physica B*, 2004, **346–347**, 397; (*b*) T. Matsumoto, T. Sugimoto, H. Aruga-Katori, S. Noguchi and T. Ishida, *Inorg. Chem.*, 2004, **43**, 3780.

3   H. Fujiwara, K. Wada, T. Hiraoka, T. Hayashi, T. Sugimoto and H. Nakazumi, *J. Low Temp. Phys.*, in press.

4   H. Fujiwara, K. Wada, T. Hiraoka, T. Hayashi, T. Sugimoto, H. Nakazumi, K. Yokogawa, M. Teramura, S. Yasuzuka, K. Murata and T. Mori, *J. Am. Chem. Soc.*, 2005, **127**, 14166.

5   M. Iwamatsu, T. Kominami, K. Ueda, T. Sugimoto, H. Fujita and T. Adachi, *Chem. Lett.*, 1999, 329.

6   (*a*) L. Binet, J. M. Fabre, C. Montginoul, K. B. Simonsen and J. Becher, *J. Chem. Soc., Perkin 1*, 1996, 783; (*b*) L. Binet, C. Montginoul, J. M. Fabre, L. Ouahab, S. Golhen and J. Becher, *Synth. Met.*, 1997, **86**, 1825.

7   Crystal data for $1 \cdot FeBr_4$: $C_{11}H_6S_7O_2FeBr_4$, $Mw = 770.05$, triclinic, P-1, $a = 7.518(1)$, $b = 9.379(2)$, $c = 16.003(3)$ Å, $\alpha = 90.14(1)$, $\beta = 86.38(1)$, $\gamma = 73.508(8)°$, $V = 1079.5(3)$ Å^3, $T = 296$ K, $Z = 2$, $R = 0.040$ and $Rw = 0.040$.

8   Crystal Data for $2_2 \cdot FeCl_4$: $C_{11}H_6Cl_2Fe_{0.5}O_3S_6$, $Mw = 477.36$, monoclinic, $C2/m$, $a = 40.7(1)$, $b = 6.92(2)$, $c = 13.13(4)$ Å, $\beta = 107.96(4)°$, $V = 3518(17)$ Å^3, $T = 293$ K, $Z = 8$, $R = 0.059$ and $Rw = 0.070$.

9   M. Teramura, K. Yokogawa, S. Yasuzuka, T. Sasaki, K. Wada, T. Hiraoka, T. Hayashi, H. Fujiwara, T. Sugimoto and K. Murata, *J. Low Temp. Phys.*, in press.

CRYSTAL STRUCTURES AND PHYSICAL PROPERTIES OF THE NEW ORGANIC CONDUCTORS (EDT-TSF)$_2$GaCl$_4$ AND (EDT-TSF)$_2$FeCl$_4$

H. Endo,[1] T. Kawamoto,[1] T. Mori,[1] T. Kakiuchi,[2] H. Sawa,[2,3] M. Kodani,[4] K. Takimiya,[4] and T. Otsubo[4]

[1]Department of Organic and Polymeric Materials, Tokyo Institute of Technology, Tokyo 152-8552, Japan
[2]Department of Materials Structure Science, The Graduate University for Advanced Studies, Ibaraki 305-0801, Japan
[3]Institute of Materials Structure Science, High Energy Accelerator Research Organization, Ibaraki 305-0801, Japan
[4]Department of Applied Chemistry, Hiroshima University, Hiroshima 739-8527, Japan

1 INTRODUCTION

Organic conductors based on an unsymmetrical π donor molecule EDT-TTF (ethylenedithiotetrathiafulvalene) with tetrahedral anions (GaCl$_4^-$, FeCl$_4^-$) have been reported.[1] The effect of the selenium substitution on the electronic properties will be interesting because the exchange of S atoms to Se atoms is regarded as the production of the 'chemical pressure'. Recently, four-selenium-substituted analogues of EDT-TTF, EDT-TSF (ethylenedithiotetraselenafulvalene), has been synthesized.[2] In this study, crystal structures and physical properties of the new organic conductors (EDT-TSF)$_2$X (X = GaCl$_4$, FeCl$_4$) are investigated.

2 METHODS AND RESULTS

2.1 Crystal Preparation

Single crystals of (EDT-TSF)$_2$X were prepared by electrochemical crystallization of EDT-TSF (2 mg) in chlorobenzene (15 ml) with n-Bu$_4$NMCl$_4$ (M = Ga or Fe ; 20 mg) under a nitrogen atmosphere. The current was kept constant at 0.2-0.3 μA for 2 weeks at 0°C.

2.2 X-ray Experiments

The crystal structural analyses of (EDT-TSF)$_2$X at room temperature were performed using a Rigaku AFC7R four-circle diffractometer equipped with a graphite monochromated Mo-$K\alpha$ X-ray source. The crystal structures were solved by the direct method SIR92 and refined by the full-matrix least-squares method. From the results of the X-ray crystal structure analyses, the extended Hückel tight-binding bands were calculated.

The crystallographic data are listed in Table 1. Figure 1 shows the crystal structure of (EDT-TSF)$_2$X (X = GaCl$_4$, FeCl$_4$) at room temperature. The crystal structural analyses reveal that (EDT-TSF)$_2$X are isostructural to each other and to the previously reported (EDT-TTF)$_2$X. These salts belong to the monoclinic system with the space group $P2_1/m$. The crystal shape is different from each other. Although the long crystal direction of the

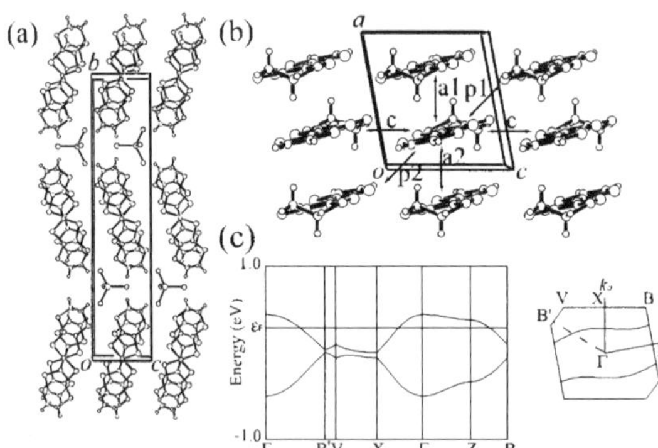

Table 1 Crystallographic data

	$(EDT\text{-}TSF)_2FeCl_4$	$(EDT\text{-}TSF)_2GaCl_4$	
Temperature (K)	293	293	8
Formula weight	1161.85	1175.72	1175.72
Space group	$P2_1/m$	$P2_1/m$	$P2_1/m$
a (Å)	7.290(3)	7.290(4)	7.1249(13)
b (Å)	30.655(4)	30.690(6)	30.3520(60)
c (Å)	6.696(2)	6.700(3)	6.6471(14)
β (°)	100.42(4)	100.38(4)	100.264(10)
V (Å^3)	1471(1)	1474(1)	1422.85(49)
Z	2	2	2
R, R_w	0.038, 0.119	0.037, 0.114	0.060, 0.228

$GaCl_4^-$ salt is the crystallographic a axis, that of the $FeCl_4^-$ salt is the c axis. Four donor molecules are located on the general positions and two anions are located on the mirror planes in the unit cell. Donor molecules form a stacked structure along the a-axis. For understanding the intermolecular interaction, the intermolecular overlap integrals of the HOMO are calculated on the basis of the extended Hückel molecular orbital calculation[3] and are listed in Table 2. All overlap modes are shown in Fig. 1(b). Overlap integrals along the stacking axis are about 6 - 18 times as large as other directions and therefore the band structure is quasi-one-dimensional (Figure 1(c)). From the calculations of the degree of dimerization along the stack, these salts have a weakly dimerized structure.

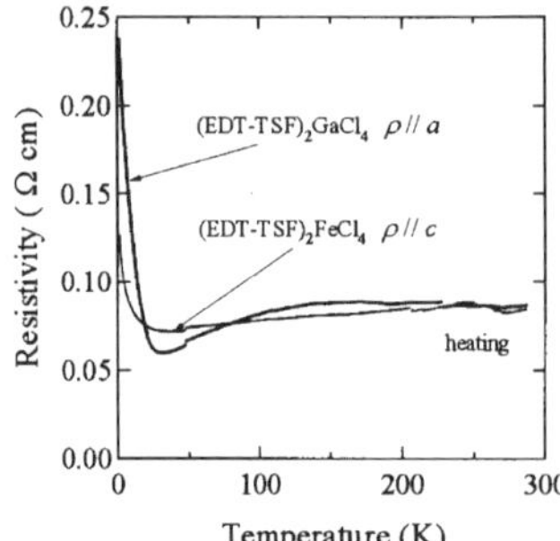

Table 2 Intermolecular overlap integrals S ($\times 10^{-3}$)

	$(EDT\text{-}TSF)_2FeCl_4$	$(EDT\text{-}TSF)_2GaCl_4$
a1	19.4	19.6
a2	21.7	21.8
S_{a2}/S_{a1}	1.12	1.11
c	1.22	1.23
p1	2.51	2.43
p2	3.15	3.16

Figure 1 (a) (b) Crystal structures and (c) energy band structure of $(EDT\text{-}TSF)_2X$.

Figure 2 Temperature dependencies of electrical resistivities of $(EDT\text{-}TSF)_2X$.

2.3 Electrical Resistivities

The electrical resistivities were measured along the direction of $\rho \parallel a$ for the $GaCl_4^-$ salt, and $\rho \parallel c$ for the $FeCl_4^-$ salt in the temperature range of 1.5-300 K at ambient pressure. The four-probe method was used with gold wires (15 μm diameter) attached to the crystal using carbon paste.

Figure 2 shows the temperature dependencies of electrical resistivities of $(EDT\text{-}TSF)_2X$. In contrast to the semiconducting $(EDT\text{-}TTF)_2X$, these salts show almost metal-like behaviour down to about 20 K and the resistivities increase slightly. The development of metallic nature is owing to the enhancement of the band width by the selenium substitution.

2.4 Magnetic Properties

The magnetic susceptibilities of $(EDT\text{-}TSF)_2X$ were measured by a SQUID magnetometer. The measurements were carried out under the fields of $H \parallel a$ for the $GaCl_4^-$ salt (1.07 mg), $H \parallel b$ and $H \parallel c$ for the $FeCl_4^-$ salt (0.455 mg).

Figure 3(a) shows the temperature dependence of the spin susceptibility of $(EDT\text{-}TSF)_2GaCl_4$ under the magnetic field of $H = 10$ kOe. The spin susceptibility was obtained by subtracting Pascal's diamagnetic contributions. The susceptibility is almost constant below room temperature, thus can be regarded as the Pauli paramagnetism. Figure 3(b) shows the magnetic susceptibility of $(EDT\text{-}TSF)_2FeCl_4$ under the magnetic field of $H = 4$ kOe ($H \parallel c$). The large susceptibility derives from the high spin state of the Fe^{3+} ions ($S = 5/2$). The magnetic susceptibilities obey the Curie-Weiss law with the Curie constant $C = 3.78$ emu K mol^{-1} ($H \parallel b$), 3.83 emu K mol^{-1} ($H \parallel c$), and the nearly zero Weiss temperatures. No magnetic transition is observed down to 1.9 K.

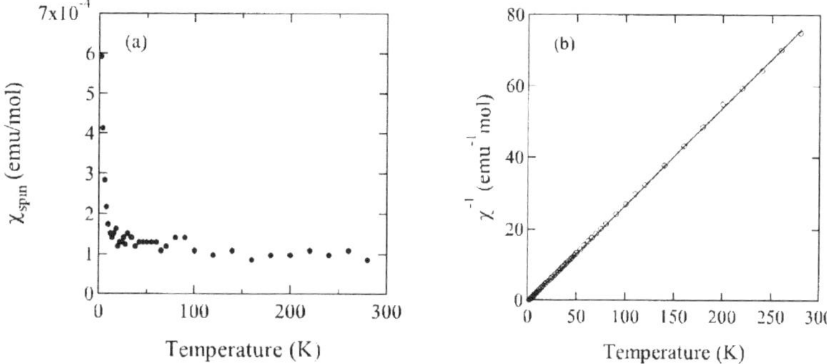

Figure 3 Magnetic susceptibility of (a) $(EDT\text{-}TSF)_2GaCl_4$ and (b) $(EDT\text{-}TSF)_2FeCl_4$. The solid line in (b) shows the Curie-Weiss law.

2.5 Low-Temperature Crystal Structure

A low-temperature crystal structural analysis of $(EDT\text{-}TSF)_2GaCl_4$ was carried out using a high-resolution synchrotron radiation X-ray. The measurements were performed by an imaging-plate system with a closed-cycle helium refrigerator at BL-1B of the Photon Factory, KEK, Tsukuba. A single crystal of $0.40 \times 0.12 \times 0.12$ mm^3 was mounted on a sapphire rod with epoxide-based adhesive. Forty six oscillation photographs were taken for crystal structural analysis at 8 K. The oscillation angle per one photograph was 5°, and the exposure time for each photograph was 2 min.

Figures 4(a) and (b) show X-ray oscillation photographs taken at 299 K and 9 K. Neither extra spot nor satellite reflection is observed at low temperatures. The crystal structural analysis at 8 K is successfully performed under the condition of the space group $P2_1/m$ (Table 1). Figure 4(c) shows the temperature dependence of the lattice parameters of $(EDT\text{-}TSF)_2GaCl_4$. All lattice parameters decrease as the temperature decreases and show no anomalous behaviour over the whole temperature range. This indicates that this salt has no structural phase transition.

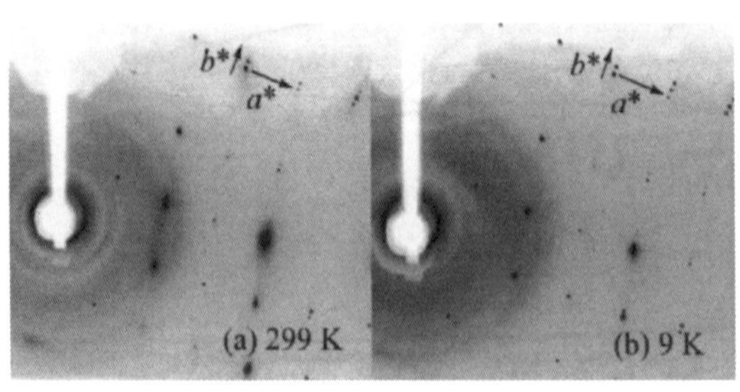

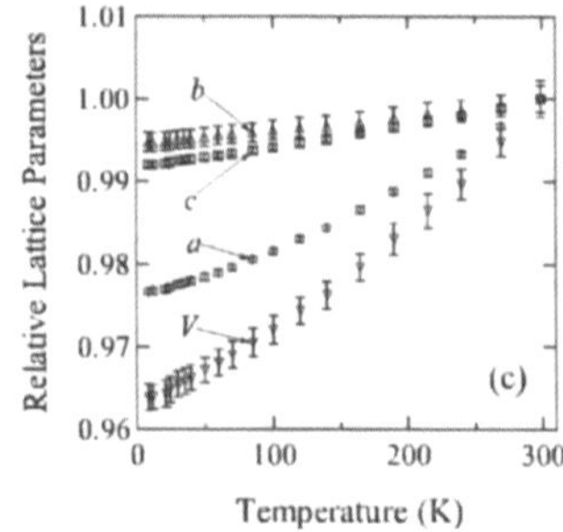

Figure 4 (a) X-ray oscillation photographs at 299 K and (b) at 9 K. (c) Temperature dependence of the lattice parameters.

3 CONCLUSION

Crystal structure analyses are carried out at room temperature for $(EDT\text{-}TSF)_2X$ and 8 K for the $GaCl_4^-$ salt. The crystal of $(EDT\text{-}TSF)_2X$ consists of weakly dimerized stacks. In the $GaCl_4^-$ salt, no structural phase transition occurs down to low temperatures. The electrical resistivities indicate the stabilization of metallic nature by the selenium substitution. $(EDT\text{-}TSF)_2X$ does not exhibit magnetic transition and the ground state is a paramagnetic insulator. In order to examine the possibility of interaction between π conduction electrons and localized Fe^{3+} magnetic moments in the $FeCl_4^-$ salt, the values of direct and indirect interactions obtained by the calculations of the Weiss temperatures are shown in Figure 5.[4] This quantitative estimation indicates π-d and d-d interactions are very weak, and the Fe^{3+} magnetic moments are almost independent of each other in consistent with the observed magnetic behaviour. Then the $FeCl_4^-$ salt is regarded as a nearly isolated spin system.

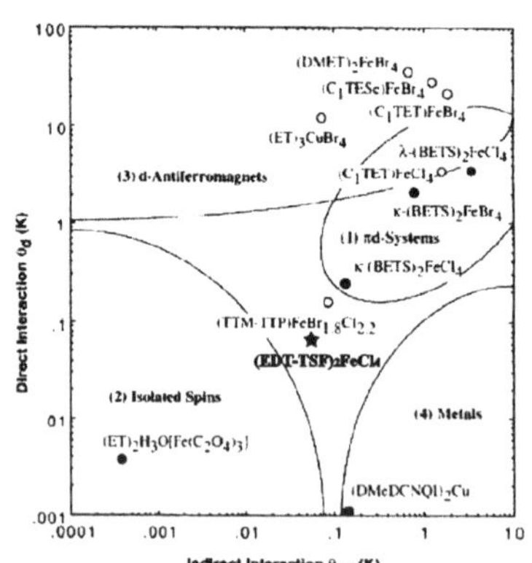

Figure 5 Calculated direct (dd) and indirect (πd) interactions.

References

1 A. Sato, E. Ojima, H. Kobayashi and A. Kobayashi, *J. Mater. Chem.*, 1999, **9**, 2365.

2 K. Takimiya, Y. Kataoka, N. Niihara, Y. Aso and T. Otsubo, *J. Org. Chem.*, 2003, **68**, 5217.

3 T. Mori, A. Kobayashi, Y. Sasaki, H. Kobayashi, G. Saito and H. Inokuchi, *Bull. Chem. Soc. Jpn.*, 1984, **57**, 627.

4 T. Mori and M. Katsuhara, *J. Phys. Soc. Jpn.*, 2002, **71**, 826.

SYNTHESES AND PROPERTIES OF TETRATHIAFULVALENE-PENDENT IRON COMPLEXES

J. Hagiwara and G. Saito

Division of Chemistry, Graduate School of Science, Kyoto University, Kyoto, 606-8502, Japan

1 INTRODUCTION

Molecule-based π-d systems in which conducting π electrons and localized spins of d electrons coexist and interplay are currently challenging targets in materials science.[1,2] Most of such π-d materials studied were radical ion salts composed of tetrathiafulvalene (TTF) derived organic cations and transition metal containing anions. These studies lead to the developments of ferromagnetic metal, $(BEDT)_3[MnCr(C_2O_4)_3]$,[3] and magnetic-field-induced superconductivity in λ-$(BETS)_2[FeCl_4]$,[4] where BEDT is bis(ethylenedithio)tetrathiafulvalene and BETS is bis(ethylenedithio)tetraselenafulvalene. Another strategy is to use monodentate or polydentate redox active ligands based on TTF derivatives functionalized by coordinating groups. Magnetic transition metal complexes with such ligands seem to be good candidate for enhancement of the interaction between conducting electrons at ligands and localized spins at metal cores through coordination bonds,[5] although only a few of radical ion salts[6,7] and no conducting complex have been reported. We report here the syntheses and properties of new tridentate ligand, *N*-(4-tetrathiafulvalenylmethyl)-*N,N*-bis(2-pyridylmethyl)amine, ttf-bpa **1**. We also report the syntheses, electrochemical and magnetic properties of iron(II) complex, $[Fe(ttf\text{-}bpa)_2](ClO_4)_2$ **2** and ligand-oxidised complex, $[Fe(ttf\text{-}bpa)_2](ClO_4)_{3.5}\cdot 2H_2O$ **3**.

2 EXPERIMENTAL

2.1 General Remarks

Absolute ethanol was prepared by a distillation over Mg/I_2. Solvents used for syntheses of Iron(II) complexes were degassed by a bubbling of Ar gas. All other solvents and reagents were purchased from commercial sources and used as received. ^{1}H-NMR spectra were recorded on a JEOL JNM-FX400 spectrometer. The electron impact mass spectrum (EI-MS) was obtained using a 70 eV electrons beam on JEOL AM-II mass spectrometer. Cyclic voltammetry measurements were carried out on the solution of 0.1 mol L^{-1} Bu$_4$NBF$_4$ in CH$_3$CN vs. Ag / AgCl by means of ALS/chi Electrochemical Analyzer Model 650A at room temperature. Voltammograms were recorded at 10 mV s^{-1} with a platinum disk for working electrode and a platinum wire for counter electrode. UV-Vis spectra were measured using the KBr disk method or MeCN solution method on SHIMADU UV-3100 spectrometer. The temperature dependence of the magnetic susceptibility from 2 K to 300 K was measured on the SQUID magnetometer of Quantum Design MPMSR2-XL applying the magnetic field of 1000 G. Magnetization data were corrected for the sample holder and for the diamagnetic contribution calculated from Pascal constants.[8] D.C. conductivity was measured with standard two-probe technique for compressed powder sample.

2.2 Syntheses of the Ligand and Fe(II) Complexes

Ttf-bpa·0.3H$_2$O (1). To a solution of 4-formyltetrathiafulvalene[9] (2.33 g, 10.0 mmol) in 200 mL of absolute ethanol was added 2,2'-dipicolylamine (2.2 g, 11 mmol), sodium cyanotrihydroborate (90 %, 0.71 g 10 mmol), and acetic acid (2.0 mL). The reaction mixture was refluxed for 6h under N$_2$ atmosphere. The mixture was then cooled and quenched with 1N HCl (100 mL). The mixture was neutralized with saturated NaHCO$_3$ and extracted with CHCl$_3$ (100 mL×3). The combined organic layer was washed with water (100 mL×3) and brine (100 mL), dried over anhydrous Na$_2$SO$_4$, filterated, and then evaporated. The residue was purified by silica gel column chromatography with at first CHCl$_3$, then successively CHCl$_3$/MeOH (20:1) as eluent and dried in vacuo to afford orange oil (yield: 3.09 g, 73 %). ^{1}H-NMR (400 MHz, CDCl$_3$, δ vs. TMS): 8.52 (d, 2H, J = 4.8 Hz), 7.73-7.59 (m, 4H), 7.18-7.16 (m, 2H), 6.31 (s, 2H), 6.18 (s, 1H), 3.83 (s, 4H), 3.49 (s, 2H). EI-MS (m/z) 415 (M$^+$), 322 ([M–CH$_2$C$_5$H$_4$N]$^+$). Anal. calcd for C$_{19}$H$_{17.6}$N$_3$O$_{0.3}$S$_4$: C 54.20; H 4.21; N 9.98; S 30.46. Found C 54.27; H 4.18; N 10.06; S 30.37.

[Fe(ttf-bpa)$_2$](ClO$_4$)$_2$ (2). All manipulations were carried out in a dry box under He atmosphere. The solution of **1** (432 mg, 1.03 mmol) in MeOH (10 mL) and the solution of Fe(ClO$_4$)$_2$·6H$_2$O (190 mg, 0.524 mmol) in MeOH (5 mL) were mixed at room temperature. The reaction mixture was added dropwise to a 50 mL of diethyl ether. The precipitate formed was collected by filteration, washed with methanol and diethyl ether and dried in vacuo to afford yellow powder (yield 352mg, 63 %). Anal. calcd. for C$_{38}$H$_{34}$N$_6$Cl$_2$FeO$_8$S$_8$: C 42.03; H 3.16; N 7.74; Cl 6.53. Found C 41.91; H 3.34; N 7.89; Cl 6.50.

[Fe(ttf-bpa)$_2$](ClO$_4$)$_{3.5}$·2H$_2$O (3). **1** (1.12 g, 2.69 mmol) and Fe(ClO$_4$)$_3$·nH$_2$O (628 mg) were treated in a similar manner as above to afford red powder (yield 732 mg, 21 %). Anal. calcd. for C$_{38}$H$_{38}$N$_6$Cl$_{3.5}$FeO$_{16}$S$_8$: C 35.90; H 3.01; N 6.61, Cl 9.76. Found C 35.93; H 2.79; N 6.64; Cl 9.85.

3   RESULTS AND DISCUSSION

3.1 Syntheses of the Ligand and Iron Complexes

TTF-pendent ligand **1** was synthesized via reductive ammination of 4-formyltetrathiafulvalene (Scheme 1). Iron complex **2** was obtained as yellow powder from **1** and Fe(ClO$_4$)$_2$·6H$_2$O. The reaction of **1** and Fe(ClO$_4$)$_3$·nH$_2$O afforded red powder of **3**. Compositions of the complexes were determined via elemental analysis. In these complexes, octahedral coordinations of two tridentate ligands to iron core were presumed (Scheme 1).

3.2 Electrochemical Property

The redox behaviors of these new ligand **1** and complex **2** were investigated by cyclic voltammetry. As seen in Figure 1, both of these compounds exhibit two quasi-reversible redox waves at 0.29 V and 0.63 V for **1**, and at 0.31 V and 0.66 V for **2** (vs. Ag / AgCl).

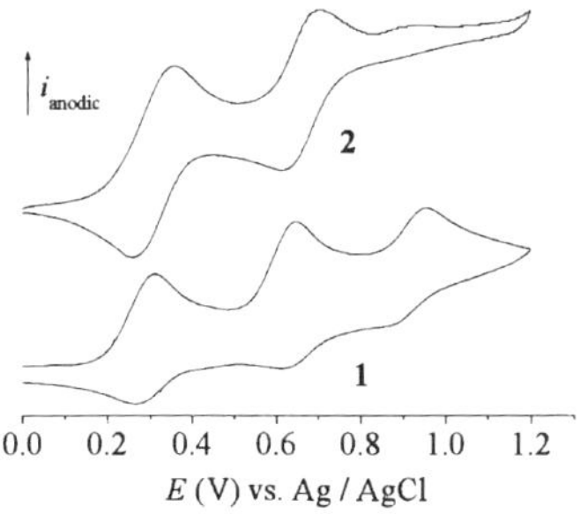

Scheme 1 *Synthetic scheme of 1-3 and proposed structure of $[Fe(ttf\text{-}bpa)_2]^{2+}$ core.*

Table 1 *Voltammetric data of 1, 2, and TTF.*

	$E^1_{1/2}$ (V)	$E^2_{1/2}$ (V)	$E^3_{1/2}$ (V)
1	0.29	0.63	0.92
2	0.31	0.66	0.93[a]
TTF	0.29	0.65	

[a] anodic potential

Figure 1 *Cyclic voltammogram of 1(lower) and 2 (upper).*

These waves correspond to the reversible waves seen in non-substituted TTF at 0.29 and 0.65 V (Table 1). This result suggests that TTF groups of **1** and **2** retain similar donor ability to non-substituted TTF. Additional quasi-reversible wave at 0.92 V for **1** may be assigned to the oxidation of bis(2-pyridylmethyl)aminomethyl group. Such a wave was hardly seen for **2** (only a very weak anodic peak was seen at 0.93 V), owing to the coordination of bis(2-pyridylmethyl)aminomethyl group to the iron center.

3.3 Electronic and Magnetic Properies

UV-Vis spectra of **1-3** in KBr and **3** in MeCN solution were shown in Figure 2. **1** and **2** exhibit similar absorption bands assigned to the TTF and pyridine chromophores. The spectra of **3** in MeCN and in KBr, on the other hand, show the absorption bands ascribable to those of TTF cation radical species in a monomer and a dimer, respectively.[10] The band at 12000 cm^{-1} in the specrum of **3** in KBr is attributed to the charge-transfer (CT) band between TTF cation radical groups, i.e. $(TTF)^{+1}(TTF)^{+1} \rightarrow (TTF)^{+2}(TTF)^0$. Such CT band does not appear in the solution spectrum of **3** because of the dissociation of dimer by solvation. These results suggest that **3** has a oxidized TTF cation radical group and a couple of the groups form a dimer in solid state. The room temperature conductivity for compacted pellet of **3** was the order of 10^{-8} Scm^{-1}, so this compound is not conductive. The temperature dependence of the magnetic susceptibility for powders of **2** and **3** showed paramagnetic behavior down to 50 K. χT products at 300 K are 3.27 and 3.04 emu K mol^{-1} for **2** and **3**, respectively. These values are in good agreement with the value of 3.00 emu K mol^{-1} for a system with $g = 2$ and $S = 2$. The decreases of χT products at low temperatures are due to intermolecular antiferromagnetic interaction and/or zero-field splitting of $S = 2$

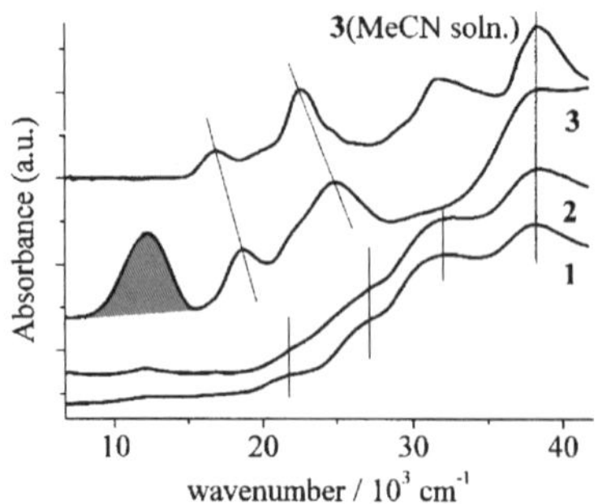

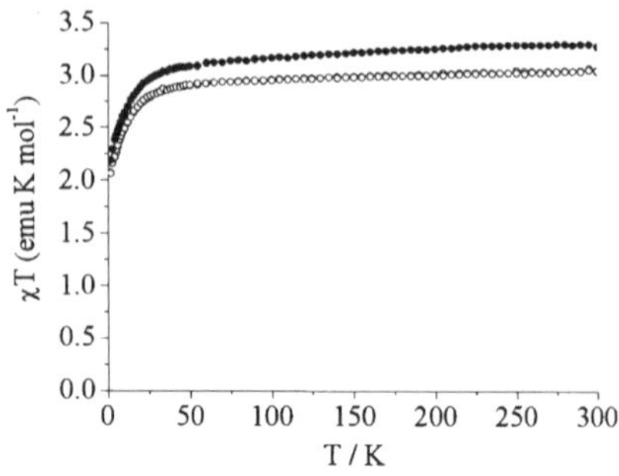

Figure 2 *UV-Vis spectra for 1-3 in KBr and 3 in MeCN*

Figure 3 Temperature dependence of susceptibility for **2**(open circle) and **3**(close circle).

center.[11] These complexes do not show any magnetic order down to 2 K. The optical and magnetic measurements suggest that compounds **2** and **3** contain high-spin Fe(II) center. In the solid of compound **3**, three fourths of the TTF groups are oxidized to form TTF cation radical groups. All of these radical groups form intermolecular dimers, which exhibit local singlet spin states through strong intradimer antiferromagnetic interaction.

4   CONCLUSION

TTF-pendent tridentate ligand **1** was newly synthesized. The reaction of iron(II) salt and **1** afforded iron(II) complex with neutral ligands **2**. On the other hand, iron(III) salt and **1** afforded iron(II) complex with oxidized ligands **3**.

Acknowledgment

This work was supported by a Grant-in-Aid (21st Century COE program on Kyoto University Alliance for Chemistry) from the Ministry of Education, Culture, Sports, Science and Technology, Japan.

References

1 E. Coronado and P Day, *Chem. Rev.*, 2004, **104**, 5419

2 T. Enoki and A Miyazaki, *Chem. Rev.*, 2004, **104** 5449

3 E. Coronado, J. R. Galán-Mascarós, C. J. Gómez-García, V. Laukhin, *Nature*, 2000, **408**, 447

4 S. Uji, H. Shinagawa, T. Terashima, T. Yakabe, Y. Terai, M. Tokumoto, A. Kobayashi, H. Tanaka, H. Kobayashi, *Nature*, 2001, **410**, 908

5 L. Ouahab, T. Enoki, *Eur. J. Inorg. Chem.*, 2004, 933

6 F. Setifi, L. Ouahab, S. Golhen, Y. Yoshida, G. Saito, *Inorg. Chem.*, 2003, **42**, 1791

7 N. Avarvari and M. Fourmigué, *Chem. Commun.*, 2004, 1300

8 C. J. O'Connor, *Prog. Inorg. Chem.*, 1982, **29**, 203

9 J. Garín, J. Orduna, S. Uriel, A. J. Moore, M. R. Bryce, S. Wegener, D. S. Yufit, J. A. K. Howard, *Synthesis*, 1994, **5**, 489

10 J. B. Torrance, B. A. Scott, B. Welber, F. B. Kaufman, and P. E. Seiden, *Phys. Rev. B.*, 1979, **19**, 730

11 O. Kahn, in *Molecular Magnetism*, VCH, 1993, ch. 2, p. 14

GIGANTIC AND ULTRAFAST PHOTORESPONSE IN MOLECULAR CHARGE ORDERING SYSTEM

M. Chollet,[1] S. Koshihara,[1,2,3] L. Guerin,[1,2,4] T. Ishikawa,[1] K. Matsuda,[5] T. Hasegawa,[6] H. Yamochi,[3,7] G. Saito,[8] R. Tazaki,[2,3] and S. Adachi,[2,3]

[1]Department of Materials Science, Tokyo Institute of Technology, 2-12-1-H61, Oh-okayama, Meguro-ku, Tokyo 152-8551, Japan. E-mail:skoshi@cms.titech.ac.jp
[2] High Energy Accelerator Research Organization (KEK)
[3] JST, ERATO
[4] Groupe Matière Condensée et Matériaux, CNRS, UM-6626, Université de Rennes 1
[5] KAST and JST, PRESTO
[6] Department of Physics, Tokyo Institute of Technology
[7] Research Center for Low Temperature and Materials Sciences, Kyoto University
[8] Division of Chemistry, Graduate School of Science, Kyoto University

1 INTRODUCTION

Organic charge transfer (CT) A_2B salts composed with cationic (or anion) molecule (molecule A) and counter anion (cation) molecule (molecule B) in the ration of 2:1 are becoming important target for materials science. Because, A_2B salts classified as typical quarter filled system show various exotic natures such as superconductivity and metal (M)-insulator (I) transition.[1-3] The appearance of charge ordering (CO) or Mott transition in quarter filled network of A molecules with 1 or 2 dimensional structure are playing a key role as a base for these exotic natures.[1,2] Recent theoretical works,[4-6] especially the one by Seo and Fukuyama have revealed that the critical balance among some physical parameters in 1/4 filled system such as bandwidth (W), on-site Coulomb interaction (U) and Nearest neighbour Coulomb interaction (V) is an origin of such exotic ground states. These previous studies stimulate us to search for the photo-induced phase transition (PIPT) in A_2B salts, because even weak photo-excitation may change the critical balance of physical parameters by switching the cooperative interaction channel leading to the macroscopic phase change.[7]

Organic donor EDO-TTF (EDO-TTF=ethylenedioxytetrathiafulvalene, see Figure 1) based molecular conductor (EDO-TTF)$_2$PF$_6$ crystal shows novel phase transition by virtue of spin-lattice-charge coupled multi-instability. This crystal shows the phase transition from M to I as decreasing temperature. In addition, this transition is accompanied with Peierls, CO and anion ordering transition at critical temperature $T_C = 280$ K as shown in Figure 1.[8] This M-I transition in (EDO-TTF)$_2$PF$_6$ can be also sensitively probed by spectroscopic method.[9] As plotted in Figure 2 by dashed lines, large spectral change really occurs around 280 K. Reflection band observed around 1.4 eV in the low temperature phase has been tentatively assigned to charge transfer excitation among EDO-TTF molecules (donor: D) in charge ordered state, i.e. excitation from D^+D^+ to $D^{2+}D^{0}$.[9]

This reflection band disappears in the high temperature phase accompanied with the melting of CO.

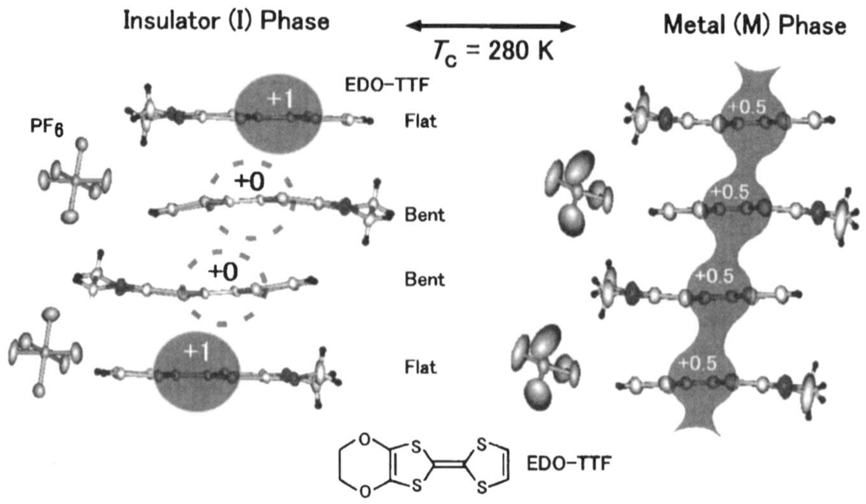

Figure 1 *Schematic view graphs of the lattice and electronic structural changes accompanied with the metal (M) to insulator (I) phase transition in (EDO-TTF)$_2$PF$_6$. Side view of EDO-TTF molecule is indicated. In the I phase, holes are localized on EDO-TTF (electron donor: D) molecules with a flat structure due to charge ordering (CO) and quasi-neutral molecules show bent structure. In the M phase, charges (holes) are delocalized.[8-10] Inset: The structure of EDO-TTF molecule.*

This crystal is an important and fascinating candidate for the study of photo-induced phase transition, because the occurrence of the photo-induced M-I transition accompanied with lattice, electronic structural and magnetic changes can be expected as discussed before. In addition, the large photo-induced reflectance change in near IR and visible region can be expected around room temperature for this material, because it is often the case that photo-induced cooperative response is largely enhanced around T_c. From this point of view, photo-response of (EDO-TTF)$_2$PF$_6$ crystal with T_c (=280 K) around room temperature is also an important target even for application. Here we review the experimental results which demonstrate the exotic photo-response observed for this crystal.

2   METHOD AND RESULTS

2.1 Experimental

To probe dynamics of the photo-induced reflectivity change, the conventional pump probe technique with a Ti:sapphire laser (1.55 eV, 120 fs, repetition rate: 1 kHz) have been utilized. The crystal of $(EDO\text{-}TTF)_2PF_6$ with a surface of 0.1 mm by 0.2 mm was excited by a femto-second pulsed laser.[10] the excitation energy was 1,55 eV, near the energy of the charge transfer band. The reflectivity change was probed utilizing a white light generated by non-linear optical method.

2.2 Femto-second laser induced spectroscopic changes in $(EDO\text{-}TTF)_2PF_6$

The reflection spectra 10 pico-second (ps) before (-10 ps) and 3ps after (+3 ps) photo-excitation observed at 180K are plotted by black and gray solid lines in Figure 2, respectively. Similar large and fast reflectivity change triggered by photo-excitation was clearly confirmed in the wide temperature range lower than 265 K and it disappeared above T_c. The obtained spectral changes indicate that optical band directly reflecting CO in I phase disappears just after photo-excitation. Therefore we can conclude that CO melting accompanied with I-to-M phase conversion occurs within 3ps just after excitation.[10]

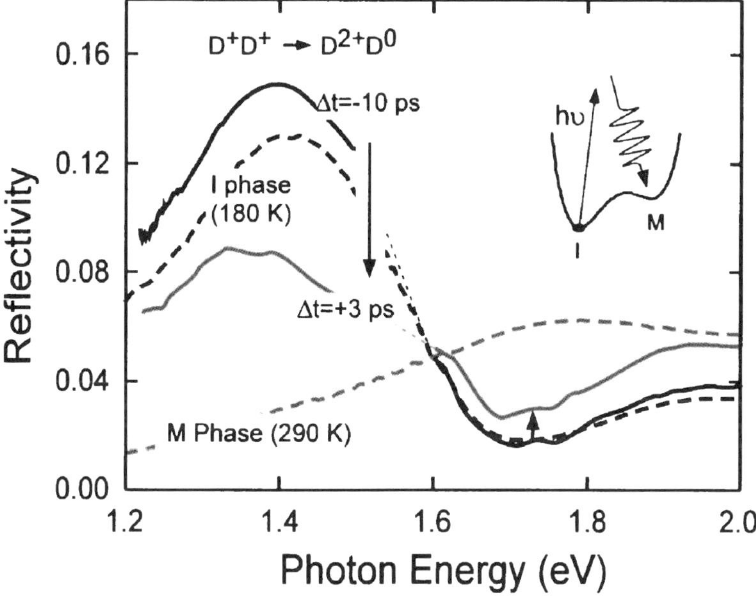

Figure 2 *Changes in the reflectivity spectra accompanied with thermally induced (T_c= 280 K) and photo-induced M-I transitions. Dashed lines: Reflection spectra in M (gray dashed line) and I (black dashed line) phase observed at 290 and 180 K, respectively. Solid lines: Reflection spectra 10 ps before (-10 ps: black solid line) and 3 ps after (+3 ps: gray solid line) photo-excitation observed at 180 K. Inset: Schematics for free-energy change accompanied with photo-induced M-I transition.[10]*

This photo-induced change in reflection spectrum is large enough to be directly observed using a CCD camera. Figure 3 shows the evolution of the reflected light coming from the sample surface observed at 10 ps before and 0.25 ps after photo-excitation.[10] The estimated photo-induced I-to-M conversion efficiency based on the reflectivity change at 3 ps after excitation comparing with the one due to thermally induced I-to-M phase change reached more than 50% with excitation intensity of 6×10^{14} photons/cm^2 (abbreviated as cm^{-2}). This excitation intensity corresponds to a single excitation photon for every 500 converted molecules based on the absorption intensity estimated from reflection spectrum by K-K transformation method. Such highly efficient conversion indicates the large contribution of strong cooperative effect in this crystal as expected.

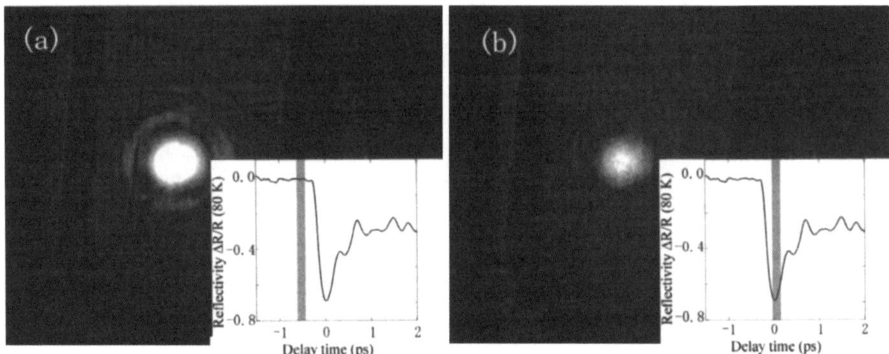

Figure 3 *Demonstration of the photo-induced reflectivity change by fs laser excitation observed by CCD camera with the resolution time of 0.2 ps.*[10] *(a): The reflected monitor light from the surface of (EDO-TTF)$_2$PF$_6$ crystal before photo-excitation. Inset shows the time profile of the photo-induced reflectivity change observed at 1.38 eV. (b): The reflected light 0.2ps after photo-excitation.*

2.3 Dynamical behaviour of the photo-induced spectroscopic changes

To make discussion on the dynamical process of CO melting, we observed the time profile of spectral changes at 1.38 and 1.72 eV as shown in Figure 4.[10] Balck and gray lines are for results at 180 K and 260 K, respectively. As indicated in this figure, the photo-induced reflectivity change ($\Delta R/R$) values have reached optimal values just after photo-excitation within resolution time (0.25 ps) and decreased until half of them within 1.5 ps. For the delay time (Δt) longer than 1.5 ps, the $\Delta R/R$ value keeps constant and the spectral change is consistent with I-to-M transition as a result of CO melting, thus we can safely conclude that the CO melting process is completed within 1.5 ps for wide temperature region utilizing the same excitation intensity. Of course, such an ultra fast photo-conversion process cannot be simply explained by the heating effect due to laser irradiation. In addition, as indicated by triangles (solid and open ones for 180 and 260 K, respectively) in Figure 4, the vibratile structure with extremely large magnitude was observed. The period of this vibration estimated from the interval between the first and second triangles was 0.45 ± 0.05 ps (74 ± 8 cm^{-1}) and 0.55 ± 0.1 ps (60 ± 10 cm^{-1}) for 180 and 260 K, respectively.

Figure 5(a) shows the Raman spectra of I phase crystal observed at 180 K. Several peaks appeared in I phase in spectral region between 30 and 150 cm^{-1}. Especially, as increasing the sample temperature, 84 cm^{-1} band (gray-colored mode) has shown the softening as plotted in Figure 5(b) by solid circles. This behaviour seems to be parallel to the temperature dependence of vibration interval estimated from time profile of $\Delta R/R$ as plotted in Figure 5(b) by solid squares. Therefore, we can conclude that the observed dynamical vibration in time domain is attributed to the coherent phonon-like vibration which corresponds to 84 cm^{-1} Raman mode. The vibrational mode calculation for a single EDO molecule shows the existence of several modes around 100 cm^{-1}, which can be characterized as the bending motion of the molecular plane. This result strongly suggest that the observed coherent vibration in the time profile is due to the strong and dynamical coupling between CO and molecular deformation.

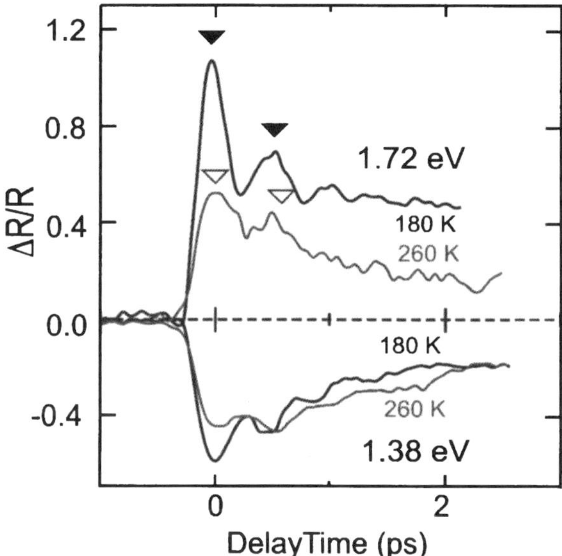

Figure 4 *Probe photon energy dependence of the time profile for the $\Delta R/R$ observed at 180 K (black lines) and 260 K (gray lines). The probe photon energy was 1.72 and 1.38 eV for the upper and lower panels, respectively. Triangles indicate the peak positions due to the vibratile structure observed at 180 K (solid triangles) and 260K (open triangles), respectively.[10]*

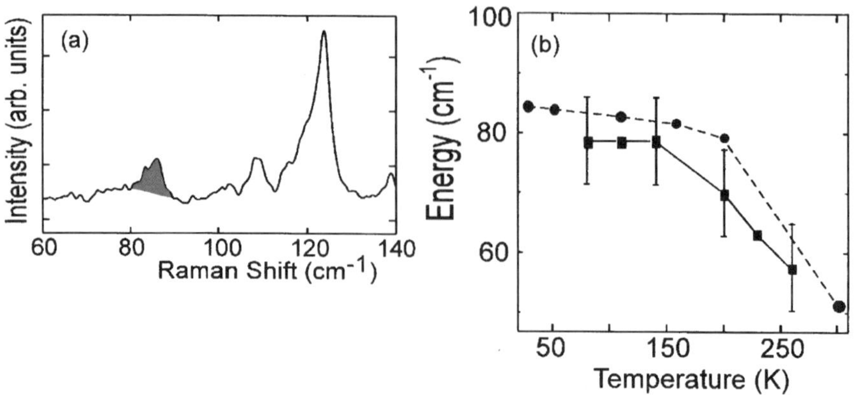

Figure 5   *(a): Raman spectrum in the low-energy region for the I phase observed at 180 K. Temperature dependence of the Raman shift energy for the gray-colored mode is plotted in (b). (b): Temperature dependence of the Raman shift (solid circles) for the gray-colored mode in (a) and that for the vibratile frequency estimated from the time profile of ΔR/R (solid squares). The solid and dashed lines serve as an eye guide.*[10]

2.4 Excitation intensity dependence of photo-reflectance change

It is one of important and attractive natures of PIPT that photo-conversion efficiency shows nonlinear and/or threshold-like dependence on the excitation intensity reflecting the role of cooperative interaction.[7] In $(EDO\text{-}TTF)_2PF_6$, indeed, the photo-reflectivity change probed with 1.38 eV monitor light at 3 ps after photo-excitation has shown nonlinear dependence on the excitation intensity as plotted in Figure 6 by solid squares. The threshold excitation intensity (I_{th}) indicated in Figure 6 is equal to 2×10^{14} /cm^2 which corresponds to one excitation photon for every 1500 EDO-TTF molecules. In addition, the estimation of conversion efficiency based on optical depth for excitation light (1.55 eV) and ΔR/R value gives the result that the one excitation photon converts 250 donor molecules. Such highly efficient and nonlinear conversion process is the result of internal cooperative interactions as expected for A_2B salts similar to other PIPT phenomena.[7] The estimated heat effect for present experimental condition based on previously reported thermo-dynamical data is only 0.1 K even for 1×10^{15} cm^{-2} excitation intensity.[11] Therefore, it also supports the idea that the observed I-to-M transition is not due to simple thermal melting of CO by light induced heating effect.

The excitation intensity dependence of ΔR/R at Δt = 0.25 ps with probe photon energy of 1.38 eV has shown scarce nonlinear behaviour in weak excitation intensity region (see solid circles in Figure 6). This result strongly suggests that the photo-induced process within time resolution includes large contribution of pure electronic (linear) excitation. In addition, the observed relaxation process in ΔR/R just after photo-excitation (Δt < 1.5 ps) seems to be reflecting the dynamical conversion from electronic to lattice relaxed excited state accompanied with intense coherent phonon vibration (see Figure 4) though its nature is not clear.

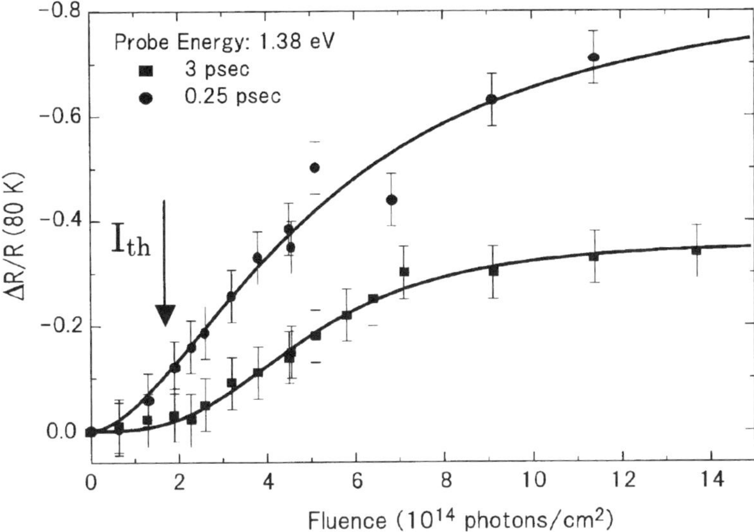

Figure 6 *Excitation intensity dependences of $\Delta R/R$ observed 3 ps (solid squares) and 0.25 ps (solid circles) after photo-excitation at 180 K. The threshold photon density was estimated at 2×10^{14} photon/cm^2 and indicated by arrow.*

3   CONCLUSION

We have reviewed that an ultra-fast photo-induced insulator to metal phase transition can be induced with very weak excitation intensity at near room temperature. The observed photo-induced spectroscopic change indicates that the PIPT process is governed by the cooperative melting of CO mediated by an optical coherent phonon. The development of combined technology for spectroscopic and X-ray structural and electron density analysis with fs and sub-fs resolution is essential not only for making clear the microscopic mechanisms of the observed exotic effect but also for molecular device oriented research utilizing strongly lattice-charge coupled system such as A_2B salts.

References

1   D.Jerome and H.J.Schulz, Advance in Physics, 2002, **51**, 293.
2   *"Organic Superconductors" 2nd ed.*, ed. by T.Ishiguro, K.Yamaji and G.Saito, Springer-Verlag, Berlin-Heiderberg (1998)
3   K.Bechgaard et.al., Solid State Commun. 1980, **33**, 1119.
4   J. E. Hirsch and D. J. Scalapino, Phys. Rev. Lett. 1983, **50**, 1168.
5   K. Penc and F. Mila, Phys Rev. B, 1994, **49**, 9670.
6   H.Seo and H.Fukuyama, J.Phys.Soc.Jpn. 1997, **66**, 1249.
7   S. Koshihara, in *"Optical Properties of Low-Dimensional Materials"*, edited by T. Ogawa and Y. Kanemitsu (World Scientific, Singapore, 1998), Vol. 2, Chap. 3, and references therein
    "Photo-induced phase transition", ed. by K.Nasu, (World Scientific, Singapore, 2004).
8   A. Ota, H. Yamochi, G. Saito, J. Mater. Chem. 2002, **12** 2600.
9   O.Drozdova, K. Yakushi, K. Yamamoto, A. Ota, H. Yamochi, G. Saito, H. Tashiro, D. B. Tanner, Phys. Rev. B, 2004, **70**, 075107.
10 M. Chollet, L. Guérin, N. Uchida, S. Fukaya, H. Shimoda, T. Ishikawa, K. Matsuda, T. Hasegawa, A. Ota, H. Yamochi, G. Saito, R. Tazaki, S. Adachi, S. Koshihara, Science, 2005 **307**, 86.
11 K.Saito, S.Ikeuchi, A.Ota, H.Yamochi and G.Saito, Chem. Phys. Lett. 2005, **401**, 76.

PHOTOCHEMICAL TRANSFORMATION OF MOLECULAR CRYSTALS INTO DEVICES

T. Naito,[1,2] H. Sugawara,[2] T. Inabe,[2] T. Miyamoto,[3] H. Niimi,[3] and K. Asakura[3]

[1] Creative Research Initiative "Sousei" (CRIS), Hokkaido University, Kita 21-jyo, Nishi 10-choume, Kita-ku, Sapporo, Hokkaido, 001-0021, Japan
[2] Division of Chemistry, School of Science, Hokkaido University, Sapporo, Japan
[3] Catalysis Research Center (CRC), Hokkaido University, Sapporo, Hokkaido, Japan

1 INTRODUCTION

All electronic devices such as transistors and diodes are based on junction structures, where a conducting solid consists of several parts with different electrical properties. Thus it is important to control the conducting properties of an arbitrary selected part of molecular materials for their applications as electronic devices. We have recently reported a novel photochemical way of fabrication of molecular devices.[1] Using a single crystal of a silver salt of an organic π–acceptor molecule 2,5-dimethyl-N,N'-dicyanoquinonediimine[2] (DM; Figure 1), $Ag(DM)_2$, we simply illuminated a part of the crystal to change the electrical conductivity. During illumination the photo-excited silver ions become strong enough oxidizers to remove electrons from neighbouring DM radical anions. Because the conduction properties are governed by unpaired π–electrons on the DM radical anions in this material, such photo-induced redox reactions result in the decrease of conductivity. Selecting the appropriate illumination conditions, one could turn the material, which has originally metallic conduction property, into a semiconducting solid with retaining its original crystal structure. It should be noted here that the conductivity changed at only the illuminated part, and the rest remained unchanged. Thus such illumination treatments should correspond to doping with spatial resolution of the excitation light diameter. The doped state survived at least for several months after the cessation of illumination. This apparent irreversibility enables us to make a junction structure starting from a single crystal. Now we are studying the detailed mechanism of this phenomenon in order to utilize it as a simple and versatile doping method for device fabrications. In this brief paper, we report the ultraviolet-visible (UV-Vis) illumination effects on the electronic and magnetic properties of $Ag(DM)_2$.

2 METHOD AND RESULTS

The single crystals of $Ag(DM)_2$ were prepared by following reported procedures and identified by spectroscopic and X-ray structural analyses.[3,4] The single crystals are fine long needles, and have the highest conduction along the needle axes. These axes coincide with the crystallographic c-axis, which are parallel with the stacking axes of the planar

DM radical anions. The single crystals were wholly illuminated with suspended and stirred in water using a Hg/Xe lamp (200-1100 nm, 200 W) equipped with a multimode optical fibre and a cold mirror. This method enabled us an easy control of temperature and homogeneous illumination on the samples. In some experiments, where only a small amount of samples were required, the single crystals were well ground to fine powder and exposed to UV-Vis light from the same light source with continued agitation. Through water or air, illumination gave the samples an approximately similar strength of light (1.5-2.5 $mWcm^{-2}$ at 200-350 nm, 40-70 $mWcm^{-2}$ at 350-550 nm, 3-15 $mWcm^{-2}$ at 550-1100 nm), but at some particular wavelengths light was characteristically strong or weak due to the light source property.

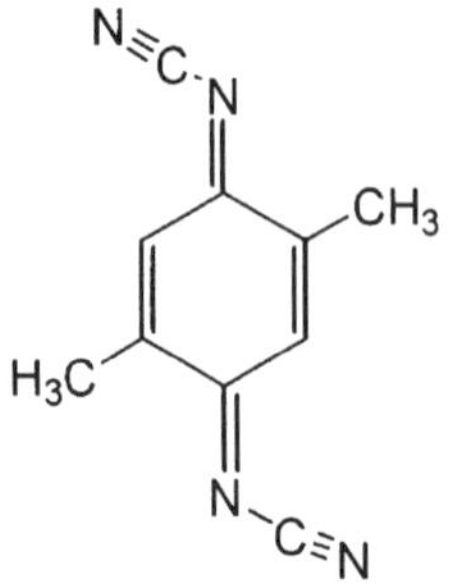

**2,5-Dimethyl-*N*,*N*'-dycyanoquinonediimine
(DM)**

Figure 1   *Structure of DM*

1.1   DM Molecules in illuminated Ag(DM)₂

As long as the sample temperature was kept around RT during illumination, the IR spectra (Figure 2) showed that the DM molecules in Ag(DM)₂ remained almost unchanged. In addition, such samples gave identical results of elemental analyses and X-ray absorption fine structure (XAFS) on Ag *L*-edge with those of the pristine samples.[5] However, Raman and X-ray photoelectron spectroscopy (RS and XPS) clearly indicated that a part of the DM molecules (up to 20-30 %), which were originally radical anion species before illumination, were oxidized toward neutral molecular species by photo-excited silver ions.[1] These experimental facts indicates that UV-Vis illumination should induce an irreversible redox reaction between the DM molecules and silver ions, but that the resultant formally neutral DM molecules should remain where it used to be in the crystal and still take part in the delocalization of the remaining unpaired π–electrons in Ag(DM)₂. Therefore one can observe only an averaged and thus intermediate oxidation state of DM species in Raman spectra.[1] In fact, powder X-ray diffraction pattern[5] indicate that the crystal structure should be largely retained as long as the redox reaction does not proceed so extensively, *i.e.* within "light-doping" stage, suggesting a certain meta-stable state after the photo-induced redox reaction. For more detailed information on the chemical state of the DM molecule, the measurements of solid state ^{13}C-NMR spectra are now under way.

1.2   Electronic Structure of Ag(DM)₂ after illumination

While the molecular and the crystal structures are largely retained at the light-doping stage, the band electronic structure after illumination could sensitively alter in accordance with the duration of illumination. We once directly observed this process by a series of electrical resistivity measurements using a partially illuminated single crystal of $Ag(DM)_2$.[1] Since electrical resistivity is generally much sensitive to extrinsic factors such

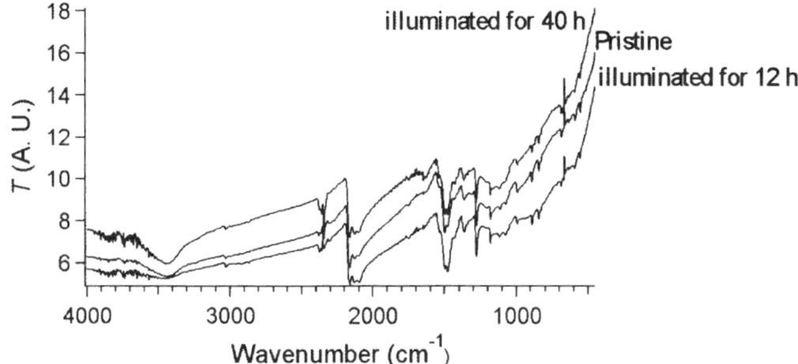

Figure 2   *IR spectra of pristine and illuminated $Ag(DM)_2$ (KBr disks)*

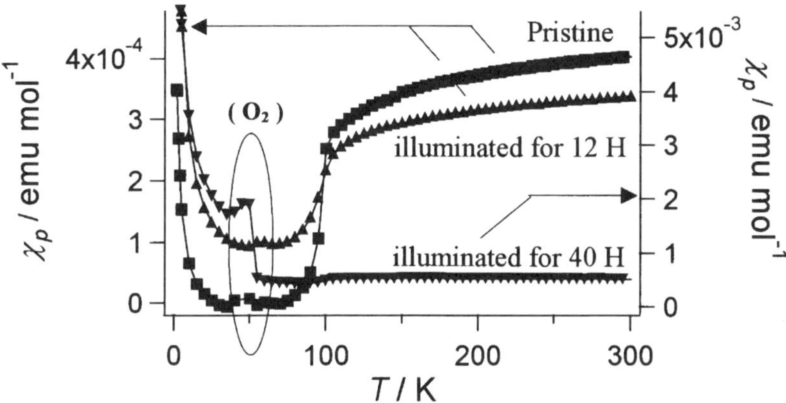

Figure 3   *Magnetic susceptibility of pristine and illuminated $Ag(DM)_2$*

as surface conditions in conjunction with electrical contacts, other measurements that require no electrical contacts are desirable to support our previous observation. Accordingly we measured magnetic susceptibility (Figure 3) and diffuse reflectance spectra (Figure 4) of the same samples used in the measurements of the IR spectra shown in Figure 2. Both of magnetic susceptibility and diffuse reflectance spectra well reflect bulk properties and electrical structures. The temperature-dependence of magnetic susceptibility of the pristine $Ag(DM)_2$ was of Pauli paramagnetic with the value of 4-5×10^{-3} (emu mol^{-1}) above 100 K, and became diamagnetic below 100 K except for a Curie-tail below ~ 30 K. Illumination clearly decreased the value of Pauli paramagnetism $(3-4\times10^{-3}$ (12 h) and 5×10^{-4} [emu mol^{-1}] (40 h)), suggesting the decrease in the number of metallic conduction electrons. On the other hand, illumination made the Curie-like behaviour at low temperatures more and more marked, while the magnetic transition at 100 K disappear. The Curie-tail, which originated from the lattice defects introduced by this doping, increased as follows; 6.9×10^{-4} (pristine), 3.0×10^{-3} (12 h), and 3.2×10^{-2} [emu K mol^{-1}] (40 h). Such magnetic behaviour could be reproduced as a sum of Curie contribution and Pauli paramagnetism. It should be noted that the transition temperature (100 K) remained constant despite the fact that the whole behaviour qualitatively changed. The results shown in Figure 3 imply that the illumination gradually and heterogeneously turns the metallic material into a semiconductor (probably from the surface), partly leaving the original metallic property of the inner sample intact. The Kubelka-Munk transformation of diffuse reflectance spectra (Figure 4) indicated that the electronic structure qualitatively changed due to the illumination.

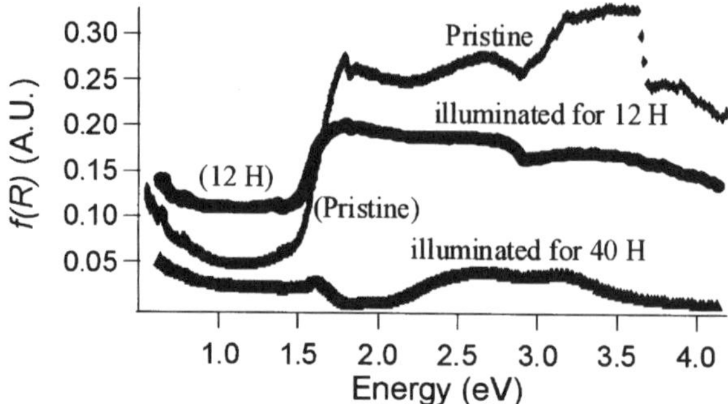

Figure 4   *Diffuse reflectance spectra of pristine and illuminated Ag(DM)$_2$*

3 CONCLUSION

UV-Vis illumination has qualitatively changed the electronic structure and magnetic properties of Ag(DM)$_2$ with retaining its crystal and molecular structure. The data show that the excited region is only at the surface of the crystals at an initial stage of illumination but finally extend to bulk part of the crystals after long enough illumination.

References

1 T. Naito, T. Inabe, H. Niimi and K. Asakura, *Adv. Mater.*, 2004, **16**, 1786.

2 A. Aumüller and S. Hünig, *Liebigs Ann. Chem.* **1986**, 142.

3 S. Hünig, M. Kemmer, H. Meixner, K. Sinzger, H. Wenner, T. Bauer, E. Tillmanns, F.R. Lux, M. Hollstein, H.-G. Groß, U. Langohr, H.-P. Werner, J.U. von Schütz and H.-C. Wolf, *Eur. J. Chem.*, 1999, 899.

4 R. Kato, H. Kobayashi, A. Kobayashi, T. Mori and H. Inokuchi, *Chem. Lett.* **1987**, 1579.

5 T. Naito, H. Sugawara, T. Inabe, Y. Kitajima, T. Miyamoto, H. Niimi and K. Asakura, submitted.

Acknowledgements

This work is partially supported by The Suhara Memorial Foundation, The Japan Securities Scholarship Foundation, Saneyoshi Scholarship Foundation, a Grant-in Aid for Scientific Research on Priority Areas of Molecular Conductors (15073207 and 15073101), and a Grant-in Aid for Scientific Research (A) (15205018).

THEORY OF OPTICAL RESPONSES OF PHOTOEXCITED HALOGEN-BRIDGED METAL COMPLEXES IN DIFFERENT INSULATING PHASES

N. Maeshima[1,2] and K. Yonemitsu[2,3]

[1]Department of Chemistry, Tohoku University, Aoba-ku, Sendai 980-8578, Japan
[2]Institute for Molecular Science, Okazaki 444-8585, Japan
[3]Graduate University for Advanced Studies, Okazaki 444-8585, Japan

1 INTRODUCTION

One-dimensional (1D) halogen-bridged metal complexes, known as MX chains, have shown various properties inherent in one-dimensional electron systems.[1] For example, the M=Ni complexes are Mott insulators owing to a strong on-site electron-electron interaction. In the Pd complexes, a relatively large electron-phonon interaction leads the systems to band insulators in a charge-density-wave (CDW) phase.

Recent experiments on their photoinduced optical responses show a distinct difference between these complexes. The Ni complexes have a Drude-like low-energy component in their optical conductivity spectra immediately after photoirradiation, which indicates the appearance of a photoinduced metallic state.[2] The Pd complexes, by contrast, remain insulating, keeping the finite charge gap without a Drude-like component, after the photoirradiation.[3]

In this study we elucidate the reason why the different photoinduced phenomena appear in the MX chains. We employ the one-dimensional (1D) ionic Hubbard model with a nearest-neighbor Coulomb interaction,[4] which is called the 1D extended ionic Hubbard model in this paper, to describe the electronic properties of the MX chains. By using the exact diagonalization method, we calculate physical quantities related to optical responses in ground and photoexcited states. The photoinduced metallic property of the Ni complexes is shown to be caused by photodoped unbound carriers, holons and doublons. In the Pd complexes, photodoped electrons and holes are bound by the nearest-neighbor interaction, resulting in the insulating property of the photoexcited state.

2 MODEL

The Hamiltonian of the 1D extended ionic Hubbard model is given by

$$H = -t\sum_{l,\sigma}(c_{l+1,\sigma}^{+}c_{l,\sigma} + c_{l,\sigma}^{+}c_{l+1,\sigma}) + U\sum_{l}n_{l,\uparrow}n_{l,\downarrow} + V\sum_{l}n_{l}n_{l+1} + \Delta\sum_{l}(-1)^{l}n_{l}, \tag{1}$$

where $c_{l,\sigma}^{+}$ ($c_{l,\sigma}$) is the creation (annihilation) operator of an electron with spin σ at site l, $n_{l,\sigma} = c_{l,\sigma}^{+} c_{l,\sigma}$, and $n_l = n_{l,\uparrow} + n_{l,\downarrow}$. The parameter t denotes the transfer integral, U the on-site repulsion strength, V the nearest-neighbor repulsion strength, and $\Delta (> 0)$ the amplitude of the alternating site potentials. In what follows, t is set to unity for simplicity. This model at half filling can describe both the Mott insulator phase and the CDW phase. Equation (1) with $\Delta = 0$ is equivalent to the Hamiltonian of the 1D extended Hubbard model, which has been used for studies of 1D Mott insulators.[5] Meanwhile, the Hamiltonian (1) with large Δ and small U has been used to describe the CDW phase of the MX chains.[6] In this study we employ the Lanczos method for diagonalization of the Hamiltonian (1). The numerical calculation is carried out for chains of length $N = 12$ under the periodic boundary condition.

To investigate the optical responses after the photoexcitation, we calculate the optical conductivity spectrum of the lowest, optically allowed, excited state $| \Psi_{1opt} \rangle$ given by[7,8]

$$\sigma_1(\omega) \equiv D_1 \delta(\omega) - \frac{1}{N\omega} \Im \left[\langle \psi_{1opt} | \hat{j} \frac{1}{\omega + i\varepsilon + E_{1opt} - H} \hat{j} | \psi_{1opt} \rangle \right]$$

$$+ \frac{1}{N\omega} \Im \left[\langle \psi_{1opt} | \hat{j} \frac{1}{\omega + i\varepsilon - E_{1opt} + H} \hat{j} | \psi_{1opt} \rangle \right], \qquad (2)$$

where D_1 is the Drude weight of $| \psi_{1opt} \rangle$ defined by

$$D_1 = -\frac{\pi}{N} \langle \psi_{1opt} | \hat{K} | \psi_{1opt} \rangle - \frac{2\pi}{N} \sum_{n \neq 1opt} \frac{\left| \langle \psi_n | \hat{j} | \psi_{1opt} \rangle \right|^2}{E_n - E_{1opt}}, \qquad (3)$$

with $\hat{K}$ being the kinetic term of the Hamiltonian (1), $\hat{j}$ the current operator defined by $\hat{j} \equiv i \sum_{l,\sigma} (c_{l+1,\sigma}^{+} c_{l,\sigma} - c_{l,\sigma}^{+} c_{l+1,\sigma})$, $| \psi_n \rangle$ the n-th excited state, and E_n the corresponding energy. The finite broadening factor ε is set to 0.1 in our calculations. It should be noted that $| \Psi_{1opt} \rangle$ corresponds to the lowest peak of the optical conductivity spectrum $\sigma(\omega)$ of the ground state, which is obtained by the Lanczos method.

3 RESULTS

We investigate the fundamental properties of the optical conductivity spectra for $V = 0$ first. Figure 1 shows $\sigma(\omega)$ and $\sigma_1(\omega)$ in the Mott insulator phase ($U = 12$, $\Delta = 0$) and those in the band insulator phase ($U = 0.1$, $\Delta = 0$). Each insulator has a charge gap $\Delta_c (\approx 8)$ and no low-energy component below Δ_c in $\sigma(\omega)$, reflecting the insulating nature. By contrast, $\sigma_1(\omega)$ has a low-energy component in both phases. In the Mott insulator phase, a large Drude component appears, suggesting the conducting property of

$\left|\psi_{1opt}\right\rangle$.[7,8] Another pronounced peak appears at $\omega \approx 1$ and a small peak at $\omega \approx 2.6$,[7] both of which correspond to the continuum band that results from two types of photodoped carriers, a holon and a doublon.[9] Hence the experimentally observed low-energy component of the Ni complexes would correspond to the Drude peak and these low-energy peaks. In the band insulator phase also, a large Drude component appears. Thus, $\left|\psi_{1opt}\right\rangle$ in the band insulator phase is also metallic due to photodoped electrons and holes.[8]

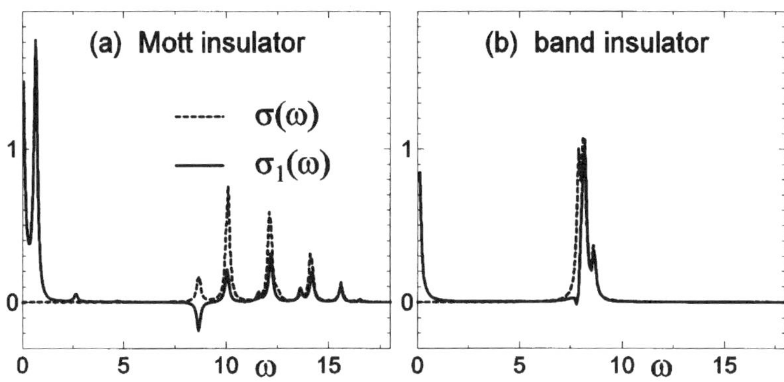

Figure 1 *Optical conductivity spectra $\sigma(\omega)$ and $\sigma_1(\omega)$, (a) in the Mott insulator phase ($U = 12$, $V = 0$, $\Delta = 0$) and (b) in the band insulator phase ($U = 0.1$, $V = 0$, $\Delta = 4$).*

The results for $V = 0$ demonstrate that the photoexcited states have the metallic Drude component in both phases. Now we focus on the effect of V. Figure 2(a) shows the V dependence of the Drude weights D_1 in the both phases. It decreases quite rapidly with increasing V in the band insulator phase, while it increases in the Mott insulator phase. This is attributed to the different tendency in binding of photodoped carriers by V. In Figure 2(b), we plot the distance R between the carriers: the holon and the doublon in the Mott insulator phase, and the hole and the electron in the band insulator phase. As V increases, R in the band insulator phase rapidly decreases and approaches 1, where the nearest-neighbor pair forms an exciton. By contrast, R in the Mott insulator is almost unchanged. These results are consistent with the known fact that the bound state appears even for infinitesimally small V in the band insulator phase,[10] and for $V > 2$ in the Mott insulator phase.[11]

4 SUMMARY

The optical conductivity spectra and the Drude weights are calculated for the ground and photoexcited states in the 1D extended ionic Hubbard model at half filling by the Lanczos diagonalization method. Although the photoexcited state in the Mott insulator phase remains metallic for small V, that in the band insulator phase loses its metallic character for infinitesimally small V owing to the binding of photodoped carriers.

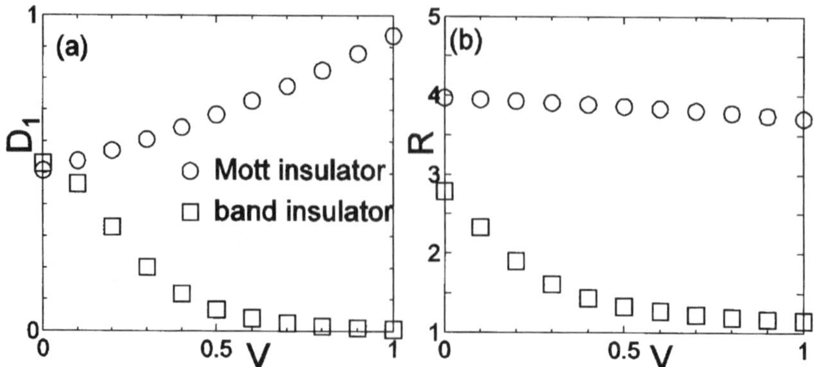

Figure 2 (a) *Drude weight of the photoexcited state* D_1 *and (b) distance between the photodoped carriers* R, *as a function of* V, *in the Mott insulator phase* ($U = 12$, $\Delta = 0$) *and in the band insulator phase* ($U = 0.1$, $\Delta = 4$).

Acknowledgements

The authors are grateful to Dr. H. Matsuzaki and Prof. H. Okamoto for showing their data prior to publication and for enlightening discussions. This work was supported by Grants-in-Aid for Creative Scientific Research (No. 15GS0216), for Scientific Research on Priority Area ``Molecular Conductors" (No. 15073224), for Scientific Research (C) (No. 15540354), and NAREGI Nanoscience Project from the Ministry of Education, Culture, Sports, Science and Technology, Japan.

References

1 H. Okamoto and M. Yamashita, *Bull. Chem. Soc. Jpn.*, 1998, **71**, 2023.
2 S. Iwai, M. Ono, A. Maeda, H. Matsuzaki, H. Kishida, H. Okamoto and Y. Tokura, *Phys. Rev. Lett.*, 2003, **91**, 057401.
3 H. Matsuzaki, PhD Thesis, University of Tokyo, 2004.
4 N. Nagaosa and J. Takimoto, *J. Phys. Soc. Jpn.*, 1986, **55**, 2745.
5 R. Neudert, M. Knupfer, M.S. Golden, J. Fink, W. Stephan, K. Penc, N. Motoyama, H. Eisaki and S. Uchida, *Phys. Rev. Lett.*, 1998, **81**, 657
6 K. Nasu, *J. Phys. Soc. Jpn.* 1983, **52**, 3865.
7 N. Maeshima and K. Yonemitsu, *J. Phys. Soc. Jpn.* 2005, **74**, 2671.
8 N. Maeshima and K. Yonemitsu, *J. Phys.: Conf. Ser.* 2005, **21**, 183
9 Y. Mizuno, K. Tsutsui, T. Tohyama, and S. Maekawa, *Phys. Rev. B*, 2000, **62**, 4769.
10 M. Ono, K. Miura, A. Maeda, H. Matsuzaki, H. Kishida, Y. Taguchi, Y. Tokura, M. Yamashita and H. Okamoto, *Phys. Rev. B*, 2004, **70**, 085101.
11 W. Stephan and K. Penc, *Phys. Rev. B*, 1996, **54**, 17269.

Fullerenes, Nanotubes and Other Related Nano Materials

ELECTRIC TRANSPORT AND MODULATED DENSITY OF STATES IN
ROTATIONAL ORDER AND DISORDER IN Na_2CsC_{60}

M. Akada[1], T. Yamamoto[1], R. Kumashiro[1], A. Hojyo[1],
H. Matsui[1], N. Toyota[1], J. P.Lu[3], and K. Tanigaki[1,2]

[1]Department of Physics, Graduate School of Science, Tohoku University
6-3 Aoba Aramaki Aoba-ku, Sendai, Miyagi 980-8578, Japan
[2]CREST-JST,
Kawaguchi Center Building, 4-1-8, Honcho, Kawaguchi-shi,
Saitama 332-0012, Japan
[3]Department of Physics, University of North Carolina, Chapel Hill
CB 3255, Phillips Hall, Chapel Hill, NC 27599-3255, U.S.A.

1 INTRODUCTION

It has become a recent common understanding that the order of orbitals, charges and spins
plays a very important role for controlling electronic states in solids, and lots of studies
have been reported regarding these issues. As for the atomic orbitals, the freedom of d-
orbitals has been shown to have strong correlations with electric conductivity and
magnetism in many manganese pervskite compounds[1-3]. In molecular solids, the freedom
of the frontier orbitals is also known to be very important in controlling ferro and antiferro

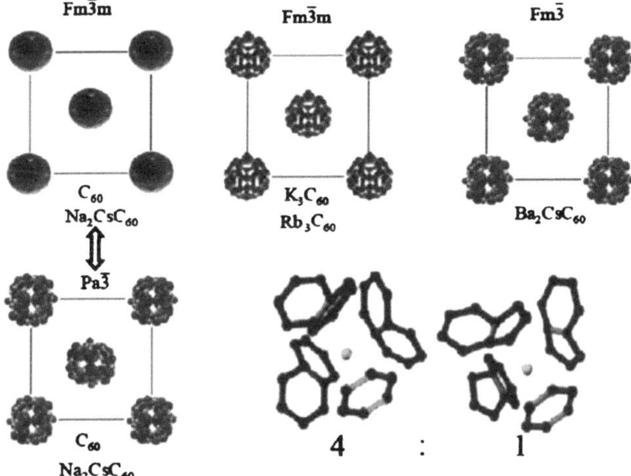

FIG. 1: Three types of rotational ordered and disordered phases have been reported so far for
fullerides. Due to the steric hindrance caused by the relatively large cation radii of K^+ and
Rb^+, K_3C_{60} and Rb_3C_{60} possess the merohedral ordering by jumping between the two
rotational positions around the $<111>$ direction with keeping the $Fm\bar{3}m$ space group[12]. The
almost free rotation at room temperature takes place when the smaller cation of Na^+ is
accommodated in the two tetrahedra sites, and examples are seen in Na_2AC_{60} (A=Cs and
Rb)[8,9]. The space same group $Fm\bar{3}m$ can also be adopted to the latter system. The third type
showing one orientational direction of C_{60} has been reported so far for Ba_2CsC_{60}[13].

magnetic spin alignment. In C_{60} solids, one of the families of molecular solids, the importance of the freedom of the C_{60} molecular orbitals is pointed out in recent reports of TDAE-C_{60} and $(NH_3)K_3C_{60}$, as for the ferro magnetic properties[4] and the antiferro-magnetic Mott insulator states[5]. Considering that the band around the Fermi level is constructed by the small orbital overlapping mostly of h_u and t_{1u} molecular levels in C_{60} solids and the C_{60} has orientational order and disorder depending on the dopants and temperature, electric transport as well as the density of states at the Fermi level N_{E_F} in C_{60} fullerides can be expected to be greatly influenced by the C_{60} rotational order and disorder[6]. Although this is very important in order to have real understanding of the C_{60} fullerides, the influence of the rotational freedom on the electronic states has only hypothetically been discussed or often ignored to be negligibly small. This is partly because C_{60} fullerides are air sensitive and are very difficult to have high quality electric transport measurements. Among the various C_{60} fullerides so far reported, Na_2CsC_{60} fulleride is the best compound[7-11] for this research, because this shows exactly the same rotational phase transition as pristine C_{60} from the rotationally ordered states in the low-temperature primitive cubic cell (low-T phase; P) to the rotationally disordered state in the high-temperature face-centered cubic (high-T phase; F) as shown in Fig.1[7,8,10]. The present paper will describe the electric transport of Na_2CsC_{60} from the low-T (P) phase to the high-T (F) one using microwave conductivity. The dynamical electric transport will clearly be shown to be greatly influenced by the C_{60} rotational order and disorder. It will also be shown that a theoretically predicted large modulation on N_{E_F} is actually induced by this rotational freedom.

2   EXPERIMENTAL

The Na_2CsC_{60} sample was prepared according to the procedure reported elsewhere[7,8]. The sample quality was checked to be sufficiently good by Xray diffraction using high energy facility at BL02B2 beam port of SPring-8. The prepared samples were introduced into a small glass tube with 3mm diameter in a glove box. Microwave conductivity σ was measured at 17 GHz by a cavity perturbation technique[14] as a function of temperature around the phase transition from the low-T (P) ordered phase to the high-T (F) disordered one. The samples put at the bottom of a glass tube were positioned in the maximum magnetic field of TE_{011} mode in a cylindrical cavity. A glass tube without samples was also measured to be used as a background signal. For high-temperature measurements of magnetic susceptibilities, an MPMS7 magnetometer (Quantum Design) was used together with a high-temperature accessory option.

3   RESULT & DISCUSSION

As the conduction band of A_3C_{60} solids can be constructed by the overlapping of the triply-degenerate t_{1u} molecular orbitals, it is very much intriguing to study how the molecular orientational order and disorder exert an influence on the electric transport. This has not been studied in detail so far. We have applied, in the beginning, a four-terminal probe DC conductivity method to the pelletized samples of Na_2CsC_{60} powder by encapsulating the samples in a special sealing cell. However, the evolution of the conductivity as a function of temperature was well fitted by the variable range hopping equation of $\rho(T) = \rho_0 \exp(T_0/T)^{1/4}$, where $k_B T_0 = 2^9/[9\pi\xi^3 N_{E_F}]$, ξ the length of electron localization. When the Fermi surface is modulated by the orientational order and disorder, ρ can be influenced even in the hopping conductivity. As shown later by the magnetic measurements, the density of states at Fermi level for the F phase is higher than that for the P phase. However, the conductivity was higher in the P phase than that in the F phase. This implies that the conductivity is controlled by the phonon scattering or the rotational disorder of C_{60} in the crystal.

Since the direct electric transport was not able to be measured, microwave measurements have been employed. The both problems of the electrical contact and the hopping among the particle grains do not become the serious problems in this measurement. The changes in both the width and the frequency of the resonance signal were recorded for the resonance of cavity with and without the sample, with decreasing temperature, the resonance frequency change directed toward a metallic shift, and the width change also increased. These behaviors indicate that the system belongs to a metallic side of a depolarization regime. In this regime, the resonance frequency change corresponds to the real part of a complex conductivity. Figure 2 shows the temperature dependence of the microwave conductivity thus obtained. A big discrete jump was observed from the high-T (F) state to the low-T (P) at 300 K. Since the temperature showing the dramatic change in the signal exactly corresponds to the rotational phase transition temperature, the observed jump of microwave conductivity is ascribed to this phase transition. Because the present microwave conductivity measurements in the maximum magnetic field mode does not give an absolute value of conductivity, the obtained data are displayed by normalizing the microwave AC conductivity at 300K as one in the inset of this figure. The conductivity was 1.30 times higher in the low-T (P) rotationally ordered phase and this change was larger than our expectations before starting measurements. It should be noted that the conductivity of ultra thin film of Au (10 nm) deposited on the C_{60} thin films shows a similar decrease in the electric transport at the P-F transition of 260K as reported earlier[15].

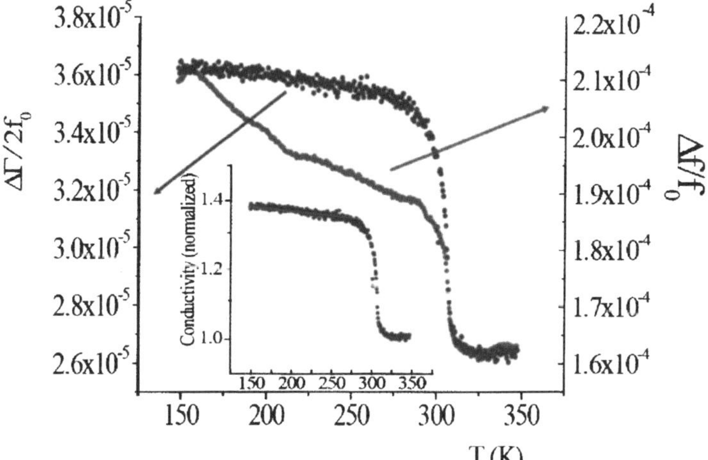

FIG. 2: The microwave conductivity of Na2CsC60. The values of $\Delta\Gamma/2f_0$ and $\Delta f/f_0$ corresponding to the real and the imaginary parts of the microwave conductivity of Na2CsC60, where $\Delta\Gamma = \Gamma - \Gamma_0$ is the linewidth difference between the empty cavity and the cavity with the sample, and $\Delta f = f - f_0$ the freqency shift between the cavity with and without the sample. The inset of the figure is the real part microwave conductivity normalized as one at 300 K.

The present direct observation is extremely large by more than two orders of magnitude compared to this report. The conductivity σ can be expressed by the followings Einstein's relation: $\sigma = 2e^2 N_{E_F} D$, where $D = v_F \ell / 3$ denotes the diffusion constant of conduction electrons with Fermi velocity v_F and mean free path ℓ. When we apply the equation to the present experimental results, $\sigma(F)/\sigma(P) = [N_{E_F}(F)D(F)]/[N_{E_F}(F)D(P)]$ can be evaluated. Using the experimental values of N_{E_F} for the F and P phases as well as the ratio of $\sigma(P)$ to $\sigma(F)$, the D(P)/D(F)=1.41 can be deduced. Therefore, the large change in the electric transport observed between the rotational order (P) and disorder (F) phases can be ascribed to the differences of the diffusion constants. Two different positions to understand the influence on the D(F) and D(P) values may be possible, one being due to the disorder phonon scattering and the other the disordered electronic states in the crystal. The electronic states at the Fermi level can also be expected to be strongly modulated by the C_{60} rotational order and disorder, since the bands are constructed by the overlapping transfer integrals of the C_{60} h_u and t_{1u} π-character orbitals. We have made local density approximation (LDA) band calculations for Na_2CsC_{60} crystals with various orientational order and disorder in the cubic lattice using the same method reported elsewhere[6]. As clearly shown in Fig. 3, the density of states (DOS) at the Fermi level is strongly modulated by the rotational order and disorder. The LDA band calculations suggest that the rotationally disordered phase (high-T; F) gives a smooth shape in the band structure, while a splitting featured band structural shape is given for the $Pa\bar{3}$ 4:1 rotationally ordered phase (low-T (P)- phase, the latter being experimentally confirmed in Na_2CsC_{60} by neutron diffraction[10]).

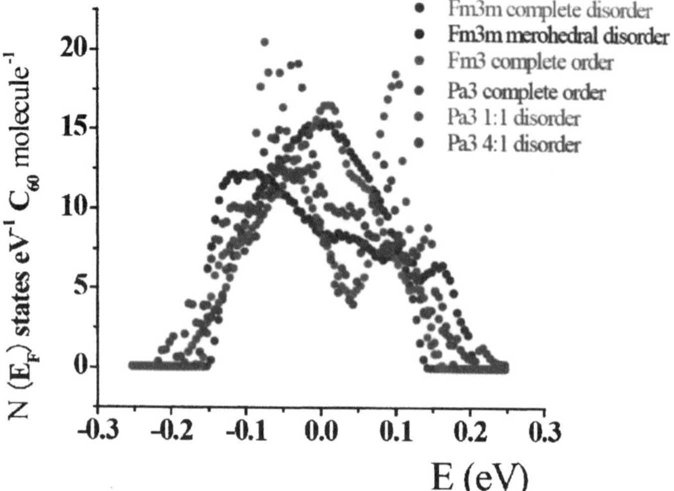

FIG. 3: The LDA band calculations of Na_2CsC_{60} for different modes of C_{60} orientations. The Fermi energy is set to zero. The two major ordered orientations in Na_2CsC_{60} are reported[10]. The tetrahedral Na atom resides above the center of a hexagonal face and the midpoints of three hexagon-hexagon fusion bonds in the major fraction and the center of a hexagonal face and the midpoints of three hexagon-pentagon fusion bonds in the minor fraction with the ratio of roughly 4 : 1. These major and minor ordered orientations are included in the present LDA band calculations for comparison to the experimental data.

Consequently, the DOS at the Fermi level N_{E_F} of 16 states eV^{-1} $(C_{60}$-mole$)^{-1}$ in the high-T (F) state becomes larger than that of 12.5 states eV^{-1} $(C_{60}$-mole$)^{-1}$in the low-T (P) state. DOS at the Fermi level can experimentally be estimated by magnetic Pauli susceptibilities. The Pauli susceptibility of Na_2CsC_{60}, therefore, has been measured using high magnetic fields. The total magnetic susceptibility was obtained as shown in Fig. 4, after the correction of the temperature-dependent Curie term arising from the defects in the lattice. In this figure, Larmor core-magnetic susceptibilities for low-T and high-T phases of C_{60} experimentally determined in a similar fashion and those of Na^+ and Cs^+ atoms in literature are included together. From these values, Pauli magnetic susceptibilities χ_{Pauli} were obtained from $\chi_{Pauli} = \chi_{total} - \chi_{core}(2Na^+ + Cs^+ + C_{60})$. The χ_{Pauli} values were 5.25×10^{-3} and 5.63×10^{-3} emu $(C_{60}\text{-mol})^{-3}$ for low-T (P) and high-T (F) phases, respectively. Using the relationship $\chi_{Pauli} = \mu_B^2 N_{E_F}$, N_{E_F} values were determined to be 16.04 and 17.42 orbitalstates eV^{-1} $(C_{60}\text{-mole})^{-1}$ for P and F phases. Although the increase in N_{E_F} from the low-T to the high-T phase by 8.6% experimentally determined is not as large as 13.3% obtained by band calculations, but this is in good agreement with each other.

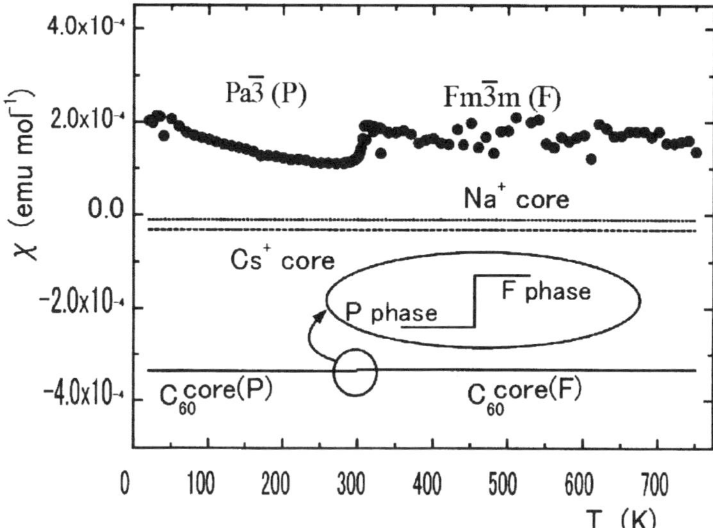

FIG. 4: Pauli susceptibilities of Na2CsC60 in the high-T (F) and low-T (P) Phases. The core diamagnetic susceptibilities are shown together.

As for the decrease in conductivity as well as the increase in N_{E_F} from the low-T (P) to the high-T (F) phase, one may presume that the lattice expansion through this phase transition would give rise to such an increase and that the influence by the orientational order and disorder will not be the important reason. Since the N_{E_F} as a function of temperature has been known from SQUID, NMR etc., the increase in N_{E_F} from low-T to high-T phase can be estimated. Using the reported experimental scaling data[16] and the lattice parameters of Na_2CsC_{60}, 1.41185 for low-T (P) and 1.41378 nm for high-T (F), the increase due to the lattice expansion is estimated to be at most 3% and this is much smaller than the observed value of 8.6%. This means that the band modulation is predominantly controlled by the rotational order and disorder of C_{60} molecules in the crystal.

It is very important now to take into account the C_{60} rotational influences on the electronic states for the real interpretations of the spin excitations from doublet to triplet observed by the NMR relaxation times[17,18]. The T1 relaxation time increases starting at 300 K. This change was explained by the following equation: $1/T_1 T = (k_B / \hbar)A^2 [\pi N_{E_F}^2 + (\chi_{se}\tau_{se})/(\mu_B^2 \hbar)]$, where, A is the scale factor and τ_{se} the relaxation time of the spin excitation from doublet to quartet. In the present experiments, N_{E_F} increases by 8.6% at the same temperature range, and therefore the real interpretation should be made considering this change in N_{E_F} originated by the C_{60} rotational order and disorder. However, compared to the change induced by the spin excitations reported by Brouet[17,18], the contribution of the change in N_{E_F} is less than 5% and is negligibly small. Therefore, the important conclusion deduced by Bouchet et al. from the NMR spectroscopy that both solids band and molecular level picturers play a very important role in the electronic states in C_{60} fullerides, is indeed correct.

4 CONCLUSION

It is important information that the rotationally disordered phase (F) at high temperatures is still metallic from the magnetic susceptibilities obtained using a high temperature accessary. The SQUID data at high temperatures have clearly shown that the Pauli temperature-independence feature continues after the phase transition and metallicity holds at least until 673 K. The results of the SQUID data are also consistent with the microwave AC conductivity experiments and NMR studies. It will still be intriguing to understand why unexpected large decrease in magnetization in high-T (F) phase can be observed by the ESR studies[19].

ACKNOWLEDGEMENTS

We are grateful to the staff members at SPring-8 (beamline BL02B2). The synchrotron radiation experiments were performed by the approval of the Japan Synchrotron Radiation Research Institute (JASRI) as Nanotechnology Support Project. This work was supported by Creation of Nanodevices and Systems Based on New Physical Phenomena and Functional Principles of CREST of JST The present work is also partially supported by the Tohoku University 21 century COE program "Particle-Matter Hierarchy" of MEXT Japan. This work was performed by a Grant-in-Aid from the Ministry of Education, Culture, Sports, Science, and Technology of Japan, No.13304031 and 17038001.

REFERENCE

1. C Renner, G. Aeppli, B. G. Kim, Y. A. Soh, S. W. Cheong, Nature **416**, 518 (2002).
2. S. Okamoto, S. Ishihara, S. Maekawa, Phys. Rev. B **61**, 14647 (2000).
3. X. J. Zhou, P. Bogdanov, S. A. Kellar, T. Noda, H. Eisaki, S. Uchida, Hussain Z, Z. X. Shen, Science **286**, 268 (1999).
4. D. Mihailovic, D. Arcon, P. Venturini et al., Science **268**, 400 (1995).
5. H. Tou, Y. Maniwa, Y. Iwasa, H. Shimoda, T. Mitani, Phys. Rev. B **62**, R775 (2000).
6. J. Lu J, L. Y. Zhang, Z. L. Cao, Mod. Phys. Letters B **10**, 1417 (1996).
7. K. Tanigaki, I. Hirosawa, T. W. Ebbesen, J. Mizuki, Y. Shimakawa, Y. Kubo, J. S. Tsai, and S. Kuroshima, Nature 356, 419 (1992).
8. K. Tanigaki, I. Hirosawa, T. W. Ebbesen, J. Mizuki, and S. Kuroshima, Chem. Phys. Letters **203**, 33 (1993).
9. K. Tanigaki, I. Hirosawa, T. Manako, J. S. Tsai, J. Mizuki and T. W. Ebbesen, Phys. Rev. B **49**, 12307 (1994).

10. K. Prassides, C. Christides, I. M. Thomas, J. Mizuki, K. Tanigaki, I. Hirosawa and T. W. Ebbesen, Science 263, 950 (1994).

11. Although both Na_2CsC_{60} and Na_2RbC_{60} show a phase transition from the low-temperature ordered primitive cubic (P) to high-temperature disordered fcc, Na_2RbC_{60} eventually undergoes polymerization in its low-T phase. Hence, among the C_{60} fullerides so far synthesized, Na_2CsC_{60} is the best material that can be used for the present study.

12. P. W. Stephens, L. Mihaly, P. L. Lee, R. L. Whetten, S. M. Huang, R. Kaner, F. Deiderich, K. Holczer, Nature **350**, 632 (1991).

13. A. C. Duggan, J. M. Fox, P. F. Henry, S. J. Heyes, D. E. Laurie, M. J. Rosseinsky, Chem. Commun. (1996) 1191.

14. O. Klein, S. Donovan, M. Dressel and G. Grunner, Int. J. Infrared Milli. Waves, **14**, 2423 (1993).

15. C. W. Chang, B. C. Regan, W. Mickelson, R. O. Ritchie, A. Zettl, Solid State Commun **128**, 359 (2003).

16. Y. Maniwa, T. Saito, A. Ohi, K. Mizoguchi, K. Kume, K. Kikuchi, I. Ikemoto, S. Suzuki, Y. Achiba, M. Kosaka, K. Tanigaki and T. W,. Ebbesen, J. Phys. Soc. Jpn. **63**, 1139 (1994).

17. V. Brouet, H. Alloul, T.-N. Le, S. Garaj, and L. Forro, Phys. Rev. Lett. **86**, 4680 (2001)

18. V. Brouet, H. Alloul, S. Garaj, L. Forro, Phys. Rev. B **66**, 155124 (2002)

19. L. Forro and L. Mihaly, Reports on Progress in Physics **64**, 649 (2001).

ULTRAVIOLET PHOTOELECTRON SPECTROSCOPY OF ENDOHEDRAL FULLERENES

Shojun Hino

Faculty of Engineering, and Graduate School of Science and Technology, Chiba University Chiba 263-8522 Japan

1 INTRODUCTION

Inner space of fullerene cages is wide enough to accommodate an atom, atoms, a molecule or clusters and they can be actually incorporated into the fullerene cages.[1-4] When entrapped atom(s) is hydrogen or helium, the interaction between the atom(s) and the fullerene cage is rather small. When metal atom(s) is entrapped, metal atom(s) gives electrons to the cage and may deform the electronic structure of the fullerene cages. This electron transfer has been clearly observed in the ultraviolet photoelectron spectra (UPS) of metallofullerenes.[5-10]

Recently formation of two yttrium atoms encapsulation into C_{82} fullerene cage has been reported and in the same cage additional two carbon atoms are also incorporated.[11, 12] As the inner diameter of such fullerene cages is around one nm, the inner space is wide enough for a single atom but could be too narrow to accommodate multiple atoms. Hence it could be expected that the interaction between multiple atoms and the cage might be different from that between a single atom and the cage.

In this article, we present UPS of multiple atoms encapsulated endohedral fullerenes and discuss the difference among the spectra in conjunction with the entrapped atomic species, cage structures and number of entrapped atoms.

2 C_{82} CAGE FULLERENES

There are nine IPR (isolated pentagon rule) satisfying fullerenes in C_{82} cage and their molecular structures are shown in Fig. 1. Numbers in the figure indicate the nomenclature given by Fowler and Manolopoulos.[13] There is only one C_{2v} cage so that cage symmetry analysis by NMR measurements alone is able to determine the cage structure. However, there are three C_s and C_2 isomers and two C_{3v} isomers, sot that symmetry estimation through NMR measurements is not necessarily adequate to determine the cage structure. Comparison between simulated spectra obtained fro theoretical MO calculation and photoelectron spectra on fullerenes has been successful to deduce the possible cage structure.[7, 9]

As for trivalent mono metal atom encapsulated metallofullerenes, two isomers having C_{2v} [14] and C_s symmetry [15] have been reported. C_{2v}-C_{82} is the major isomer and C_s is the minor one and its ratio is approximately 9 : 1. Their absorption spectra differ so much, which indicates their different electronic structures. As for divalent mono metal atom encapsulated fullerenes, $M^{2+}@C_{82}{}^{2-}$ (M denotes entrapped metal atom), more than three isomers have been isolated. Four isomers have been isolated for Ca@C_{82} and one of them (isomer III) has C_2 symmetry, [16] and three isomers have been isolated for Tm@C_{82} and their symmetry has been determined as C_s for isomer I, C_2 for II and C_{2v} for III. [17]

When multiple atoms are incorporated into the cage, situation is a little bit unclear since the amounts of isolated endohedral fullerenes are rather poor and/or the variety of such endohedral fullerenes is not so much for the moment. It is still progressing to obtain these fullerenes with more abundant quantity. Among these endohedral fullerenes, three $Y_2C_2@C_{82}$ isomers have been isolated and their symmetry has been determine by NMR to be C_s (No. 6) for isomer I, C_{2v} (No. 9) for II and C_{3v} (No. 7 or 8). [11, 12] Two yttrium atoms entrapped metallofullerene has also been isolated and its cage symmetry was determined to be C_{3v} (No. 7 or 8). [11,12]

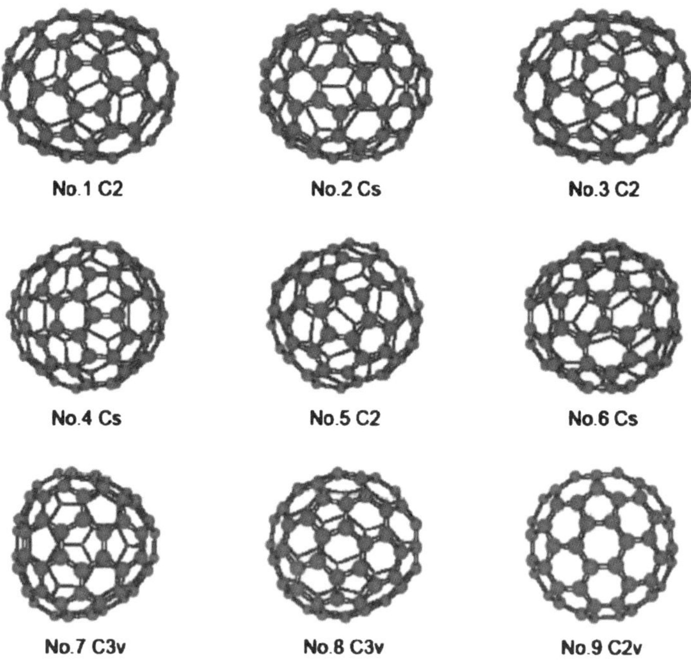

Figure 1   *Isolated Pentagon Rule (IPR) satisfying cages for C_{82}. Symmetry of each cage is given together with numbering nomenclature given by Fowler and Manolopoulos [13].*

3 EXPERIMENTAL

Metallofullerenes were synthesized by arc heating of metal oxides and graphite composite rods and isolated by multiple stages high performance liquid chromatography (HPLC). The details of syntheses and isolations are described elsewhere.[1, 11, 12, 14-18]

Photoelectron spectra were measured by a photoelectron spectrometer at BL8B2 of UVSOR (Ultraviolet Synchrotron Orbital Radiation facility), Institute for Molecular Science, Okazaki, Japan. Resolution of the spectra varied from 150 meV to 100 meV due to replacement of the electron energy analyser. Spectra referenced against the Fermi level which was determined by measuring the Fermi level of gold deposited films.

Specimens for the photoelectron measurements were thin metallofullerene films sublimed from resistive heating quartz crucibles. They were prepared in a sample preparation chamber attached to the photoelectron spectrometer. Sublimation temperature was 550 – 700 °C and it varied from one metallofullerene to the other ones. It seemed that the amount of metallofullerenes was one of the crucial factors for their sublimation temperature. At the first stage of deposition of metallofullerenes heavy degassing was always observed, which might be due to evaporation of solvent and its residue used at the HPLC separation processes.

Pressure of the sample preparation chamber during the sublimation went up to 2×10^{-5} Pa from base pressure of less than 1×10^{-7} Pa. Base pressure of the photoelectron spectrometer was 2×10^{-8} Pa, and the pressure during the measurements was around 4×10^{-8} Pa.

4 UPS OF MONO METAL ATOM ENCAPSUALTED ENDOHEDRAL FULLERENES

Ultraviolet photoelectron spectra of metallofullerenes, La@C_{82} (major C_{2v} isomer containing less than 10 % minor C_s isomer which is too little to influence the UPS critically),[5] Tb@C_{82} (major C_{2v} isomer),[8] Tm@C_{82} (isomer III, C_{2v} symmetry),[9] and Pr@C_{82} (minor C_s isomer) are shown in Fig. 2. Spectral resolution of the UPS of La@C_{82} was poorer than that of other metallofullerenes. When the same major and minor isomer mixed La@C_{82} was measured with higher resolution, the first structure denoted as N in the figure was clearly observed as a distinct structure[19] and its UPS are almost identical to those of Tb@C_{82}.

The UPS of fullerenes can be principally divided into two parts; one is between 0 and 4.5 eV binding energy region and the other is the region deeper than 4.5 eV. The former is due to π-electrons that delocalized on the fullerene cage and the latter is mainly due to σ-electrons that construct the backbone of skeletal structure of fullerenes.

The electron configuration of La@C_{82}, La@C_{82} Tb@C_{82} and Pr@C_{82} is $M^{3+}@C_{82}^{3-}$ but that of Tm@C_{82} is $M^{2+}@C_{82}^{2-}$. The spectral onset of Tb@C_{82} or La@C_{82} is much smaller than that of Tm@C_{82}, which derives from the difference in the amounts of transferred electrons; Tb or La donates three electrons to the cage whereas Tm donates only two electrons. Donation of odd electrons brings to form an open shell molecule so that the highest molecular orbital (HOMO) becomes unstable and its binding energy becomes shallow. Transfer of three electrons makes two levels that are specified by structure N and not clearly discernible one at around 1.2 eV, while that of two electrons induce only structure N. Taking into account this difference, structures B, C and ones of deeper binding energy region of La@C_{82}, Tb@C_{82} and Tm@C_{82} quite resemble. On the

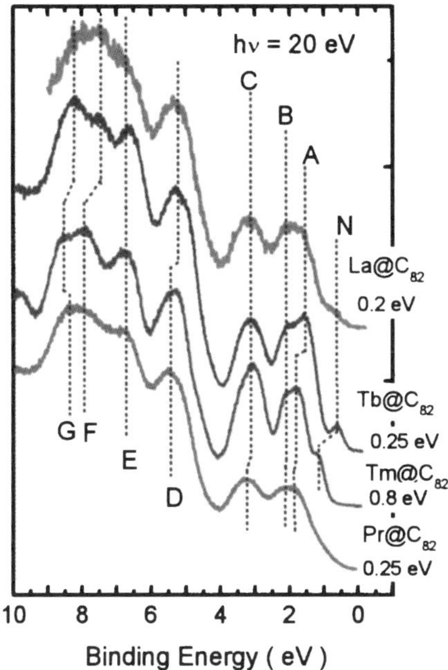

Figure 2   *Ultraviolet photoelectron spectra of C_{2v} symmetry metallofullerenes, $La@C_{82}$, $Tb@C_{82}$ and $Tm@C_{82}$ (III) and C_s symmetry $Pr@C_{82}$.*

other hand, the UPS of C_s-$Pr@C_{82}$ differs so much from those of other metallofullerenes. There is no correspondence among the UPS in both upper valence π-electron band region and deeper σ-electron band region. Deeper valence σ-electron band region is diffuse compared with that of other metallofullerenes. This diffuse structure might be due to wide distribution of σ-electron molecular orbitals derived from poor cage symmetry. These findings bring a conclusion that the cage structure is the dominating factor to specify the electronic structure of metallofullerenes and entrapped species do not affect the electronic structure so much.

The UPS of $Ca@C_{82}$ (III)[7] and $Tm@C_{82}$ (II)[9] support this conclusion. These two metallofullerenes have the same C_2 symmetry and their UPS are almost identical, which is an indication of their isoelectronic structures. Therefore, we could conclude that the cage structure is the decisive factor for the electronic structure of mono metal atom entrapped metallofullerenes and entrapped metal species do not have significant effect to it.

5 UPS OF MULTIPLE ATOMS ENCAPSULATED ENDOHEDRAL FULLERENES

5.1 $Y_2C_2@C_{82}$ (III) and $Y_2@C_{82}$

As symmetry of $Y_2C_2@C_{82}$ (III) and $Y_2@C_{82}$ is the same C_{3v} and their absorption spectra are analogous, it was suggested that the electronic structures of these two endohedral fullerenes are similar.[11] Their UPS are shown in Fig. 3. These two spectra resemble each other and slight difference can be observed in the π-electron valence band region. Spectral onset of $Y_2@C_{82}$ is smaller than that of $Y_2C_2@C_{82}$ (III) by about 0.4 eV and troughs between the first structures of $Y_2@C_{82}$ are shallow. These points imply that these two endohedral fullerenes have analogous electronic structure and the difference was induced by encapsulation of additional two carbon atoms.

To clarify the difference between them, the UPS of $Y_2C_2@C_{82}$ (III) was subtracted from that of $Y_2@C_{82}$ and their difference spectrum is shown in Fig. 3. The difference spectrum reveals a small peak at around 0.8 eV and its area corresponds to two electrons. That is, it is highly plausible that $Y_2@C_{82}$ bears additional two electrons compared with $Y_2C_2@C_{82}$ (III). The UPS of $Y_2C_2@C_{82}$ (III) can be well reproduced by a simulated spectrum obtained by molecular orbital calculation assuming C_{3v} (No. 8) geometry with additional four electrons on the cage.[10] This indicates that the electronic configuration of these metallofullerenes are $(Y_2C_2)^{4+}@C_{82}{}^{4-}$ and $(Y_2)^{6+}@C_{82}{}^{6-}$. Additional two electrons do not deform the electronic structure of C_{82} cage drastically. In these endohedral fullerenes the cage geometry seems to dominate the electronic structure as mono metal encapsulated metallofullerenes. However, an existence of the small peak in the spectral onset region of $Y_2@C_{82}$ indicates that there is a clear difference in the electronic structures of these two endohedral fullerenes. The resemblance of their absorption spectra could be accidental.

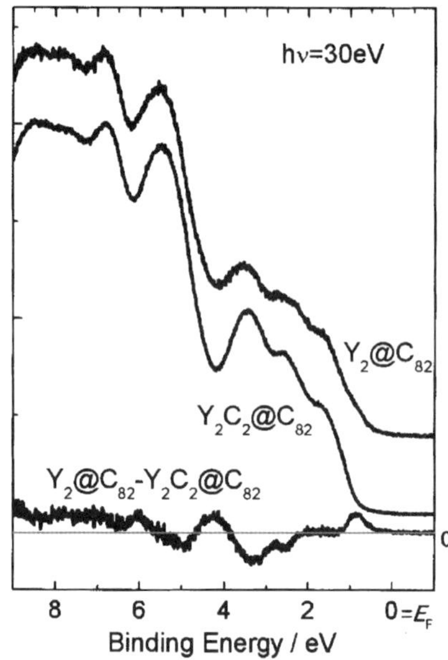

Figure 3   *Ultraviolet photoelectron spectra of $Y_2C_2@C_{82}$ (III) and $Y_2@C_{82}$ obtained with 30 eV photon energy, and their difference spectrum.*

5.2 Three $Y_2C_2@C_{82}$ Isomers And $Lu_2C_2@C_{82}$ (III)

Three $Y_2C_2@C_{82}$ isomers have been isolated and their absorption spectra are different each other, suggesting their different electronic structures.[12] Among them $Y_2C_2@C_{82}$ (II) has the longest wavelength optical absorption edge and it extends to more than2000 nm whereas those of the other two isomers are 1400 nm (isomer I) and 1000 nm (isomer III). This suggests that $Y_2C_2@C_{82}$ (II) has the smallest HOMO – LUMO (lowest unoccupied molecular orbital) gap and/or band gap among the three isomers. Their UPS are shown in Fig.4. The UPS spectral onsets of the isomers are 0.9 eV for isomer I, 0.65 eV for II and 0.8 eV for III. Isomer II has the smallest onset value, which might relate to optical absorption gap. However, the relations in isomers I and III in the absorption spectral edges and the UPS spectral onsets do not correspond each other. Again, this is an indication that optical absorption spectra do not always reflect the electronic structure of endohedral fullerenes.

The upper valence band UPS of these three isomers differ so much, which indicates their completely different π-electronic structures. We do not go into detailed discussion on these spectra, but this observation supports that the cage structure dominates the electronic structure of fullerenes even multiple atoms are encapsulated as was found in mono metal atom entrapped fullerenes.

The spectrum of $Lu_2C_2@C_{82}$ is also shown in Fig. 4. This $Lu_2C_2@C_{82}$ has C_{2v} symmetry, which is the same as that of $Y_2C_2@C_{82}$ (II). If the encapsulated metal species

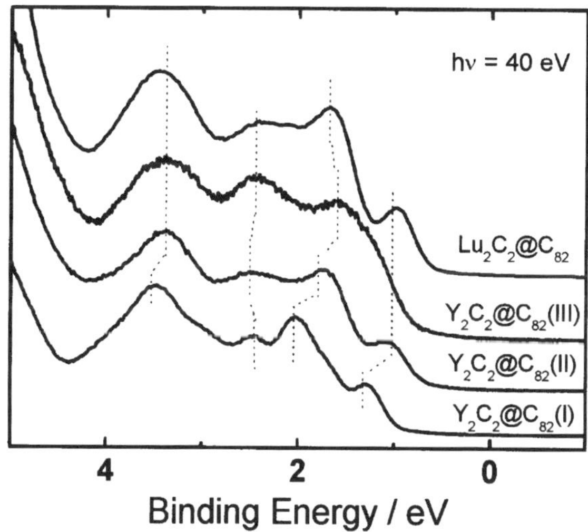

Figure 4   *Upper valence band UPS spectra of three $Y_2C_2@C_{82}$ isomers and C_{2v}-$Lu_2C_2@C_{82}$.*

affect the electronic structure of endohedral fullerenes, the UPS of $Lu_2C_2@C_{82}$ and $Y_2C_2@C_{82}$ (II) could have been different. However, these two spectra are almost identical. This is another support that the electronic structure of endohedral fullerenes are dominated by the cage structure itself.

5.3 Other Multiple Atoms Encapsulated Endohedral Fullerenes

Up to now the UPS of $Ti_2C_2@C_{78}$ [20] (it was thought to be $Ti_2@C_{80}$ but now this formula is decisive and have D_3 (5) symmetry, numerics is after nomenclature of Fowler and Manolopoulos) have been reported. $La_2@C_{78}$ has been also isolated and D_3 (5) structure was proposed for its cage structure. [21] The UPS of $La_2@C_{78}$ have been measured (not published yet), but its UPS are completely different from those of $Ti_2C_2@C_{78}$. From the conclusion described above, the origin of this difference was considered to be derived from the difference in their cage structure and validity of the symmetry analysis was questioned. Later, it has been found that theoretical calculation using the same D_3 (5) cage geometry with encapsulation of two Ti atoms and two C atoms or two La atoms in the cage is able to reproduce the UPS quite well. [22] These facts cast doubt on the universal validity on the conclusion that the cage structure itself dominates the electronic structure of endohedral fullerenes.

Analogous findings are also observed in the UPS of $Ti_2C_2@C_{82}$. The upper valence band UPS of $Ti_2C_2@C_{82}$ (C_{3v} symmetry) and $Y_2C_2@C_{82}$ (III) differ so much (Note that deeper σ-electron band regions are in good agreement). As Ti atoms usually form carbide at high temperature such as fullerene synthesis is performed, encapsulated Ti atoms may have strong bond with encapsulated carbon atoms as well as the cage carbon atoms. Hence, the electronic structures of $Ti_2C_2@C_{78}$ or $Ti_2C_2@C_{82}$ could be under strong influence of Ti – C bonding. That is, they could be exception to the conclusion. For the moment, it cannot be fully concluded that cage structure is the dominant factor that govern the electronic structure of multiple atoms encapsulated fullerenes. Further acquisition of UPS data of multiple atoms entrapped fullerenes is required to have a clear understanding of the electronic structure of these fullerenes.

Acknowledgements

This study was conducted as part of a Joint Research Program of UVSOR, Institute for Molecular Science. This work was supported by a Grant-in-Aid for Scientific Research in the Priority Area Molecular Conductors (No. 15073203) from the Ministry of Education, Science, Sports, and Culture, Japan.

References

1 for example, H. Shinohara, *Rep. Prog. Phys.* 2000, **63**, 843.

2 K. Komatsu, M. Murata and Y. Murata, *Science*, 2005, **307**, 238.

3 M. Saunders, R. J. Cross, H. A. Jimenez-Vazquez, R. Shimshi, and A. Khong, *Science*, 1996, **271**, 1693.

4 S. Stevenson, G. Rice, T. Glass, K. Harich, F. Cromer, M. R. Jordan, J. Craft, E. Hadju, R. Bible, M. M. Olmstead, K. Maitra, A. J. Fisher, A. L. Balch and H. C. Dorn, *Nature*, 1999, **401**, 55.

5   S. Hino, H. Takahashi, K. Iwasaki, K. Matsumoto, T. Miyazaki, S. Hasegawa, K. Kikuchi, Y. Achiba, *Phys. Rev. Lett.*, 1993, **71**, 4261.

6   S. Hino, K. Umishita, K. Iwasaki, T. Miyamae, M. Inakuma and H. Shinohara, *Chem. Phys. Lett.*, **300**, 1999, 145.

7   S. Hino, K. Iwasaki, K. Umishita, M. Aoki, K. Kobayashi, S. Nagase, T. John S. Dennis, T. Nakane and H. Shinohara, *Chem. Phys. Lett.*, **337**, 2001, 65.

8   K. Iwasaki, N. Wanita, S. Hino, D. Yoshimura, T. Okazaki, H. Shinohara, *Chem. Phys. Lett.*, 2004, **398**, 389.

9   S. Hino, K. Iwasaki, N. Wanita, D. Yoshimura, N. Ozawa, T. Kodama, K. Sakaguchi, H. Nishikawa, I. Ikemoto, K. Kikuchi, *Chem. Phys. Lett.*, 2005, **402**, 217.

10 S. Hino, N. Wanita, K. Iwasaki, D. Yoshimura, T. Akachi, T. Inoue, Y. Ito, T. Sugai, and H. Shinohara, *Phys. Rev. B, in press*.

11 T. Inoue, T. Tomiyama, T, Sugai, and H.Shinohara, *Chem. Phys. Lett.*, 2003, **392**, 226.

12 T. Inoue, T. Tomiyama, T. Sugai, T. Okazaki, T. Suematsu, N.Fujii, H. Utsumi, K. Nojima, and H. Shinohara, *J. Phys. Chem. B*, 2004, **108**, 7573.

13 P. W Fowler, D. E. Manolopoulos, An Atlas of Fullerenes, Oxford Press, Oxford, 1995.

14 T. Akasaka, T. Wakahara, S. Nagase, K. Kobayashi, M. Waelchli, K. Yamamoto, M. Kondo, S. Shirakura, S. Okubo, Y. Maeda, T. Kato, M. Kako, Y. Nakadaira, R. Nagahata, X. Gao, E. Van Caemelbecke, and K.M. Kadish, *J. Am. Chem. Soc.*, 2000, **122**, 9316.

15 T. Akasaka, T. Wakahara, S. Nagase, K. Kobayashi, M. Waelchli, K. Yamamoto, M. Kondo, S. Shirakura, Y. Maeda, T. Kato, M. Kako, Y. Nakadaira, X. Gao, E. Van Caemelbecke, and K.M. Kadish, *J. Phys. Chem. B*, 2001, **105**, 2971.

16 Z. Xu, T. Nakane and H. Shinohara, *J. Am. Chem. Soc.*, 1996, **118**, 11309.

17 T. Kodama, N, Ozawa, Y, Miyake, K, Sakaguchi, H, Nishikawa, I, Ikemoto, K, Kikuchi, and Y, Achiba, *J. Am. Chem. Soc.*, 2002, **124**, 1452.

18 Z. Shi, T. Okazaki, T. Shimada, T. Sugai, K. Suenaga, and H. Shinohara, *J. Phys. Chem. B*, 2003, **107**, 2485.

19 D. M. Poirier, M. Knupfer, J. H. Weaver, W. Andreoni, K. Laasonen, M. Parrinello, D. S. Bethune, K. Kikuchi, and Y. Achiba, *Phys. Rev. B.*, 1994, **49**, 17403.

20 K. Iwasaki, S. Hino, D. Yoshimura, B. Cao, T. Okazaki, and H. Shinohara, *Chem. Phys. Lett.*, 2004, **397**, 169.

21 B. Cao, T.Wakahara, T. Tsuchiya, M. Kondo, Y. Maeda, G. M Aminur Rahman, T. Akasaka, K. Kobayashi, S. Nagase, and K. Yamamoto, *J. Am. Chem. Soc.*, 2004, **126**, 9164.

22 M. Otani, S. Okada, and A. Oshiyama, The 28th Fullerene-Nanotubes General Symposium, Meijou University, Jan. 7-9, 2005, Nagoya, Japan. And private communication from S. Okada.

MANIPUTLATION AND ASSEMBLY OF CARBON NANOTUBES BY
DIELECTROPHORESIS

H.Z. Geng[1], J. Tang[2,3], G. Yang[2], and O. Zhou[1,2].

[1] Curriculum in Applied and Materials Sciences, University of North Carolina
Chapel Hill, NC 27599, USA
[2] Dept of Physics and Astronomy, University of North Carolina,
Chapel Hill, NC 27599, USA
[3] National Institute for Materials Science
Tsukuba, Ibaraki 305-0047, Japan

1 INTRODUCTION

Carbon nanotubes (CNTs) have become an important class of materials for both
fundamental studies and technological applications[1]. Considerable progress has been made
recently in understanding the formation mechanism and in developing synthesis routes for
fabrication of homogeneous CNTs with controlled structure and morphology.
Experimental studies have demonstrated unique properties and potentials for a wide range
of applications such as sensors, electron field emitters and microelectronics[1]. Eventual
utilization of these novel materials however requires new processes for controlled
assembly and integration, which is current lacking.

Dielectrophoresis is a versatile technique often used to manipulate and separate
colloidal particles based on the interaction between the alternating-current (AC) field and
the induced dipole moments[2,3]. It has been widely applied in biological systems[4]. Recently
it has been utilized to orient, purify and align CNTs,[5-7] and to make CNT interconnects.[8-10]

Here we report the results from a systematic study of the experimental conditions
including the viscosity and ionic conductivity of the medium, and the frequency and
amplitude of the AC field on CNT assembly. The feasibility of utilizing dielectrophoresis
to fabricate CNT fibrils and atomic force microscope (AFM) probes are demonstrated.

2 EXPEREMENTS AND RESULTS

2.1 Dielectrophoresis

When a neutral particle is placed in an electrical field that is spatially
inhomogeneous, equal and opposite charges are induced on the particle by the electrical
field. Because the field strength is different on the two sides of the particle, a net force acts
on the particle. This is termed as dielectrophoresis. The dielectrophoretic force $F(t)$ on a
particle is given by the following general expression[11]:

$$\vec{F}_{DEP} = (\vec{m}(t) \bullet \nabla)\vec{E}(t), \tag{1}$$

where m (t) is the polarizability of the particle and E (t) is the electrical field. A CNT can be considered as a rod with very large aspect ratio. When a rod shaped particle is placed in the electrical field, it will experience a torque, which tends to align one of the particle's axes with the field. The rod will align so that all components of torque are zero. The magnitude and orientation of the torque depend on the frequency of the applied field and the properties of the particle and the medium. The orientation of a rod shaped particle can be controlled by the conductivity of the medium and the frequency of the applied field[3]. The dielectrophoretic force on a rod whose major axis is parallel to the electric field is given by equation (2)[11].

$$\vec{F}_{DEP} = \frac{2\pi abc}{3}\varepsilon_m \, \mathrm{Re}\left\{\frac{\varepsilon_p^* - \varepsilon_m^*}{\varepsilon_m^*}\right\}\nabla \vec{E}^2, \tag{2}$$

where a,b,c are the dimensions and ε_p^* is the permittivity of the rod. ε_m^* is the permittivity of the medium. ε^* is given by equation (3)

$$\varepsilon^* = \varepsilon - j(\sigma/\omega), \tag{3}$$

where σ is conductivity and ϖ is the frequency of the AC field. From these equations, permittivity, conductivity of the suspending medium and the frequency of AC field are shown to be critical parameters as the properties of CNTs are not variables in our experiments. Single walled CNTs were used in this study. Because semiconducting and metallic tubes tend to bundle together, the whole bundle appears to be metallic.

2.2   General Study on Manipulation of CNTs by Dielectrophoresis

A systematic study was conducted to investigate the effects of parameters such as the AC frequency, conductivitys and viscosity on dielectrophoresis of CNTs. Gold electrode arrays were patterned on glass substrates by photo-lithography and thermal evaporation. The patterned electrode array was connected with a function generator. The experimental setup for in situ manipulation and observation is shown in Figure 1.

SWNTs synthesized by the laser ablation method at UNC were first purified, then etched into desired length (0.5 to 2 µm) using a strong acid. They were then dispersed in

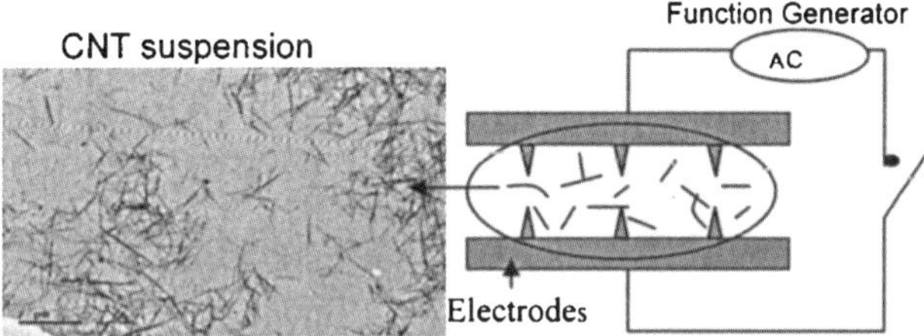

Figure. 1. *Manipulation of CNTs on a patterned gold electrode array by dielectrophoresis*

deionized water at the concentration of 0.1mg/ml. A droplet of the CNT suspension was applied on top of the electrode array. The AC field of 200V/CM was then applied. The movement of the CNTs was recorded by a CCD camera attached to an optical microscope.

Figure 2(a) shows several frames from a recorded movie revealing the CNTs at different stages of the assembly process under the applied AC field. In figure 2(b) a schematic is drawn according to the optical images.

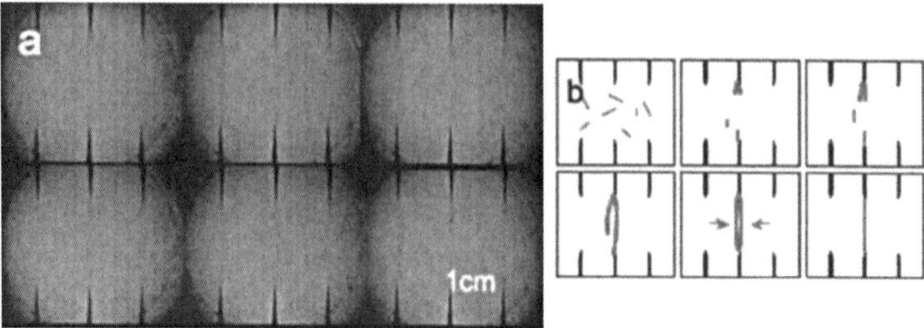

Figure 2   *The process of assembling CNT wires between two opposing electrodes.*

2.2.1 Effects of the AC Field Frequency. The experiment was conducted by applying a 2V AC field across two opposing metal electrodes that were in contact with a droplet of the CNT suspension. The AC field was turned on till the CNTs were assembled into a stable structure. The experiment was repeated using different AC frequencies varying from 20Hz to 2MHz. For consistence CNT suspensions from the same batch were used in all experiments. The results are shown in Figure 3. It was found that under low frequency (20Hz) the CNTs moved to the electrodes rapidly without aligning to the field direction and formed irregular patterns around the electrodes. When the frequency was increased to 2MHz, fine CNT structures were aligned along the AC field direction. In the middle frequency range (20KHz to 200KHz), the CNTs were found to have very low mobility and especially in the frequency range of 20K to 150K. No CNT structure was assembled even after the field was on for extended period of time.

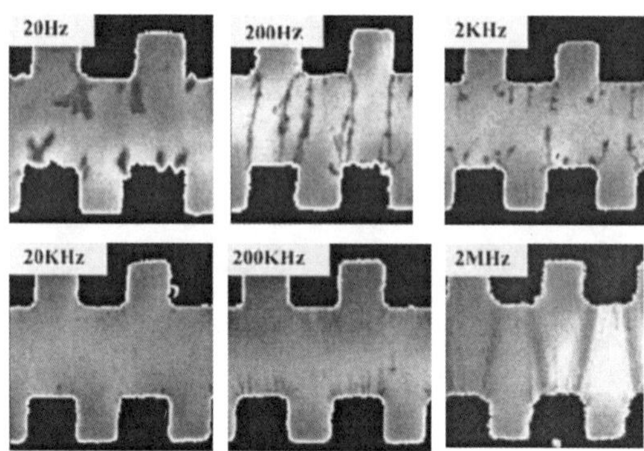

Figure 3   *A series of optical images showing different CNTs structures formed under different frequency AC field.*

2.2.2 Effects of the Ionic Conductivity of the Medium. The conductivity of CNT suspension was tuned by adding NaCl/water solution into the suspension. Experimental results showed that highly ordered CNT structures were fabricated in the suspending medium with low conductivity. However, when the conductivity of the medium increased the CNT structures formed became less ordered and when it reached 2S/m no CNT structures was observed. Experimental results are recorded in Figure 4.

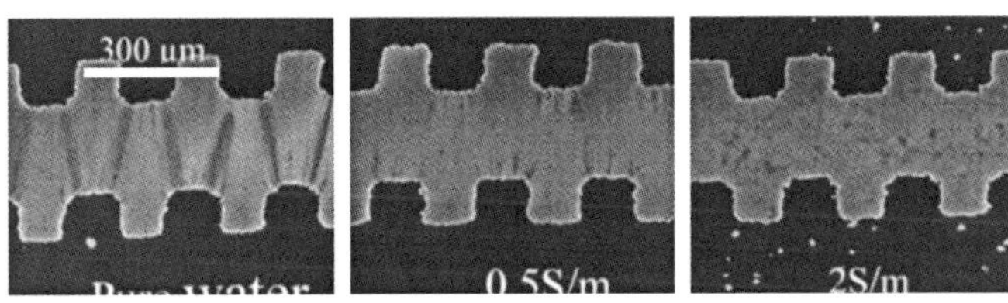

Figure 4   *A series of optical images showing different structures CNTs formed under the same AC field but different ionic conductivity of the medium.*

2.2.3 Viscosity of the Suspension. The viscosity of the medium was changed by adding glycerol into the suspension. Glycerol was mixed with the standard suspension at 1:1 ratio which gave the medium twice the viscosity of pure water. The frequency dependent experiment was repeated using this more viscous medium. Figure 5 shows the experimental result. When a CNT bundle in the suspension moves under the dielectrophoretic force to form certain structures, it also encounters a resistant force, which is the shearing force between the surface of the moving CNT bundle and the surrounding medium. When the viscosity of the medium is increased this resistant force is increased accordingly, which decreases the net force applied on the CNT bundle and slows down the movement of the CNT bundle. Experimental results show that under low frequency range (below 200 Hz), the resistant force prevented CNT bundles from moving rapidly and coagulating around edges of electrodes, therefore more ordered CNT structures were former. While under frequency higher than 200 Hz, no CNT structure was observed after the field was turned on for extended period of time. One explanation is that the slow movement of CNT bundles won't result in observable motions during the short pitch time when the AC field frequency is high.

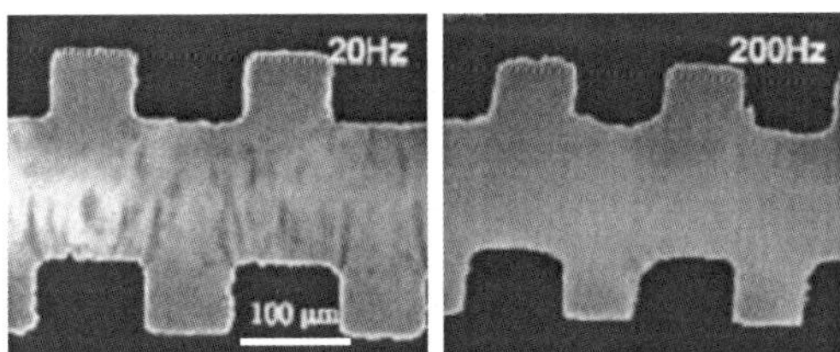

Figure 5   *Optical images showing CNT structures form in viscous medium under different frequencies.*

2.2.4 Different CNT Structures Fabricated by Dielectrophoresis The dielectrophoresis technique provides a convenient and efficient way to transfer CNTs from the liquid to a solid surface at predetermined locations with considerable control and accuracy without the use of photo or shadow mask. The technique can be used to fabricate micro-wires connecting pre-fabricated electrodes and oriented films. Figure 6 shows two examples of the CNT structures fabricated using this method.

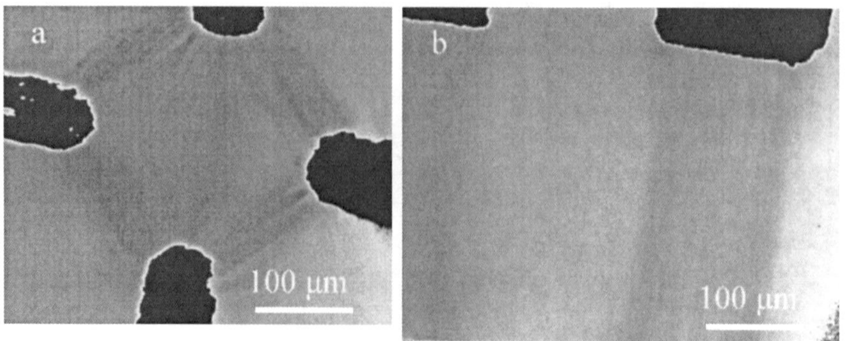

Figure 6   *Different CNT structures fabricated by using different electrode array configurations.*

2.3   Fabrication of Functional CNT Structures

Among all the CNT structures assemble under AC field, thin fibrils assembled on sharp probes were found to be of great interest[13-15]. Figure 7 shows the CNT fibrils attached to an etched tungsten probe[13] and a commercial AFM probe[15] by dielectrophoresis. This method gives fine control over the length and diameter of the fibril, and the capability of parallel processing. Field emission properties of the CNT fibrils assembled on tungsten probes have also been studied.[14] Their capability of delivering currents exceeding 10^5 A/cm^2 in density and their stability and lifetime make them promising for potential applications as field emission electron sources for precision vacuum electronic instruments such as electron microscopes. CNT fibrils attached to the commercial AFM probes were also studied and were found to provide the commercial Si probes with improved lateral resolutions and probing depth.[15]

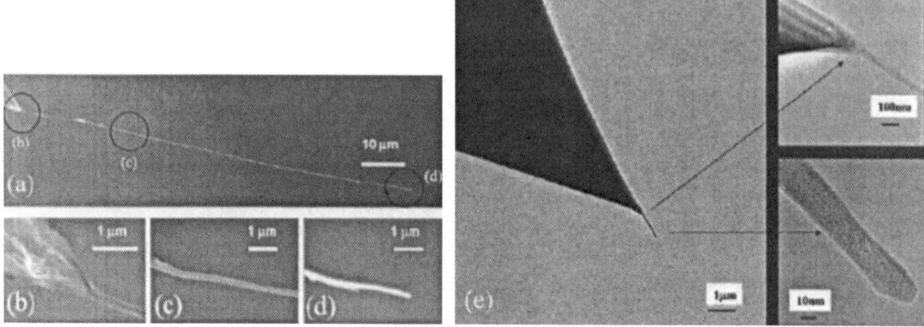

Figure 7   *SEM images of CNT fibrils attached to a tungsten probe and a CNT fibril attached to a commercial AFM probe.*

3 CONCLUSION

We showed that functional CNT structures can be assembled with considerable control and ease by dielectrophoresis. The interaction between the CNTs and AC field was found to depend sensitively on several parameters including the AC frequency, and the ionic conductivity and viscosity of the medium.

References

1 M.S. Dresselhaus, G. Dresselhaus, and P. Avouris, *Carbon Nanotubes: Synthesis, Structure, Properties, and Applications*. Topics in Applied Physics. Vol. 80. 2000, Springer-Verlag: Heidelberg.

2 H.A. Pohl, *Dielectrophoresis*. 1978, Cambridge: Cambridge University Press.

3 T.B. Jones, *electromechanics of Particles*. 1995, Cambridge: Cambridge Univ. Press.

4 M.P. Hughes, Electrophoresis, 2002(23): 2569-2582.

5 K. Yamamoto, S. Akita, and Y. Nakayama, J. Phys. D: Appl. Phys, 1998. **31**: L34.

6 X.Q. Chen, T. Saito, H. Yamada, and K. Matsushige, Appl. Phys. Lett., 2001. **78**(23): 3714-3716.

7 M.Diehl, S. N. Yaliraki, R. A. Beckmann, M. Barahona, J. R. Angew. Chem. 2002, 114, 363-366

8 L. A. Nagahara, I. Amlani, J. Lewenstein, R. K. Tsui, Appl. Phys. Lett. 2002, 80, 3826-3828

9 R. Krupke, F. Hennrich, H.b. Weber, D. Beckmann, O. Hampe, S. Malik, M. M. Kappes, H.v. Lohneysen, Appl. Phys. A 2003, 76, 397-400

10 R. Krupke, F. Hennrich, H. B. Weber, M. M. Kappes, H.v. Lohneysen, Nano Letters, 2003, 3, 1019-1023

11 H. Morgan, N. G. Green. Journal of Electrostatics 42, 1997, 279-293

12 O. Zhou, H. Shimoda, B. Gao, S.J. Oh, L. Fleming, G.Z. Yue, Acc. Chem. Res. Vol. 35, 2002, 1045-1053.

13 J. Tang, B. Gao, H. Geng, O. D. Velev, LC. Qin, O. Zhou. Adv. Mater. 2003, 15, No.16, 1352-1355.

14 J. Zhang, J. Tang, G. Yang, Q. Qiu, LC. Qin, O. Zhou, Adv. Mater. 2004, 16, No. 14, 1219-1222

15 J. Tang, G. Yang, Q. Zhang, A. Parhat, B. Maynor, J. Liu, LC. Qin, O. Zhou. Nano. Lett. 2005, 5, No.1, 11-14.

Acknowledgement

This work was partially supported by a NASA URETI on Bio Inspired Materials.

SUPERCONDUCTING PHASE MADE FROM C_{60} DOPED WITH LANTHANUM

M. Akada[1], T. Hirai[2], J. Takeuchi[2], N. Hiroshiba[1], R. Kumashiro[1], T. Yamamoto[1] and K. Tanigaki[1]

[1]Department of Physics, Graduate School of Science, Tohoku University, and CREST-JST 6-3 Aoba Aramaki Aoba-ku, Sendai, Miyagi 980-8578, Japan
[2]Department of Materials Science, Graduate School of Science, Osaka City University, 3-3-138 Sugimoto, Sumiyoshi 558-8585, Japan

1 INTRODUCTION

After finding of the first reports on conductivity[1] and superconductivity[2] in C_{60} fullerides, a variety of alkali and alkaline-earth metal-doped C_{60} fullerides have been found to be conducing and superconducting.[3] The studies of C_{60} fullerides have been extended to ones with divalent rare-earth elements like Eu,[4] Yb,[5] and Sm[6] and the details of their structure and physical properties have eventually been clarified. Later, C_{60} was also reported to show superconductivity when La was doped by the pioneering work of Kortan *et al.*[7] in 1996. This was the first situation that superconductivity was found when the trivalent atom was used. Furthermore, the superconducting transition temperature T_c of 12 K was relatively high among rare-earth metal C_{60} fullerides, the T_cs of which generally range from 6 to 8 K.[8–10] However, the fraction of this superconducting phase was too small to determine the real superconducting phase at that time as was reported and many years have passed without elucidating what actually occurs, since preparations of the superconducting phase were very difficult.

Because it is very intriguing and essential in science to understand the details of the compound showing superconductivity when C_{60} is doped by the trivalent La, we have decided to revisit on this system. Better quality samples with high-resolution x-ray diffraction (XRD) studies have led us to the conclusion that the real superconducting phase with T_c of 12 K made from La- C_{60} should be ascribed to a carbide but not to any C_{60} fullerides. This is the first situation encountered as a fact that the superconductivity found from C_{60} is not coming from the C_{60} fullerides.

2 EXPERIMENTAL

A various methods were employed to synthesize La-C_{60} compounds. The first method was using thermal decomposition of La azide. Although the heating temperature can be lowered, crystallinity was not improved. An ammoniation method was also tried, but La was not dissolved in liquid ammonium. The best samples were prepared by mixing and pelletizing various stoichiometric quantities of C_{60} and La in an Ar glove box. Especially pelletization and grinding after heating were essential to have high-quality samples. The prepared pellets were then introduced into a tantalum cell followed by being sealed in a quartz tube in *ca.* 400 Torr under He atmosphere. The La- C_{60} compounds were then heated at various temperatures of 873 to 943 K for the purpose of an increasing superconducting fraction by suppressing the formation of by-product compounds. Because some other existing crystal phases may be contaminated to show superconductivity, the conventional carbide phases made from La and C reported in the literature were also prepared. The prepared samples were subjected to the magnetic susceptibility measurements for determining superconductivity using MPMS (Quantum Design) and to high resolution XRD measurements for crystal structural analyses at the BL02B2 beamline port in the high-energy facility of SPring-8.

3 RESULT & DISCUSSION

When La is doped to C_{60}, superconductivity appears and this can clearly be detected by the superconducting diamagnetic susceptibilities as shown in Fig. 1. The highest onset temperature of 12 K was the same value that was first reported by Kortan *et al.*[7] After repeating experiments many times, we have realized that the signal is composed of mainly two superconducting critical temperatures, the intensities of which change depending both on the stoichiometries of La to C_{60} and on the preparation temperatures. It should be noted here that the two superconducting phases were possible from these reactions as was described later, and therefore the fitting using two phases was very reasonable. One of the typical situations can be viewed in Fig. 1. The observed magnetic susceptibility curve as a function of temperature was deconvoluted into two signals, one of which corresponds to the lower-temperature superconducting phase with an onset of 6 K and the other the higher-temperature superconducting phase at 12 K.

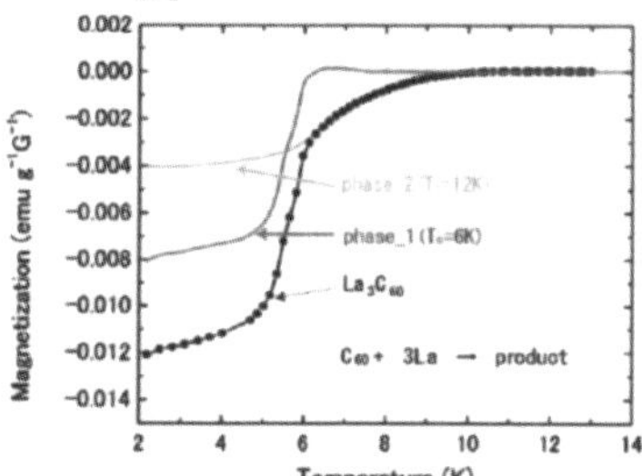

Figure 1. *Superconducting diamagnetic susceptibilities as a function of temperature for the samples with nominal compositions of La$_3$C$_{60}$ in the feed. The observed signal can be deconvoluted into the two phases labeled as phase1 T_c=12 K and phase 2 T_c=6 K.*

After many trials to make samples showing a large superconducting fraction from C_{60} and La, we have focused on the two key samples, and have made more detailed analyses on them. It is interesting to point out that the carbides of both La_2C_3 and the unknown one increased as intercalation reactions proceed and the starting material of C_{60} decreases, as shown in Fig. 2. Corresponding to the increase in the signals of these carbides, the total superconducting fractions also increased. The disappearance of C_{60} and the appearance of both carbides have a good correlation with the heating temperature and the heating time. Although the superconducting fraction can be increased to *ca.* 10% in total (the phase with T_c=12 K plus the one with T_c=6 K), the heating continuation has not led to a sufficient amount of the superconducting phase of T_c=12 K because reactions to amorphous states also proceed as the heating time increases.

The known existing crystal phase of pristine C_{60}, La, and La_2C_3 were removed after three-phase Rietveld fittings from the whole XRD spectrum. As indicated by the downward arrows in Fig.2, some unknown peaks still remained. These remaining peaks

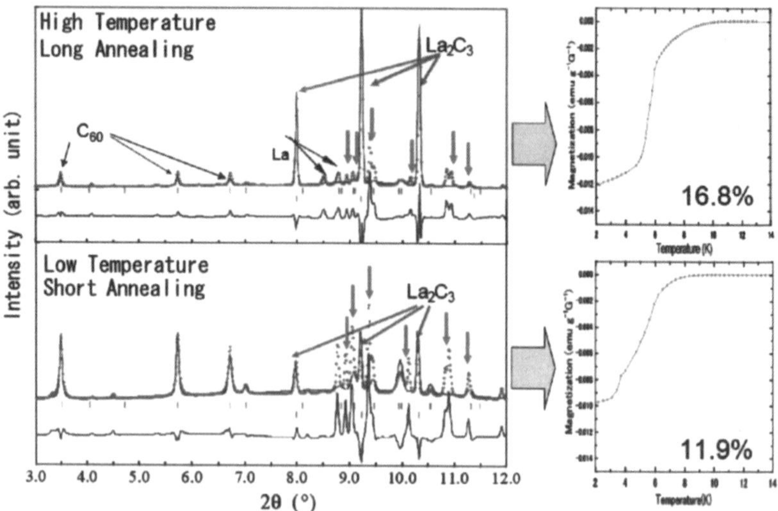

Figure 2. *XRD spectra and Superconducting diamagnetic susceptibilities for the compounds prepared from 3 La and 1 C_{60} in the feed. The wavelength of x-ray is 0.05010 nm. Two preparation conditions were employed: the high-temperature 943 K with a long annealing time 9 h and the low temperature 823 K with a short annealing time 6 h.*

will most likely result from some other carbides, considering the positions of the diffraction peaks. When the ratio of the intensity of La_2C_3 to that for the remaining carbide increases by changing the heating conditions from low temperatures to high temperatures, the superconducting fraction of T_c=12 K also increased, as can be seen in this figure. This seems to imply that the superconductivity of T_c=12 K shall be ascribed to La_2C_3 and that of T_c=6 K may come from the unknown carbides.

While repeating experiments, we have also realized that exactly the same situation can surprisingly be obtained when the reactions start from amorphous carbon with La. As shown in Fig.3, the reactions of amorphous carbon with La can also produce the superconductivity both at 6 and 12 K. It is emphasized that we have never obtained any superconductors when the La carbides do not form in all of our experiments. This is

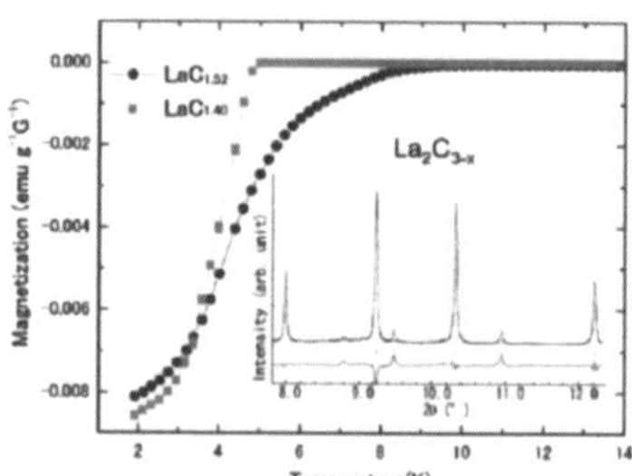

Figure 3. *Superconducting diamagnetic susceptibilities of La carbides prepared from La and carbon powder: $LaC_{1.52}$; heated at 1273 K for 72 h and $LaC_{1.40}$; heated at 1273 K for 8 h. The inset is the XRD measured using the wavelength of 0.05011 nm.*

completely different from the situations so far encountered for the superconducting C_{60} fullerides.

In literature searches, we have found three important studies as for the La-C carbide compounds. Giorgi *et al.*[11] first reported in 1968 that the superconductivity was observed in La_2C_3 with T_c ranging from 5.9 to 11.0 K. In 1980, Novokshonov *et al.*[12] also reported that La_2C_3 superconducts with a T_c of 11 K. Afterward, Reinhard *et al.*[13,14] reported in 2004 that La_2C_3 superconducts at 13.2 K, and that another unknown carbide showing superconductivity at T_c=5.5 K is produced when the carbon content is reduced from this ideal stoichiometry. Our experimental data look like similar to the report by Reinhard *et al.*, i.e., the superconductivity at 12 K is ascribed to the La_2C_3 carbide phase and the one at T_c=6 K to another unknown carbide.

The stoichiometry of $LaC_{1.52}$ in the feed gave a sufficient amount of superconducting fraction for the La_2C_3 crystal phase with T_c=12 K. The Rietveld fitting employing the crystal data of La_2C_3 in the literature[15] clearly indicates that La_2C_3 is produced, as shown in the inset of Fig.3. The refinement parameters were a=0.8803 nm with Rp factor of 11.11% in the space group $I\bar{4}3d$.[16] When the less carbon contents were used, a single superconducting transition of T_c=6 K was still observed with a residue of the superconductivity of La. On the other hand, larger contents of carbon in the feed resulted in the appearance of the higher transition signal of T_c=12 K. These results imply that the superconducting phase with T_c=6 K is most likely ascribed to a similar structure of La_2C_3 with certain deficiencies.

4 CONCLUSION

The superconductivity with T_c of 12 K reported in the reactions of C_{60} with La was confirmed, but it was found that the true superconducting phase is not ascribed to any C_{60} fullerides. [17] Instead, the detailed studies apparently showed that the superconductivity results from the La carbides that are most likely related to the La_2C_3 phases.

References

1   R. C. Haddon, A. F. Hebard et al., Nature **350**, 320 (1991).
2   A. F. Hebard, M. J. Rosseinsky et al., Nature **350**, 600 (1991).
3   K. Tanigaki, I. Hirosawa et al., Nature **356**, 419 (1992).
4   H. Yoshikawa, S. Kuroshima et al., Chem. Phys. Lett. **239**, 103 (1995).
5   B. Xia, M. W. Ruckman, and M. Strongin, Phys. Rev. B **48**, 14623 (1993).
6   D. Claves, Y. Ksari et al., Solid State Commun. **99**, 359 (1996).
7   A. R. Kortan, N. Kopylov, and E. Ozdas, *Proceedings of ECS Meeting, Fullerenes*, 1996, Vol. 3, pp. 423–428.
8   B. Gogia and K. Tanigaki, Phys. Soc. Meeting, Japan, 1998.
9   E. Ozdas, A. R. Kortan et al., Nature **375**, 126 (1995).
10   X. H. Chen and G. Roth, Phys. Rev. B **52**, 15534 (1995).
11   A. L. Giorgi, E. G. Szklarz et al., J. Less-Common Met. **17**, 121 (1969).
12   V. I. Novokshonov, E. P. Khlybov et al., Russ. Metall. **3**, 167 (1980).
13   K. Reinhard, R. K. Kremer, and A. Simon, Curr. Appl. Phys. **4**, 563 (2004).
14   X. Wang, I. Loa, K. Syassen et al., cond-mat 0503567, 2005.
15   Cerius2 package program database.
16   F. Izumi and T. Ikeda, Mater. Sci. Forum **198**, 321 (2000).
17   Misaho Akada, Toshinari Hirai et al., Phys. Rev. B, **72**, 132505 (2005).

ELECTRIC TRANSPORT PROPERTIES OF DOPED C_{60} NANOWHISKERS

R. Kumashiro[1], K. Tanigaki[1,2], N. Miyata[1], H. Ohashi[1], T. Fujiki[1], Y. Gao[3], Y. Murata[3], K. Komatsu[3], T. Akasaka[4], J. Minato[5] and K. Miyazawa[5]

[1] Department of Physics, Graduate School of Science, Tohoku University, 6-3 Aoba Aramaki Aoba-ku, Sendai, Miyagi 980-8578, Japan
[2] CREST, Japan Science and Technology Agency, 4-1-8 Honcho, Kawaguchi, Saitama 322-0012, Japan
[3] Institute for Chemical Research, Kyoto University, Kyoto 611-0011, Japan
[4] Center for Tsukuba Advanced Research Alliance, University of Tsukuba, Tsukuba, Ibaraki 305-8577, Japan
[5] Advanced Materials Laboratory, National Institute for Material Science, 1-1 Namiki, Tsukuba, Ibaraki 305-0044, Japan

1 INTRODUCTION

Fullerenes, carbon nanotubes and the related materials are greatly expected as new materials for electronic devices using nano technology. They have specific structures in the nano-scale size, and show a variety of electronic properties depending on the crystal structure as well as their chemical compositions. In addition, these materials will provide feasibility for controlling accurate structure in nano scale. Being motivated by these situations, many researchers have been making intensive efforts for investigation and elucidation of physical and chemical properties of these materials.

Recently, it was reported that crystallization of C_{60} using the interfacial boundary in solution of two kind of solvents, such as toluene and isopropyl alcohol, yields needle like precipitates consisting of C_{60}[1,2]. The structural observation using an electron microscope reveals that the precipitates have quasi one-dimensionality and a variety of needles with different diameters can be produced. These materials are so called C_{60} nanowhiskers[3-6] and their electronic properties are worth to investigate. So far, electronic properties of fullerene solids and the related materials upon chemical and physical doping have been studied by many researchers. It has been understood that the C_{60} solid is semiconducting and that the resistivity ρ of the pristine C_{60} ranges from 10^8 to 10^{14} $\Omega\,$cm in solids and thin films.

In this paper, we present the electronic properties of C_{60} nanowhiskers when they are intercalated by I_2.

2   METHOD AND RESULTS

2.1 Samples and Measurements

C_{60} nanowhisker samples were grown by a liquid-liquid interfacial precipitation method using toluene and isopropyl alcohol[3,4]. The obtained precipitates were thoroughly washed, and dried at room temperature. C_{60} bulk crystals were removed carefully, and only C_{60} nanowhisker crystals were used for measurements. The C_{60} nanowhisker sample was pelletized (ca.1.0x0.5x0.02mm) under hydrostatic pressure of 100MPa, and Au electrodes were fabricated on the pellet surface by an Au evaporation method. The pellet sample was then set into a glass cell for electric conductivity measurements, succeedingly degassed at 393K in vacuo.

After these pretreatments, small amounts of I_2 solid were set into the glass cell under an inert atmosphere, and I_2 doping in vapour phase was carried out at 300K and 473K, respectively.

Two-probe electric measurements were carried out at temperatures from 80K to 300K, using an ultra-high resistance meter (ADVANTEST R8340A). Conductivity of the samples was monitored during the temperature change.

The samples prepared in the same way are stuffed into glass capillary tubes to be subjected to the Xray measurements. Pretreatments and I_2 doping procedures were carried out in the same manner for resistivity measurements. The capillary tubes were set to a large Debye-Scherrer Camera at SPring-8 BL02B2. The $1.0\,\text{Å}$ wavelength was used for achieving high angular resolution. XRD spectra were collected in the temperature range from 92K to 390K.

2.2 Temperature Dependence of Resistivity

The nanowhiskers were exposed to I_2 vapour for five minutes at room temperature and the resistivity has been monitored at different temperatures. Figure 1 shows a temperature dependence of the electric resistivity of this sample. The initial resistivity was $5.4\text{x}10^7$ Ω cm and greatly increased with the exposure time to I_2 vapour. It seems that the gradient of ρ vs. T changes at three temperature points of 190, 220, and 240K as shown in Fig.1, although it is difficult to judge at moment whether these changes are intrinsic or not.

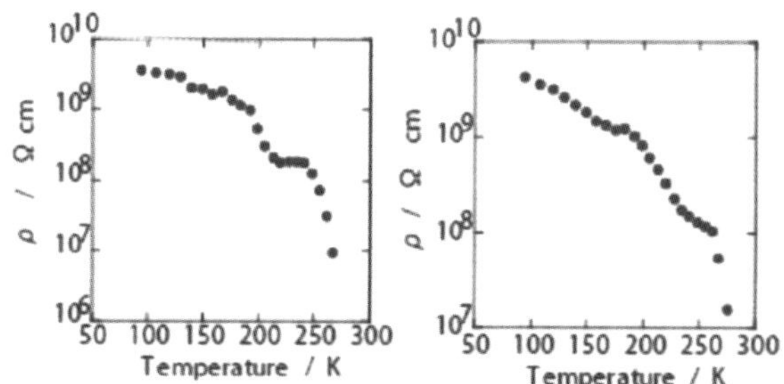

Figure 1   *Resistivity as a function of temperature of the C_{60} nanowhisker; exposed to I_2 at room temperature for 5 minutes (left) and 60 minutes (right).*

Another similar experiment has been carried out for the C_{60} nanowhisker sample exposed to the I_2 vapour for 60 minutes at room temperature. The resistivity became to be 7.0×10^6 Ω cm and this was lower than that of the sample with 5min exposure time. Importantly, similar changes in the ρ -T curve were detected, although the temperatures showing the changes in the gradient of the curve were slightly different from those observed for the 60 min exposed sample. They were 180, 240, and 260K as shown in the right of Fig.1.

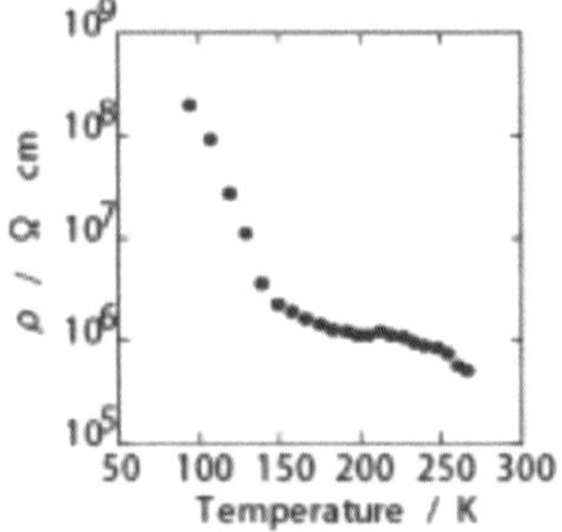

Figure 2   *Resistivity as a function of temperature of the C_{60} nanowhisker; doped with I_2 at 473K.*

From the two experiments described above, the doping level seems to exert a certain influence on the temperature dependence of resistivity. Therefore, the third experiment has been carried out at a higher temperature of 473K. Figure 2 shows the temperature dependence of electric resistivity of this sample. The value of ρ measured at room temperature was 2.1×10^6 Ω cm and this is the lowest among the three samples described in the present work. Although the resistivity decreased, the temperature dependence shows that this is not metallic but is still semiconducting. As seen in this figure, the temperature showing a change in the gradient at the lower temperatures seems to shift further to a low temperature with holding the change at 260K and the change observed around at 200K seems to have smeared out.

2.3 XRD Results; Doping Level and Temperature Dependence

To understand the temperature dependence of resistivity in C_{60} nanowhiskers in detail upon I_2 vapour exposure, the crystal structure of the I_2 doped nanowhisker has been studied as a function of temperature. Figure 3 shows XRD patterns of the pristine and I_2 doped C_{60} nanowhisker samples measured at various temperatures ranging from 92 to 390K. Three major peaks were observed at the low 2θ region (6, 10, 12°) for the pristine C_{60} nanowhisker sample, which are similar to those of the C_{60} crystal lattice. However, in addition to these diffraction peaks at low angles, new peaks appeared in the relatively higher 2θ region.

Taking into account the development of the peak intensities with increasing I_2 doping level, it can be considered that these new peaks are attributed to the structural changes of C_{60} nanowhiskers.   The lattice of the C_{60} nanowhiskers would be deformed by the intercalation of I_2 molecules. Actually, at 92K five sharp peaks can additionally be observed (15, 16, 17, 17.5, 18°). When temperature increases to 150K, the relative intensity ratio of these

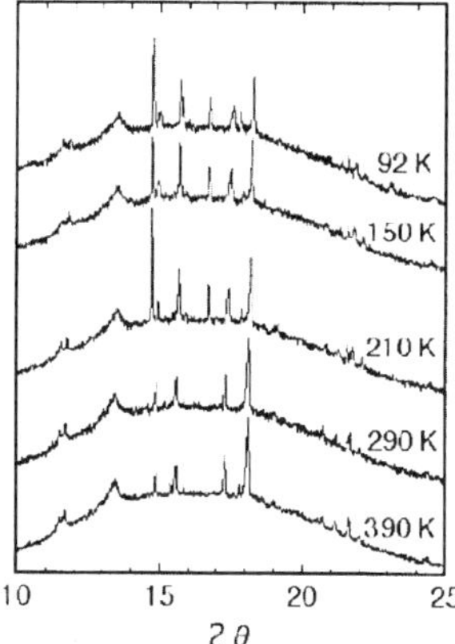

Figure 3 *XRD spectra of I_2-doped C_{60} nanowhisker as a function of temperature.*

peaks slightly changes.   By increasing temperature further to 210K, the intensity ratio again returns to the state originally observed at 92K. With raising the temperature, diffraction pattern showed a drastic change (Fig.3). A peak positioned at 17 degree in 2θ disappeared and an intense peak at 15 degree reduced in its intensity. The latter change is most likely associated with a low temperature ordered phase (sc, space group Pa3) to a high temperature disordered one (fcc, space group fm3m).

3 CONCLUSION

Electric transport of the C_{60} nanowhiskers was studied as a function of temperature. When the C_{60} nanowhiskers were exposed to I_2 vapor, the conductivity was greatly increased, but its temperature dependence indicated its semiconductive nature. Temperature dependence of the conductivity showed that there are two or three temperatures relation changes and they depend on I_2 doping level. Because Xray diffraction pattern upon temperature changes, the observed change in the gradient of electric transport as a function of temperature could be ascribed to the crystal phase transitions. Further detailed studies on the other related compounds will be helpful for having a more detailed understanding.

References

1 K. Miyazawa, A. Obayashi, and M. Kuwabara, *J. Am. Ceram. Soc.* **84**, 3037 (2001).
2 K. Miyazawa, Y. Kuwasaki, A. Obayashi, and M. Kuwabara, *J. Mater. Res.* **17**, 83 (2002).
3 K. Miyazawa and K. Hamamoto, *J. Mater. Res.* **17**, 2205 (2002).
4 K. Miyazawa, *J. Am. Ceram. Soc.* **85**, 1297 (2002).
5 K. Miyazawa, M. Akaishi, Y. Kuwasaki, and T. Suga, *J. Mater. Res.* **18**, 166 (2003).
6 M. Tachibana, K. Kobayashi, T. Uchida, K. Kojima, M. Tanimura, and K. Miyazawa, *Chem. Phys. Lett.* **374**, 279 (2003).

Variable range hopping conduction and magnetic anomalies in ordered mesoporous carbon materials CMK-n

S. Kasamatsu[1], E. Negishi[1], N. Toyota[1], R. Ryoo[2], O. Terasaki[3]

[1] Physics Department, Graduate School of Science, Tohoku University, Sendai, Japan
[2] National Creative Research Initiative Center for Functional Nanomaterials and Department of Chemistry, Korea Advanced Institute of Science and Technology, Daejeon, South Korea
[3] Structural Chemistry, Arrhenius Laboratory, Stockholm University, Stockholm, Sweden

1 INTRODUCTION

Ordered mesoporous carbon materials synthesized using a mesoporous silica as a template have attracted much attention for the development of new catalysts, adsorbents, and electrodes because of their various pore structures. Since the physical properties have been little known for these materials, we have measured the electrical resistance, the magnetoresistance and the magnetic susceptibility of CMK-n (n = 3, 3G, 5), where all the CMK-n are the ordered mesoporous carbon materials synthesized by using SBA-15 (2-dimentional hexagonal, *p6mm*) as a template.[1] CMK-3 consists of the carbon rods with amorphous carbons arranged in a hexagonal formation. We have measured three kinds of CMK-3. We will call them CMK-3-a, CMK-3-b and CMK-3-c, whose diameters of a carbon rod are 9 nm, 6 nm and 5 nm, respectively. CMK-3G also consists of carbon rods with discoid graphen sheets stacked along the rods. The local structure of CMK-3G is more ordered than that of CMK-3. The diameter of a CMK-3G carbon rod is 9 nm. CMK-5 is formed by carbon pipes arranged in a hexagonal formation. Each carbon rod and pipe is interconnected to maintain the hexagonal structure. The details of measured samples are shown in Table 1.

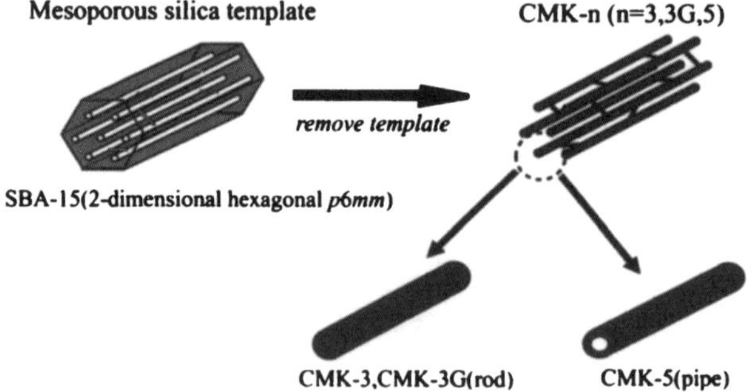

Figure 1 *Schematic illustration of the structure of CMK-3, CMK-3G and CMK-5*

Table 1 *Outline of measured samples*

CMK-n	CMK-3-a	CMK-3-b	CMK-3-c	CMK-3G	CMK-5
Template	SBA-15	SBA-15	SBA-15	SBA-15	SBA-15
Carbon source	sucrose	sucrose	sucrose	acenaphthene	furfuryl alcohol
Type	rod	rod	rod	rod	pipe
Diameter [nm]	9	6	5	9	9

2 EXPERIMENTAL

All CMK-n samples are the fine powder. Therefore, we have sintered all samples at 573 K for four hours in the air to measure dc electrical resistivity by a four-probe method. This sintering condition is determined by the thermogravimetric curves and X-ray diffractions before and after pressurization.[2] The magnetoresistance has been measured with magnetic field H up to 8 T applied transversely to the electrical currunt direction. The magnetic susceptibility measurements have been carried out with a commercial SQUID magnetometer (Quantum Design, Inc.) on both non-degassed samples and degassed sample. The non-degassed sample was sintered under the same condition as one for the electrical resistivity measurement. Then it was kept in the gelatin capsule for the measurement. The degassed sample was covered by the commercial wrap in the He atmosphere after vacuuming the sample at 573 K for a day.

3 RESULTS AND DISCUSSIONS

3.1 Electrical Resistance and Magnetoresistance

The electrical resistance in all the CMK-n follows not a thermal activation-type dependence but a variable range hopping (VRH) conduction.[3] It is expressed as

$$\rho = \rho_0 \exp[(T_0/T)^{1/(n+1)}] \tag{1}$$

where n represents the dimension of a given electronic system. We have analysed the electrical resistivity using VRH model. In the present analysis, we concentrate on only lower temperature region than the kink temperature. The best fit is obtained with $T^{-1/3}$ as shown in Figure 2, which means $n = 2$.

Next the magnetoresistance (MR) is shown in Figure 3 for CMK-3-a as an example. All the other samples show very similar magnetoresistance. The wavefunction of the electrons shrinks due to the external magnetic field. Therefore, the hopping probability decreases because the localization radius gets smaller than that in absence of magnetic field, resulting in a positive magnetoresistance. We have analyzed the magnetoresistance using VRH model. In VRH model, the resistivity at low magnetic field obeys the following equation,

$$\ln \frac{\rho(T,B)}{\rho(T,0)} = \gamma_n B^2 \tag{2}$$

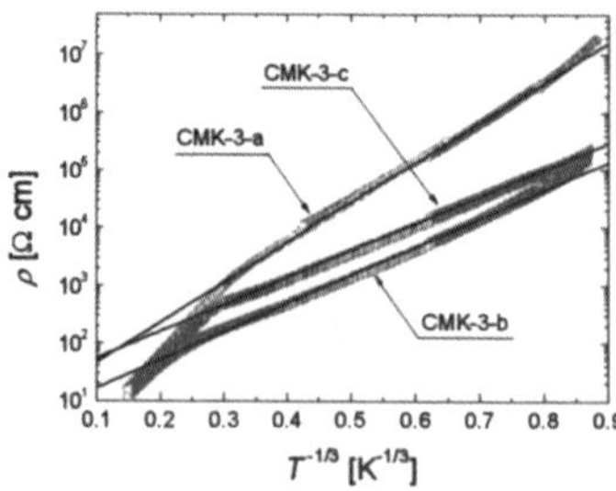
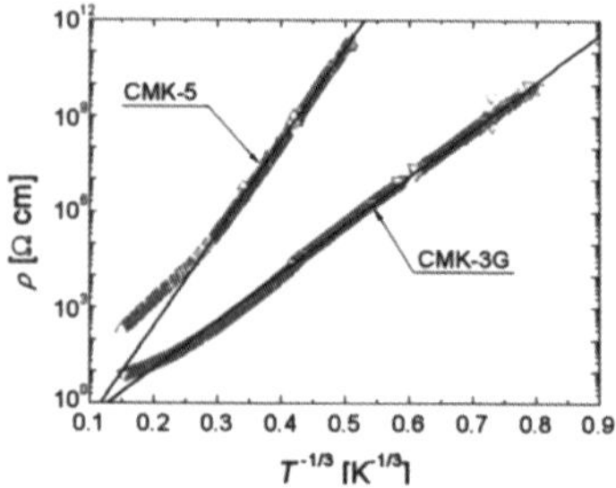

Figure 2 *Temperature dependence of the electrical resistivity for each samples. The points show the experimental data and the straight line shows the fitting results using VRH model with n=2*

where

$$\gamma_n = t_n \frac{e^2 \alpha^4}{\eta^2} \left(\frac{T_0}{T}\right)^{3/(n+1)} \tag{3}$$

Here, α is a localization radius of charge carriers. For a two dimensional system,[4] $t_2 = 2.8 \times 10^{-3}$. Figure 3 shows the data of magnetoresistance for CMK-3-a in comparison with Eq. (2). At high temperatures 4.2K and 10K, the B^2 dependence holds up to 8T, while the data above 4T at 1.6K deviate downward from the B^2 dependence, indicating a saturation behavior. The saturation-like behavior at high fields and low temperatures may be attributed to Kamimura type spin-dependent VRH taking the effect of conduction electron spins into account.[5]

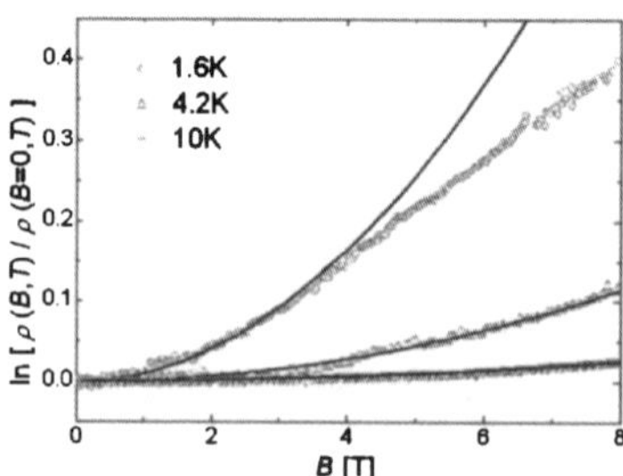

Figure 3 *Analysis of magnetoresistance of CMK-3-a at various temperatures. The straight line shows the fitting result with the function proportional to B^2*

We have estimated some physical parameters from data fitting for all CMK-n.[4,6] As an example, we describe the parameters of CMK-3-a at 4.2K. The localization radius of conduction electrons α is about 4 nm. A mean hopping length of the conduction electrons is about 8.8 nm. A density of localized state at Fermi level is about 2.3×10^{18} eV^{-1} m^2. The hopping length becomes longer when temperature lowers. The details of these parameters will be obtained in the paper we will submit in the future. We have concluded the hopping conduction between carbon rods is dominant at low temperature.

3.2 Magnetic Susceptibility

Figure 4 shows the temperature dependence of magnetic susceptibilities, χ, for non-degassed CMK-3-a. The cusp around 70K has been observed in all non-degassed CMK-n with increasing temperature under both zero field cool (ZFC) and field cool (FC) condition. In addition, the cusp is enhanced under FC. However, the obvious cusp cannot be observed with decreasing temperature. Here, it is noted that a solid oxygen has a structural phase transition from γ- to β-phase around 43K.[7] Therefore, we prepared the degassed samples to remove oxygen effect. In the degassed samples, the cusp has been smaller. The magnetization curves of all the degassed samples can be fitted by a Brillouin function with $S = 1/2$, which means the spin comes from the dangling bonds or localized π-electrons of CMK-n, not from oxygen. We have estimated the spin density N_s. For example, N_s of CMK-3-a, CMK-3G and CMK-5 are 2×10^{19} g^{-1}, 1.5×10^{19} g^{-1} and 6×10^{19} g^{-1}, respectively.

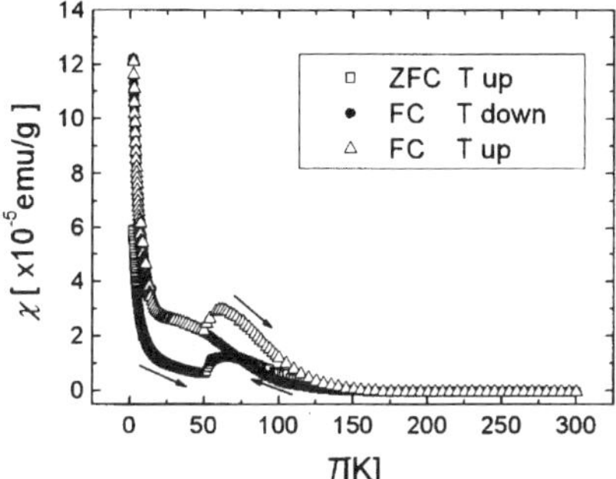

Figure 4 *Temperature dependence of magnetic susceptibilities for non-degassed CMK-3-a. Open squares and triangles means ZFC and FC data with increasing temperature. Solid circles indicate FC data with lowering temperature*

References

1   For review, Ryong Ryoo, *Mesoporous crystals and related nano-structured materials*, edited by O. Terasaki, Elsevier B.V., 2004
2   Tae-Wan Kim, In-Soo Park and Ryong Ryoo, *Angew. Chem. Int. Ed.*, 2003, **42**, 4375
3   N.F. Mott, *Metal-Insulator Transition*, Taylor and Francis, 1974
4   K.G. Lisunov, B. Raquet, H. Rakoto, J.M. Broto, E. Arushanov, X.Z. Xu, H. El Alami and C. Deville Cavellin, *J. Appl. Phys.*, 2003, **94**, 5912
5   A. Kurobe and H. Kamimura, *J. Phys. Soc. Jpn.*, 1982, **51**, 1904
6   B.I. Shklovskii and A.L. Efros, *Electronic Properties of Doped Semiconductors*, Springer solid-state science 45
7   U. Köbler and R. Marx, *Phys. Rev. B*, 1987, **35**, 9809

MAGNETIC PROPERTIES OF HYDROGENATED AND FLUORINATED SURFACE LAYER OF DIAMOND NANOPARTICLE

M.V. Baidakova [1], V.Yu.Osipov [1,3], C. Katsuyama [1], K. Takai [1], T. Enoki [1], A. Yonemoto [2], H. Touhara [2], A.Ya. Vul'[3]

[1] Department of Chemistry, Tokyo Institute of Technology, Japan
[2] Department of Materials Chemistry, Shinshu University, Ueda, Nagano, Japan
[3] Ioffe Physico-Technical Institute, St.Petersburg, 194021 Russia

1 INTRODUCTION

Nanodiamond (ND) produced today in large amounts by the detonation method is one of the new kinds of nanocarbons that have been intensively studied since the middle of the 90s.[1] A unusual feature of detonation ND is a sharp distribution in size, with the size of most particles varying between 4 and 5 nm.[2,3] Every ND particle consists of a diamond sp^3-core of about 5 nm and a sp^2- hybridized (graphene) carbon-shell around the core. The shell structure and its thickness depend on the detonation conditions and chemical treatments of the non-diamond forms of carbon on the ND surface.[4] The large specific surface area and the possibility of chemical modification of the ND surface make it possible to use detonation ND as a starting material for many applications in solid state and molecular nanoelectronics.[5,6] Detonation ND particles can also be used to produce onion-like carbon and nanographite by heat treatment.[7]

The surface of ND pristine particles is subjected to a structural reconstruction which produce a mixture of sp^2/sp^3 hybridized carbon atoms on surface.[8] Therefore, the removal of this reconstructed layer is a prerequisite for a chemical modification. It was shown that hydrothermal treatment (HyTT) etches off both the sp^2- and sp^3-hybridized carbon phases from the surface.[9] Fluorine is expected to play a role similar to that of hydrogen in the chemical modification of the ND surface. But since fluorine is chemically so active that it breaks graphitic π-bonds,[10] we may expect structural changes in the near-surface region of the diamond core due to the chemical modification of ND.

In this work we produced ND particles with their surface modified chemically with hydrogen and fluorine, and analyzed their structural, electronic and magnetic properties using XRD, SAXS (small-angle X-ray scattering), high-resolution TEM, FTIR, thermogravimetric analysis (TGA), chemical analysis, and magnetic susceptibility measurements.

2 RESULTS AND DISCUSSION

2.1 Sample preparation and structural features

Detonation ND samples were used as the starting material. To avoid the possible effect of magnetic impurities on the experiments, the ND powder was subjected to acid treatment.[7] HyTT was carried out for hydrogen termination of the ND surface. The ND powder was sealed in a gold capsule filled with distilled water in a molar ratio of $ND:H_2O = 1 : 3$ and treated at $650^{\circ}C$ and 100 MPa for 1 - 94 hours.[9] Hydrothermally treated nanodiamond (HyTTND) samples thus produced were dried in vacuum condition at room temperature. Fluorination was carried out by the direct reaction between fluorine gas and ND powder in a passivated nickel tube under a fluorine pressure of 1 atm at 500 $^{\circ}C$ for 1 day (for FND1 sample) or 3 days (for FND2 sample). Fluorinated sample (FND3) was prepared also by fluorination of a HyT-treated sample in the same condition for 3 days.

FTIR, XRD and SAXS measurements, chemical analysis were performed for all samples prepared. For the FND samples, the thermal decomposition process was tracked using TGA measurements up to $900^{\circ}C$. The magnetic susceptibility and magnetization measurements were made using a SQUID magnetometer in the temperature range of 2 – 380 K in a field up to 5.5 T. The samples for SQUID measurements were vacuum-sealed in tubes after heating at $400^{\circ}C$ for several hours at a vacuum level of about 1×10^{-6} Torr.

Figure 1 shows typical FTIR spectra from the ND, HyTTND and FND samples. One can see that the pristine ND contains a large amount of various oxygen-containing groups in the near-surface region of the diamond core. This is indicated by the presence of wide bands at 1000 - 1500 cm^{-1} and near 1750 cm^{-1} that are assigned to C-O stretches in ether-type structures as well as to OH bending modes in water and to the different type of carbonyl species (C=O stretches), respectively. One should also note strong bands at 1650 and 3400 cm^{-1} (not shown) normally attributed to water adsorbed by the sample and weak bands in the C-H stretching region at 2800 - 3000 cm^{-1}.[11]

The HyTT leads to changes in the FTIR spectra: the 1750 cm^{-1} line disappears, the intensities of the 1650 cm^{-1} and 3400 cm^{-1} lines (not shown) decrease, the band at 1000-1500 cm^{-1} becomes less intense, split into several regions with well-defined peaks. An additional narrow line at 1330 cm^{-1} appears for the HyTTND treated for time more that 6 hrs and becomes the most predominant band in the 1000 - 1500 cm^{-1} region after HyT treatment for 94 hours. This line is due to the C-C stretching vibrations of the diamond crystal lattice with the same frequency of the Raman line. In general, this band is an active mode for Raman spectroscopy, but is an inactive mode for IR spectroscopy. In this case, it can arise owing to lack of symmetry in the C-C bonds of the diamond crystal in the interface region.[12]

Figure 2a illustrates the evolution of the total absorption band (A) in the C-H stretching region at 2800 - 3000 cm^{-1} as a function of the HyTT time. It shows an increase in A and, hence, in the number of C-H bonds as the treatment time increases to 40 hours. The treatment above 40 hours leads to a large drop in the A value, and there is no noticeable change in the number of C-H bonds after a long HyTT time (94 hours).

The fluorine treatment leads to similar changes in the FTIR spectra as shown on Figure 1c: the 1750 cm^{-1} line disappears, the intensities of the 1650 cm^{-1} and 3400 cm^{-1} lines (not shown) sharply decrease, and the band at 1000-1500 cm^{-1} also becomes less intense. In the FND samples, strong lines for C-F surface groups appear at 1096, 1251 and 1347 cm^{-1}.[12] It should be emphasized that these features persist in the spectra of FND heated to $500^{\circ}C$. Only annealing at $700^{\circ}C$ can change the FTIR spectra considerably, which indicate a changes in the coverage of chemisorbed fluorine.[11] This is interpreted as being due to disintegration of the C-F groups (see Figure 1d). These FTIR data agree well with the TGA data: the FND sample is stable on heating up to $500^{\circ}C$, at which the mass loss is estimated as only 3.5 wt %.

The high-resolution TEM and XRD data show that the scatterers in the pristine ND and those subjected to chemical treatments (HyTND and FND) are diamond grains of

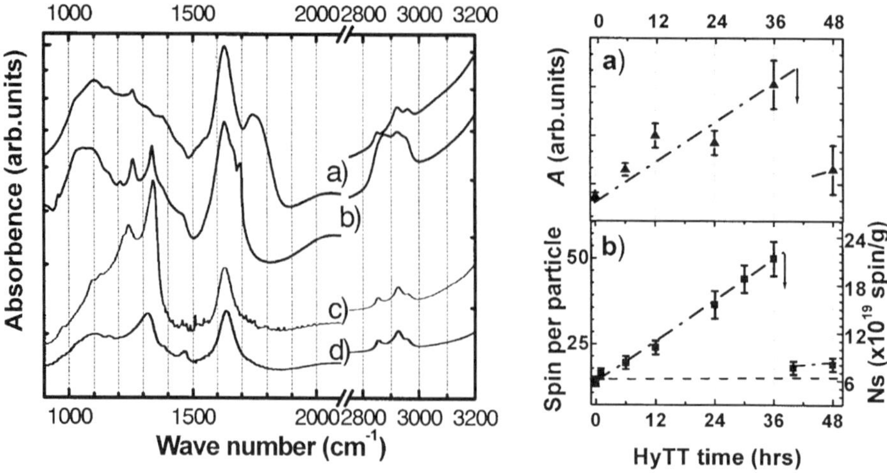

Figure 1 (right) FTIR spectra of pristine ND sample (a), sample after HyTT for 36 hours (b), and FND2 samples before (c) and after heart-treatment at 700°C (d).

Figure 2 (left) The HyTT time dependence of the area of the total absorption band A in the C-H stretching region at 2800 - 3000 cm^{-1} (a), and the number of spins per particle and spin concentration N_s (b). The horizontal dashed line for N_s= 6.3×10^{19} spins/g is the guide for the eyes.

ca.4.4 nm in size. It has been shown that the pristine ND sample consists of loosely aggregated ND particles with a narrow size distribution.[13] In this case the ND structure can be described in terms of fractal geometry.[4] The SAXS data suggest that pristine ND behaves as a scatterer with a size of L = 6.8 nm and volumetric fractal dimension D = 2.5. The SAXS theory states that D is equal to 3 in uniform bulk clusters. Therefore, the deviation from D = 3 as well as the apparent difference in size between the scatterer and the diamond core is due to a fractal-like structure of the ND shell. It can be accounted for by different electronic density distributions or different atoms arrangement between the fractal shell and the core. In the HyTT, L and D depend strongly on the treatment time. In the early stage (12 hours), a scatterer has L = 5.8 nm and volumetric D = 2.85, which is characteristic of particles with smeared boundaries. A scatterer after a long-term HyTT (24 hours) represents only a diamond core, because L becomes the same to the core size 4.4 nm and D = 2.93 approaches 3. The slight deviation of the volumetric fractal dimensions D = 2.93 from D = 3 indicates the presence of stretch or a lattice misfit between the interface region and the central part of diamond core. The tendency in L and D with increasing HyTT time is accounted for by the dissolution of the carbon transient layer (sp^2- hybridized carbon-shell and interface sp^3-region) covering the ND core under the HyT condition.

So, HyTT is capable to remove transient layer. In this process, the most defective sites of the interface structure are etched selectively to form transient layer with the fractal structure, that is also indicated by presence of a line at 1330 cm^{-1} in IR spectra of HyTTND samples after HyTT for 94 hours. In our model the diamond core surface is assumed to be the inner part of the transient layer immediately adjacent to the core, or so called interface region. It should be noted that the core of ND particles has octahedral shape (all of its 8 facets are diamond (111) surfaces) or cubooctahedral one, as suggested by TEM.[5] Naturally, we should not think that surfaces faceting diamond core as perfect as the surface of bulk diamond. In particular, the presence of a lattice misfit between the interface region and the central part of diamond core is indicated by the appearance of a 1330 cm^{-1} line in the FTIR spectrum and follows from SAXS data.

Fluorination also changes the structure of the transient layer, so that FND2 particle scatters X-ray as a volumetric fractal with $L = 5.5$ nm and $D = 2.84$. In other words, we can suggest only a partial "dissolution" of the transient layer, which occurs due to the evolution of gaseous fluorinated carbon compounds. The difference in the sizes of the scatterer and the ND core allows the estimation of the transient layer thickness as c.a. 0.6 nm. The comprehensive analysis of SAXS data with data of chemical analysis and TGA allow estimation of the F/C atomic ratio in the interface region for FND samples. Fluorination of the pristine ND leads to the fact that up to 35% of the surface atoms become chemically bonded to fluorine (FND2 sample) and that the sample obtained by fluorination of HyT treated ND (FND3 sample) has F/C $\cong$ 50%.

2.2 Magnetic properties

The magnetic properties of the ND particles have been studied by several workers.[6-7, 14-15] The results on the magnetic properties of ND show that the magnetic phenomena are associated only with localized defects with $S = \frac{1}{2}$ and the concentration of localized spins does not exceed $N_s \sim 10^{20}$ spins/g. The concentration of intrinsic localized paramagnetic centers estimated as $N_s = 6.3 \times 10^{19}$ spins/g is found to be about 12 spins per ND particle,[15] and there is no noticeable exchange coupling between them.[14, 15] The low concentration of localized spins, as compared with the total number of surface atoms in a particle, indicates that the surface bonds of the ND core are almost totally terminated and the radical-like paramagnetic centers are due to the point-like structural defects inside/on the core.[7, 14]

The magnetic susceptibility χ in ND and chemically-modified NDs is described in terms of two contributions; the temperature-dependent term and the temperature-independent term. The temperature dependent term follows well, or is close to, the Curie law. The detailed behavior of the susceptibility for the HyTTND samples is given in our previous work.[15] In this work was shown that the HyTT induced additional spins on the surface. As shown in Figure 2b the number of these excess spins is linearly dependent on the HyTT time, reaching ca. 23 per particle at 24 hours. Here we assume that the actual number of surface spins per particle can be evaluated as the difference between the total number of spins for the HyTT samples and untreated pristine one. Above 40 hours, the dependence of concentration of surface spins vs. HyTT time shows a discontinuous drop, which correlates well with the change in the number of C-H bonds, as shown in Figure 2a.

Interestingly, the temperature-dependent susceptibility significantly deviates from the Curie law below 200 K in the HyTTND, in contrast to that in the pristine ND. This suggests the presence of a strong antiferromagnetic coupling between the surface spins. The susceptibility behavior of the surface spins is explained in terms of a model of a 2D spin S=$\frac{1}{2}$ clusters of various nuclearity ($N = 2-5$) which are predominantly localized on the particle facets having octahedral shape and also on the crystal ribs.[15] We found that the 24 hrs HyT treated sample has ca.7 dimers per particle and the exchange constant is approximately $J \sim -50$ K if we assume that most of the spin clusters are dimers. The most dramatic change in the magnetic properties of the HyTTND surface takes place in the 36 hours HyTT, where the astounding amount of surface spins (ca. 40 per particles) is produced in addition to a large increase in the positive temperature-independent paramagnetic contribution (χ_0 = 2.0×10^{-6} emu/g). We suppose that the evolution of metallic-like features at T > 100 K occurs before/at the beginning of the surface reconstruction, when multiple alternating chains –C=C-C=C- arise as forerunners of the surface reconstruction.[15]

Here is a question on the nature of the defect-related spins emerging from the modification of the ND core surface and determining its magnetic properties. The nature of hydrogen-related paramagnetic defects arising on ND particles has been studied in detail with reference to polycrystalline CVD diamond films.[17,18] This defect is created by a hydrogen atom entering a stretched C-C bond at a grain boundary, allowing the carbons to relax back, one bounding to the hydrogen and the other with an unpaired electron predominately localized in its dangling bonds. Basically the structure of ND particles is the same both for present detonation ND samples and for CVD diamond films in the former of which. Since the interface layer can be described as stretched region from the SAXS data. Note that in the detonation ND particle hydrogen occur primarily in the transient layer, in contrast to the case of CVD diamond films. This fact and a small size of HyTTND particles account for the observed specific magnetic properties of the hydrogenated surface layer in the HyTTND particles

So, in the present HyTTND particles, the correlation between the defect spin concentration and the total C-H infrared absorption is consistent with the hydrogen-related defects being located in hydrogen-decorated interface layer.[18] We have pointed out that a ND particle in the samples subjected to a HyTT for over 24 hours is an octahedron, so one may expect that the concentration of dangling bonds will be higher for the octahedron edges (ribs). The dangling bonds defects, which interact with each other as collective defects, account for the observed specific magnetic properties of the HyTTND particles, namely, the appearance of antiferromagnetic spin fluctuations in the short HyTT time region. When the HyTT time is prolonged over 30 hours, the large temperature-independent susceptibility reminiscent of Pauli paramagnetic contribution happens to appear. Taking into account the occurrence of surface reconstruction evidenced by the decrease in the spin concentration and the hydrogen content, the unstable surface structure becomes stabilized to form zigzag alternating chains –C=C-C=C-C-, as in polyacetylene, after metallic fluctuations inherited to the electronic instability.[15,16]

The magnetic properties of the FND samples have features different from those of the HyTTND particles, although the susceptibility can be described with a combination of the Curie term and the temperature-independent term as shown in Figure 3. All the FND samples are characterized by the positive temperature-independent paramagnetic contribution χ_0 ~ 0.2 – 1 ×10^{-6} emu/g. It is seen that fluorination produces additional spins. In contrast to the HyTTND samples, no noticeable exchange coupling between the spins has been detected. The concentration of fluorine-induced defect spins remains nearly the same (23-25 spins per particle) in the FND samples, no matter whether the initial material is the ND with a well-developed transient region (made up of ca.1.2 nm) or HyT treated ND for 18 hours (Figure 3b). In addition, the number of the defect-related spins created upon fluorination is practically independent of the fluorination time (1 or 3 days). Since the ND particle possesses a complex structure, we subjected the FND2 sample to a graded annealing in order to clarify the origin of the additional spins. The spin concentration is presented in Figure 3b for the FND particles and their defluorinated products obtained at annealing temperatures of 250-700°C. The characteristic behavior of the spin concentration vs annealing temperature curve is the presence of a maximum and minimum at T ~ 250 and 500°C, respectively. After the temperature of the minimum the spin concentration shows a steep increase above 650°C. Note that the higher spin concentration for the sample

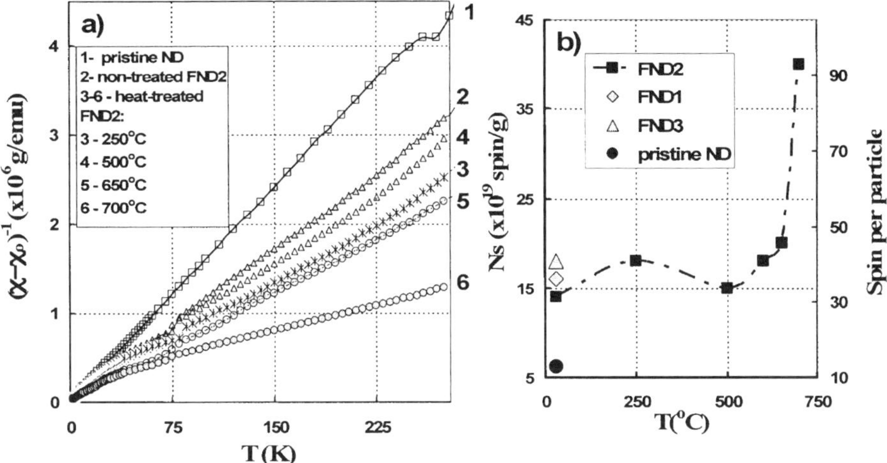

Figure 3 (a) Temperature dependence of $1/(\chi-\chi_0)$ for the pristine ND (1) and FND2 samples: non-treated (2), treated at 250 (3), 500 (4), 650 (5) and 700°C (6). (b) Comparison of the number of spins per particle and spin concentration N_s for ND, FND1, FND2 and FND3 samples and the N_s vs. temperature of annealing (T) for FND2 samples (solid squares).

annealed above 700°C does not entail any noticeable coupling between the spins in contrast to the HyTTND samples.

The observed behavior of the localized spins can be explained in relation to the chemical activity of fluorine. As fluorine has strong chemical activity, it is expected to react not only with the transient surface carbon layer but also with the ND core. In general, fluorine reacts with the carbon atoms of the transient layer, which consists of mainly sp^2-graphene structures, in two ways. Namely, in reaction with the edge carbon atoms of graphene structures, the termination of the carbon atoms with fluorine gives rise to non-magnetic structure. In another reaction associated with the carbon atoms in the interior of the graphene sheets, the formation of bonds with fluorine atoms creates dangling bonds on the carbon atoms adjacent to those bonded to fluorine atoms.[10] This enhances magnetism in the transient layer. The competition and cooperation of these two types of reactions govern the magnetism of the transient layer of the FND particles. As the SAXS data indicate that fluorine treatment only partially removes the transient carbon layer, the spin concentration of the as-reacted FND particles, which is higher than that of the pristine ND, suggests that dangling bonds created in the reaction of the interior carbon atoms predominates in the magnetism of the transient layers.

The reaction of fluorine with the ND core is characterized with the termination of fluorine atoms to the surface carbon atoms, where the fluorine termination reduces the number of magnetically active dangling bonds, degrading the magnetism. The total number of surface atomic sites that can form C-F bonds is ca. 1,300, assuming the ND particle to be octahedral.[16] Then the composition ratio of the fully surface fluorine-terminated ND is calculated to be F/C $\approx$ 50%. According the chemical analysis data, the FND3 sample has F/C $\approx$ 50%, so we can make conclusion that in this case the diamond surface appears to be practically fully modified with fluorine atoms and non-magnetic.

The change in the magnetism upon defluorination can be explained along the story same to that in the fluorination of the ND particle. In the first stage of the defluorination taking place in the low annealing temperature, the dangling bond spins are created through the dissociation of C-F bonds, where the spin concentration becomes the maximum when a half of the carbon atoms are free from fluorine.[10] Complete defluorination in the transient layer makes the fluorine-induced spins entirely disappear as a result of the formation of non-fluroianted graphene structures. After deflurorination of transient layer is completed around 500°C, the dissociation of fluorine atoms from the ND core produces dangling bond spins on the ND core surface as evidenced by the steep increase in the spin concentration above 700°C.

3 CONCLUSION

We investigated the structures and magnetic properties of the ND particles with their surfaces chemically modified with hydrogen and fluorine. The pristine ND has transient (sp^2- hybridized carbon-shell and interface sp^3-region) layer covering the ND cores. HyTT works to dissolve the transient layers, making the surface carbon atoms terminated with hydrogen. The positive correlation between the hydrogen concentration and the concentration of the localized spins created on the surface demonstrates that the hydrogen termination works to increase the dangling bond spins on the surface. The magnetic susceptibility investigation suggests the presence of strong antiferromagnetic interaction between the surface spins. This implies the importance of collective magnetic defects, in which the hydrogen atoms play an essential role. In the prolonged HyTT time above 30 hours, a structural reconstruction takes place, resulting in the reduction of magnetic defects.

Fluorine atoms react with the carbon atoms not only on the surface of the ND core but also in the transient layer. Fluorination creates localized spins in the transient surface layer as a result of the formation of incompletely fluorinated graphitic structures. Defluronation from the transient layer first increases the spin concentration and then reduces it after all the fluorine atoms disappear from the transient layer. At elevated annealing temperatures above 700°C, dissociation of fluorine atoms from the surface of the ND core creates dangling bond spins inherited to the diamond surface structures.

Acknowledgement

The present work is supported by the Grant-in-Aid for Scientific Research No. 15105005 from the Japanese Society for the Promotion of Science (JSPS) and partly supported by the "NEDO International Joint Research Grant Program" for 2004 (Nanocarbon team, Project 04IT4) A.Ya.Vul's research was supported in 2005 by the Russian Ministry of Education and Science (contracts 02.435.11.2013 and 02.434.11.7108) V.Yu.O. and M.V.B. thanks JSPS for the fellowship and financial support. The authors express their thanks to Dr. M. Ishikawa (Tokyo Institute of Technology) for chemical analysis of ND samples and to Mr. H. Arikai (Department of Material Chemistry, Shinshu University) for preparation of fluorinated samples.

References

1 N. R. Greiner, D. S. Phillips, J. D. Johnson, and F. Volk, *Nature*, 1988, **333**, 440.
2 V.L. Kuznetsov, A.L. Chuvilin, E.M. Moroz, V.N. Kolomiichuk, S.K. Shaikhutdinov, Y.V. Butenko, *Carbon*, 1994, **32**, 873.

3 M. Yoshikawa, Y. Mori, H. Obata, M. Maegawa, G. Katagiri, H. Ishida, and A. Ishitani, *Appl. Phys. Lett.*, 1995, **67**, 694.

4 M.V. Baidakova, A.Ya. Vul', V.I. Siklitski, *Chaos, Solitons & Fractals*, 1999, **10**, 2153.

5 V.L. Kuznetsov and Yu.V. Butenko, in *Nanostructured Materials and Coatings for Biomedical and Sensor Applications,* ed. Y.G. Gogotsi and I.V. Uvarova, Kluwer Academic Publishers, Netherlands, 2003, p.187.

6 B.V. Spitsyn, M.N.Gradoboev, T.B.Galushko, T.A.Karpukhina, N.V.Serebryakova, I.I.Kulakova, and N.N.Melnik, in *Synthesis, Properties and Applications of Ultrananocrystalline Diamond*, ed. Dieter M. Gruen, Olga A. Shenderova, Alexander Ya. Vul', Springer, Dordrecht - Berlin, Heidelberg, New York, 2005, NATO Science Series. Series II: Mathematics Physics and Chemistry, vol. 192, p. 241.

7 B.L.V. Prasad, H. Sato, T. Enoki, Y. Hishiyama, Y. Kaburagi, A.M. Rao, P.C. Eklund, K. Oshida and M. Endo, *Phys. Rev. B*, 2000, **62**, 11 209.

8. J.-Y.Raty, G. Galli, C. Bostedt, T.W. van Buuren and L.J. Terminello, *Phys.Rev.Lett.*, 2003, **90**, 037401.

9 Y. Gogotsi, T. Kraft, K.G. Nickel, M.E. Zvanut, *Diamond and Related Materials*, 1998, **7**, 1459.

10 K. Takai, H. Sato, T. Enoki, N. Yoshida, F. Okuno, H. Touhara, and M. Endo, *J.Phys.Soc.Jpn*, 2001, **70**, 175.

11 T Jang and K.Xu, *Carbon*, 1995, **33**, 1663.

12 T. Ando, K. Yamamoto, M. Kamo, Y. Sato, Y.Takamatsu, S. Kawasaki, F. Okino and H. Touhara, *J.Chem.Soc.Faraday Trans.*, 1995, **91**, 3209.

13 A.M. Affoune, B.L.V. Prasad, H. Sato, T. Enoki, *Langmuir*, 2001, **17**, 547.

14 A.I. Shames, A.M. Panich, W. Kempinski, A.E. Alexenskii, M.V. Baidakova, A. T. Dideikin, V.Yu. Osipov, V.I. Siklitski, E. Osawa, M. Ozawa and A.Ya. Vul', *Journ. Phys. Chem. Solids*, 2002, **63**, 1993.

15 V. Osipov, M. Baidakova, K. Takai, T. Enoki, A. Vul', in *Defects in Advanced High-κ Dielectric Nano-Electronic Semiconductor Devices*, ed. E.Gusev, Springer, Netherlands, 2006, p.445.

16 V.Yu. Osipov, M.V. Baidakova, K. Takai, T. Enoki, A.Ya. Vul', to be published.

17 X. Zhou, G.D. Watkins, K.M.McN. Rutledge, R.P. Messmer, S. Chawla, *Phys.Rev.B.*, 1996, **54**, 7881.

18. D.F. Talbot-Ponsondy, M.E. Newton, J.M. Baker, G.A. Scarsbrook, R.S. Sussmann, A.J. Whitehead, S. Pfenninger, *Phys.Rev.B*, 1998, **57**, 264.

MODELING AND DESIGNING OF NANO-CARBON MAGNETS BY A NEW DENSITY-MATRIX FUNCTIONAL THEORY

K. Kusakabe, M. Takahashi, Y. Minamigawa and N. Suzuki

Graduate School of Engineering Science, Osaka University, Toyonaka, Osaka 560-8531, Japan

1 INTRODUCTION

We have performed the first-principles study on the magnetic nanocarbon structures.[1,2] This class of nanometer scale carbon materials contains graphite fragments, finite carbon nanotubes, and various graphitic nanostructures. Theoretical investigation of this topic has been successfully done, by the usage of a combined approach, i.e. designing with simplified models of many electron systems and certifying the designed magnetism by the first-principles calculations. The latter method is often applied to design of not only the carbon-based magnets but also inorganic compounds. However, successful examples of the combined approach are mostly found in the magnetic molecules and carbon based magnets This may partly because of a long history of magnetic molecules. In this field, chemistry researchers have utilized design technique based on quantum mechanical consideration on simplified electron models. Thus, we may say that our approach is firstly based on the Longuet-Higgins rule[3] and the Ovchinnikov rule[4] for the magnetic molecules. Our key ingredient is the edge state found in zigzag edged nanographite.[5,6] At the graphite zigzag edge, there appears non-bonding degenerate surface states. This state was thought to cause magnetic instability in graphitic structures. However, from the knowledge of the Ovchinnikov counting rule, zigzag graphite ribbons should not show ferromagnetic order. One of the authors (K.K.) and Maruyama have designed magnetic nanographite.[1,2] This design was realized by finding of realistic Klein's edge,[7] in addition to the zigzag edge. Combination of these two types of the edge is a key. The structures naturally gain sublattice structures with different numbers of sublattice cites as π networks. The certification was given by DFT-LSDA calculations, which is thought to be reliable for extending magnetic systems containing magnetic nanocarbon structures.

In this paper, we reconsider the design of magnetic nanographite from two points. Our previous approach was applied to several possible structures. Satisfying the Ovchinnikov rule, the structures are found to be magnetic. However, if we consider synthesis of these magnetic structures, preparation of zigzag edges may not be easy. Recently, Kobayashi *et al.* have found existence of edge states at several defective arm-chair edges of graphite.[8] As a feasible example, we consider the defective armchair edges of graphitic structures. Another point left for future development is the electron model itself. So far, the first-principles technique based on the density functional theory was not able to derive the LDA+U type functional from the first principles. Recently, Kusakabe succeeded in formulating the modelling of effective Hamiltonians containing the short range interactions explicitly. This rigorous approach is an extension of DFT. We discuss this new technique and future prospects of the first-principles approach for modelling and designing the nanocarbon magnets.

2 DESIGN OF MAGNETIC NANOCARBON MATERIALS

2.1 Magnetic nanographite and magnetic nanotubes

Our successful examples of design contain magnetic nanographite, magnetic nanotubes, and graphite-diamond hybrid structures. These examples show that design technique using the combined approach works well for the π conjugated systems. This is because π networks and its variants are rather well described by the tight-binding model of the electron systems. Including the short range repulsive interaction, we obtain naturally the Hubbard model. By the knowledge of the magnetism in the Hubbard model, we can rather freely design magnetic structures. Interestingly, DFT-LSDA calculations result the same magnetism, if the graphitic structure maintains planner configuration. If the graphitic structure has a curved surface, as for the nanotubes, DFT-LSDA may conclude differently. Actually, for the magnetic nanotubes, DFT-LSDA solutions have smaller magnetic moments than those expected by the Ovchinnikov counting rule. This is because the electronic structure of the nanotube is not pure π systems. Interestingly, however, the edge state counting rule seems to hold instead as discussed by Okada and Oshiyama.[9] In addition to this theoretical limit, we face at difficulty in creation of these magnetic nanocarbon structures as they are.

2.2 Magnetic defective armchair nanographite ribbons

Kobayashi *et al.* have found that the armchair edge is stably created at cleaved edges of graphite.[8] At the top of graphite layers, they observed a single graphene layer with a well-defined hydrogenated edge. At some edges, there exist lacks of carbon atoms. Thus, we can expect existence of defective armchair edges. Interestingly, at some defects, there appears the edge state. This observation suggests that we would have edge-state based magnetism even in these defects in the armchair edge. To model the structure, we considered armchair graphene ribbons. In Figure 1, we show a first example. This structure is obtained by removing 5 carbon atoms at the armchair edge. Due to the Ovchinnikov counting rule, we expect existence of magnetism in the π system. To keep the flat structure, however, there is a dangling bond at one carbon atom. This also results in a magnetic moment. Figure 2 shows spin density of the system by DFT-LSDA calculation. Spin density profile of this defective armchair ribbon shows that a dangling bond at a carbon atom gives strong contribution in the spin density. At the zigzag edge, we see spin density. We also considered another ribbon with 3 carbon atoms removed in Figure 3. The spin density in Figure 4 shows that the edge state originated spin density appears at the short zigzag edge. This simple example suggests that fabrication of magnetic edges starting from the armchair edges is expected. The spin moment is accumulated at the defect. Thus, we may hope that continual creation of defects at the armchair will make aligned localized magnetic moments at the graphene stripes.

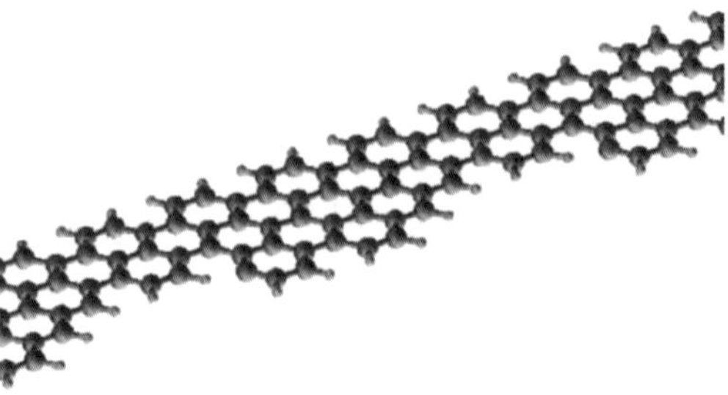

Figure 1 *Structure of a grapheme ribbon with defected armchair edge. The structure is obtained by removing 5 atoms from a perfect armchair ribbon. Two edges are hydrogenated.*

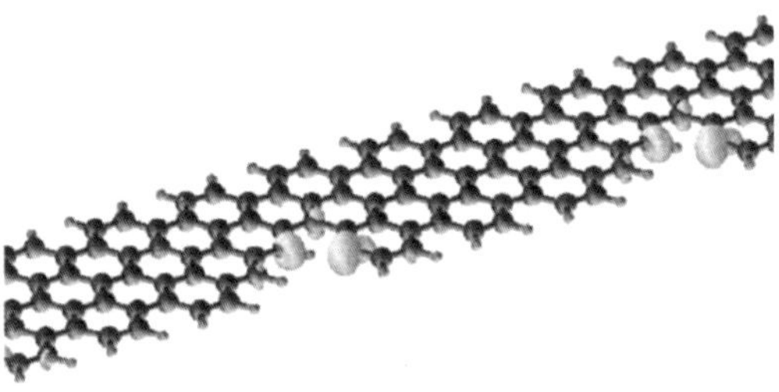

Figure 2 Spin density profile of the defective armchair ribbon in Figure 1. A dangling bond at a carbon atom shows strong contribution. At the zigzag edge, we see spin density.

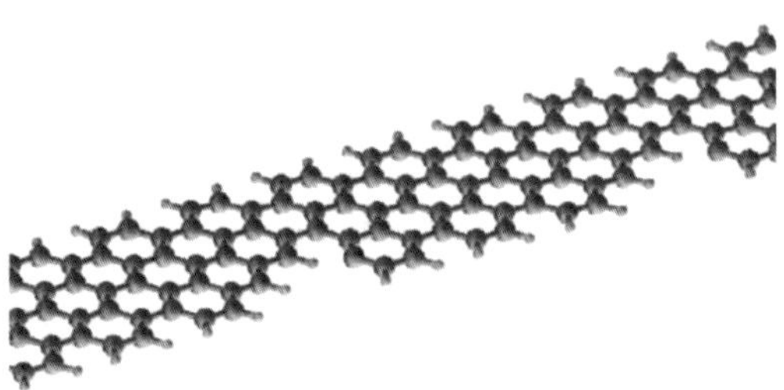

Figure 3 *Structure of a grapheme ribbon with defected armchair edge. The structure is obtained by removing 3 atoms from a perfect armchair ribbon. Two edges are hydrogenated.*

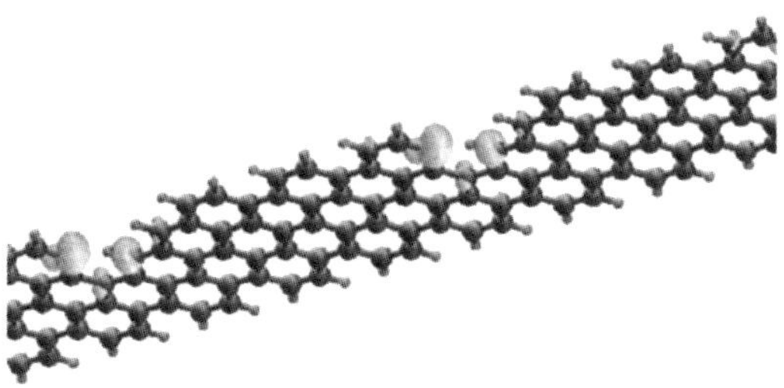

Figure 4 Spin density profile of the defective armchair ribbon in Figure 3. A dangling
bond at a carbon atom shows strong contribution. At the zigzag edge, we see
spin density originated clearly from the edge state.

3 DESIGN OF MODELS FOR NANOMAGNETS

3.1 Density-matrix functional theory

Recently, Kusakabe has found a way to determine an effective short-range interaction in an
extension of the density-functional theory.[10,11] Requiring that a density-density
correlation on a localized orbital is reproduced in a model, the interaction parameter U is
uniquely determined, if it exists. Based on this new theory in the density functional theory,
we can make a connection between a Coulombic system and an effective model rigorously.

The basic theorem of our extension of the Kohn-sham scheme is Theorem 3 in the
reference [11]. This theorem tells us that the so-called Hubbard interaction is uniquely
determined, if the resulting model reproduces a density-density correlation function. This
conclusion is rather natural, because DFT has wide flexibility. If we consider the multi-
reference DFT, the extension of Kohn-Sham scheme and its artificial wavefunction can
reproduce various correlation functions as well as the single particle density. This is a kind
of modelling of the effective Hamiltonian from the first principles. Since the method is for
determination of U as a functional of the density-density correlation, the theory itself is
regarded as a density-matrix functional theory.

3.1 Approximation of LDA+U and the Hubbard model

The theory of a density-matrix functional (T-DMF) is a theoretical framework. It gives a
principle on the selection of the localized orbital, at which the U-term is required. Once the
model is settled, the design of the magnetic nanocarbon systems will be much reliable.

To perform realistic calculation, we have to introduce approximations for the energy
functional. After decoupling the kinetic energy functional and the 2-body effective
interaction part, the exchange-correlation energy functional is identical to that of the usual
Kohn-Sham DFT. Using LSDA, we have the LDA+U model. Since the original extended
Kohn-Sham scheme is a self-consistent approach, the LDA+U model should be
approximately treated as a self-consistent calculation. More crudely, if we utilize the given
one-body part as a tight-binding Hamiltonian, the Hubbard model is naturally obtained.

Thus, T-DMF connects the Hubbard model and DFT calculation in a first-principles
manner. We may say that the combined approach utilized so far has got a foundation. If the
result of the previous approach shows consistency, the approximate treatment is justified.

If one needs a more accurate unified approach, one can utilize T-DMF to determine the U-term and the residual exchange-correlation energy functional. This effective model is written as MR-DFT, the effective model naturally be a many-body model. Implicit inclusion of the U-term will allow us to apply the known accurate evaluation of the quantum phase transition to this first-principles effective theory. Design of functional materials with various order becomes possible due to T-DMF in principle.

4 CONCLUSION

We have shown some examples of feasible magnetic nanocarbon structures. Especially, the defected armchair nanographite ribbon is possible to fabricate in real experiments, since application of the external pressure via the STM tip will remove some atoms from the structure.

To make the prediction more accurate, we have also developed the first-principles Hubbard model. When we utilize this theory, we hope that the design of magnetic carbon systems would be much reliable and flexible. In addition to the static calculation, many developments were made on the dynamic calculation. Combining the effective many-body theory and the dynamical simulation will allow us to realize the quantum device simulator, which can do ab-initio simulations of nano-scale devices in operation.

Acknowledgment

K.K. is grateful for fruitful discussions with Prof. T. Enoki and Dr. Y. Kobayashi on nanographite. He is also grateful for collaboration of Prof. K. Yamaguchi and Prof. S. Yamanaka on the density functional theory. This work was supported by a Grant-in-Aid for Scientific Research in Priority Areas (Nos. 17064006 and 17064013) from MEXT and a Grant-in-Aid for Scientific Research (Nos. 15310086 and 15GS0213) from JPS. It is also supported by 21th COE program.

References

[1] K. Kusakabe and M. Maruyama, Phys. Rev. B 2003, **67**, 092406.
[2] M. Maruyama and K. Kusakabe, J. Phys. Soc. Jpn. 2004, **73**, 656.
[3] H.C. Longuet-Higgins, J. Chem. Phys. 1950, **18**, 265.
[4] A.A. Ovchinnikov, Theor. Chem. Acta 1978, **47** 297.
[5] M. Fujita, K. Wakabayashi, K. Nakada and K. Kusakabe, J. Phys. Soc. Jpn. 1996, **65**, 1920.
[6] K. Nakada, M. Fujita, G. Dresselhaus and M.S. Dresselhaus, Phys. Rev. B, 1996, **54**, 17954.
[7] D.J. Klein, Chem. Phys. Lett. 1994, **217**, 261.
[8] Y. Kobayashi, K.-i. Fukui, T. Enoki, K. Kusakabe, and Y. Kaburagi, Phys. Rev. B 2005, **71**, 193406.
[9] S. Okada and A. Oshiyama, J. Phys. Soc. Jpn. 2003, **72**, 1510.
[10] K. Kusakabe, J. Phys. Soc. Jpn. 2001, **70**, 2038.
[11] K. Kusakabe, cond-mat/0505703.

Integration and Functionalities of Molecular Materials

EIGHT UNIMOLECULAR RECTIFIERS

R. M. Metzger

Department of Chemistry, Box 870336, University of Alabama, Tuscaloosa, AL 35487-0336, USA

1 INTRODUCTION

Eight unimolecular rectifiers are discussed (Figure 1): γ-hexadecylquinolinium tricyanoquinodimethanide, **1**; its tetrafluoro analog **2**; 2,6-di[dibutylamino-phenylvinyl]-1-butylpyridinium iodide, **3**; dimethylanilino-aza[C_{60}]-fullerene, **4**; fullerene-bis-[4-diphe-nylamino-4''-(N-ethyl-N-2'''-ethyl)-amino-1,4-diphenyl-1,3-butadiene] malonate, **5**; *N*-(10-nonadecyl)-*N*-(1-pyrenyl-methyl)perylene-3,4,9,10-bis(dicarboximide) **6**; N-(10-nona-decyl)-N-(2-ferrocenyl-ethyl)-pyrenyl-methyl)perylene-3,4,9,10-bis(dicarboxyimide),   **7**; and   4,5-dipentyl-5'-methyltetrathiafulvalen-4'-methyloxy-2,4,5-trinitro-9-dicyanomethyl-enefluorene-7-(3-sulfonyl-propionate), **8**. These molecules have electron donor regions or moieties (D), and electron acceptor moieties (A). Langmuir-Blodgett (LB) or Langmuir-Schaefer (LS) monolayers of these molecules show asymmetric electrical conductivity between Au or Al electrodes. When the films are compact, and if there is an intramolecular charge transfer (ICT, or intervalence transfer, IVT) band, then the rectification is persistent.

These rectifiers are part of "unimolecular electronics" (UE), the study of electrical processes on the scale of a single molecule[1]. The first UE serious proposal came when Aviram and Ratner (AR) proposed electrical rectification by a single molecule[2].

Electroactive molecules (sizes between 0.5 and 3 nm) may be the *reductio ad absurdum* of integrated electronics. Since the mid 1960's the computing power has doubled, by now every 18 months[3], because the "design rules" (DR) (i.e. the smallest distance between adjacent components) get smaller and smaller[4]: DR < 100 nm is now common in commerce, and DR ≈ 50 nm in research; however, DR cannot shrink *ad infinitum*. When DR < 20 nm, huge technical hurdles may face silicon-based electronics, and molecules may provide a competitive alternative.

We must touch individual molecules with metal electrodes; we want to exploit their chemical structure to control the flow of electricity[5-9]. Here we discuss four published rectifiers (**1**, **3**, **4**, **5**) and four unpublished ones (**2**, **6**, **7**, **8**).

2 THE MEASUREMENT OF RECTIFICATION

One can probe the electrical conductivity of individual molecules by scanning tunneling

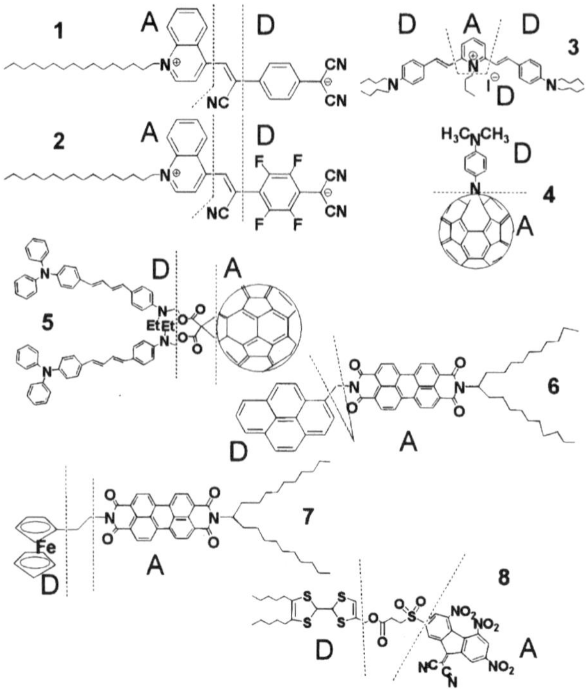

Figure 1. *The eight rectifying molecules. The dotted lines within each structure delineate approximately the "boundaries" of the donor (D) and acceptor (A) moieties.*

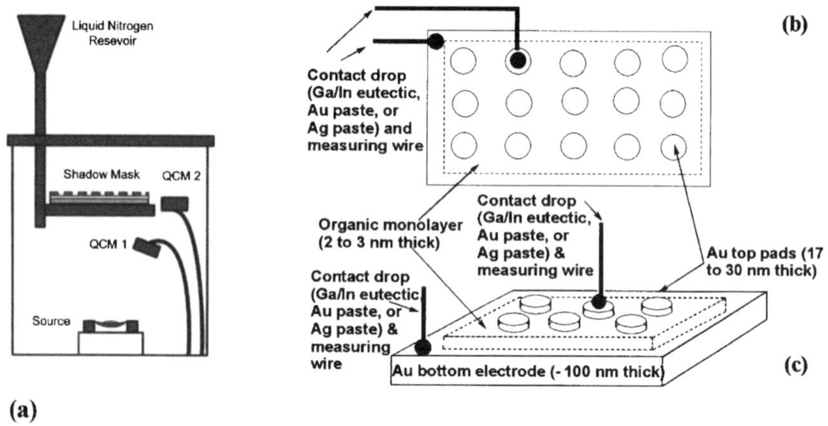

Figure 2. *(a) Evaporator with dual thickness monitors QCM1 and QCM2; (b) top view and (c) projected view of the Au | monolayer | Au pads.*

microscopy (STM). A single molecule can be touched on an electrically conducting surface by an STM tip, by a conducting-tip atomic force microscopy (AFM) tip, or when bonded covalently to two Au "break junction" tips[10]. The current vs. voltage (IV) characteristics can be measured for a single molecule by scanning tunneling spectroscopy (STS).

One can also assemble monolayers of molecules transferred atop a metal electrode by the Langmuir-Blodgett (LB) method, by the Langmuir-Schaefer (LS) method, or as "self-assembled monolayers" (SAMs) covalently bonded to an inorganic bulk substrate (e.g. organothiols on gold). The electrical measurements are made using electrical contacts to "metal | organic | metal" (MOM) sandwiches of a single organic monolayer. Top metal "pads" have areas from 1 μm^2 to 1 mm^2, but the monolayer is only 2 to 3 nm thick beneath each pad. Thus $\sim 10^{12}$ molecules in a monolayer are measured in parallel.

One must fabricate the MOM pads while not over-heating the organic layer. We first deposit organic monolayers on macroscopic and hydrophilic Al | Al_2O_3[11], or very fresh and hydrophilic Au substrates[12] by the LB or LS method. Then we place the substrate plus monolayer on a platform cryocooled to 77 K by an external supply of liquid N_2, in an evaporator, and deposit on top 20 to 50 small pads (areas 0.265 to 2 mm^2 each) of the same metal by evaporation. This works well for Al (+ Al_2O_3)[11], but with Au, one must also pre-cool the hot Au vapor by collisions with Ar atoms, before the Au atoms descend on the substrate ("cold gold" method)[12]. Figure 2 shows the evaporation chamber and the MOM pads.

3 THEORY

AR proposed[2] a D-σ-A molecule connecting an electron donor moiety (D) to an electron acceptor moiety (A) through an insulating saturated "sigma" bridge σ; the mechanism involves inelastic tunneling through the molecule from its first highly polar electronic excited state D^+-σ-A^- to the less polar ground state D^0-σ-A^0 [2]. It may be assisted by an ICT or IVT mixing of the donor and acceptor states, i.e. the existence of an ICT or IVT absorption band. If the two moieties are too far apart (the bridge is too long), then they will not communicate, and no rectification will occur. If they are too close, then a new single mixed ground state will form, and the molecule will not rectify.

The metal-molecule interfaces are important: (1) Do we arrange the molecules by physisorption or by chemisorption? (2) Are there image charges, image dipoles, and Schottky barriers? (3) Do the molecules rearrange or disintegrate during electrical measurement? (4) Can we avoid compressing the molecules? (5) How do we get a miniumum-resistance metal-molecule contact?

There are three processes for asymmetrical conduction in MOM assemblies: (1) with Schottky ("S") barriers[13-15]; (2) if the "chromophore" is placed asymmetrically ("A") within a MOM sandwich, e.g. because of the presence of a long alkyl "tail"[16-18]; (3) when the current involves electron transfers between molecular orbitals, whose dominant probability amplitudes are asymmetrically placed within the chromophore: this third process is true unimolecular ("U") rectification[16]. Molecules may exhibit combinations of these effects; pure "U" rectifiers are rare[17].

The tunneling current I across a single molecule is approximately exponential with potential V (its logarithm is linear with some power of V): a sigmoidal IV curve is seen, symmetrical between the first and third quadrants. When the conduction involves a very large number of channels ($N \rightarrow \infty$), or if scattering occurs during the electron transport[19], then Ohm's linear law holds.

The rectification ratio (RR) is defined as the current I at a positive bias V divided by the absolute value of the current at the corresponding negative bias -V:

$$RR(V) = I(V) \, / \, | \, I(-V) \, | \tag{1}$$

4 MONOLAYERS OF 1

Instead of an AR D^0-σ-A^0 system, the first rectifier **1**[11,12,20-22] was a ground-state zwitterion D^+-π-A^-, connected by a twisted "pi" bridge, where tunneling from the lower-polarity excited state D^0-π-A^0 to the higher-polarity ground state D^+-π-A^- is possible[23].

Molecule **1** is slightly soluble in polar solvents; its ground-state electric dipole moment is $\mu_{GS} = 43 \pm 8$ Debyes[11]. Its absorption spectrum in solution shows a narrow. hypsochromic band, peaked between 600 and 900 nm; this band is an IVT band[11,23], with emission in the near IR region[23]. Its excited-state moment is $\mu_{ES} = 3$ to 9 Debyes[23]. Since **1** is hypsochromic, its ground state must be D^+-π-A^-, and its first electronically excited state is D^0-π-A^0. Molecule **1** forms an amphiphilic Pockels-Langmuir (PL) monolayer at the air-water interface, with a collapse pressure of 34 mN m^{-1} at 20°C[11]; this monolayer transfers on the upstroke only (Z-type) onto hydrophilic glass, quartz, or Al[11,26] or fresh hydrophilic Au[12], but poorly on the downstroke onto graphite[26]. A peak at 570 nm was seen in an LB multilayer[11], but this peak occurs only soon after film collapse, when molecules cluster[24]; it is polarized <u>in the plane</u> of the film, and must be due to some intermolecular aggregation[24]. If LB films are transferred before film collapse, the true IVT band appears, polarized <u>normal to the film</u>, with peak at 530 nm[24]. The monolayer thickness of **1** is 2.3 nm[11] or 2.9 nm[12] (X-ray diffraction), 2.3 nm (spectroscopic ellipsometry, SE)[12], 2.2 nm (surface plasmon resonance, SPR) [11,27], and 2.5 nm (X-ray photoelectron spectrometry, XPS)[27]. Using 2.3 nm and a calculated molecular length of 3.3 nm, yields a tilt angle of 46° from the surface normal[11]: this is confirmed by grazing-angle FTIR studies of **1**[11,27]. The XPS spectrum of a monolayer of **1** on Au shows two N(1s) peaks[27]. Angle-resolved XPS shows that the cyano Ns lie closer to A, than the quinolinium N[27]. Since a monolayer survives in the ultra-high vacuum of an XPS instrument, it must adhere to it by its CN terminations. When an LB monolayer of **1** is placed between oxide-covered Al electrodes, an asymmetric current is seen, with RR = 26 at 1.5 Volts[11], and a small current of 0.33 electrons molecule^{-1} s^{-1} (using a molecular area of 50 Å^2 from the pressure-area isotherm) [11]. The RRs of **1** vary from pad to pad: in these measurements, all electrical resistances (Al, Ga/In or Ag paste, oxides, wires, etc.) are in series. As high potentials are cycled, the RRs decrease: 1.5 Volts across a monolayer of thickness 2.3 nm creates a field of 0.65 GV m^{-1}, and molecules may turn around end-over-end[11]. The rectification of **1** is independent of temperature in the range 105 K < T < 390 K[22].

With oxide-free Au electrodes the current through the pads rises to a maximum of 90,400 electrons molecule^{-1} s^{-1}, and is asymmetrical[12]: the same asymmetric conduction occurs with Au or Al electrodes[12]. The maximum RR at 2.2 Volts is 27.53[12]. Figure 3(a) shows how RR decreases[12]. Z-type 30-layer films of **1** rectify between Au electrodes[25], with currents three orders of magnitude smaller than for the monolayer.

The IV measurements outlined above are done with electrodes of the same materials on both sides of the monolayer. However, oxides on the Ga/In eutectic drops,

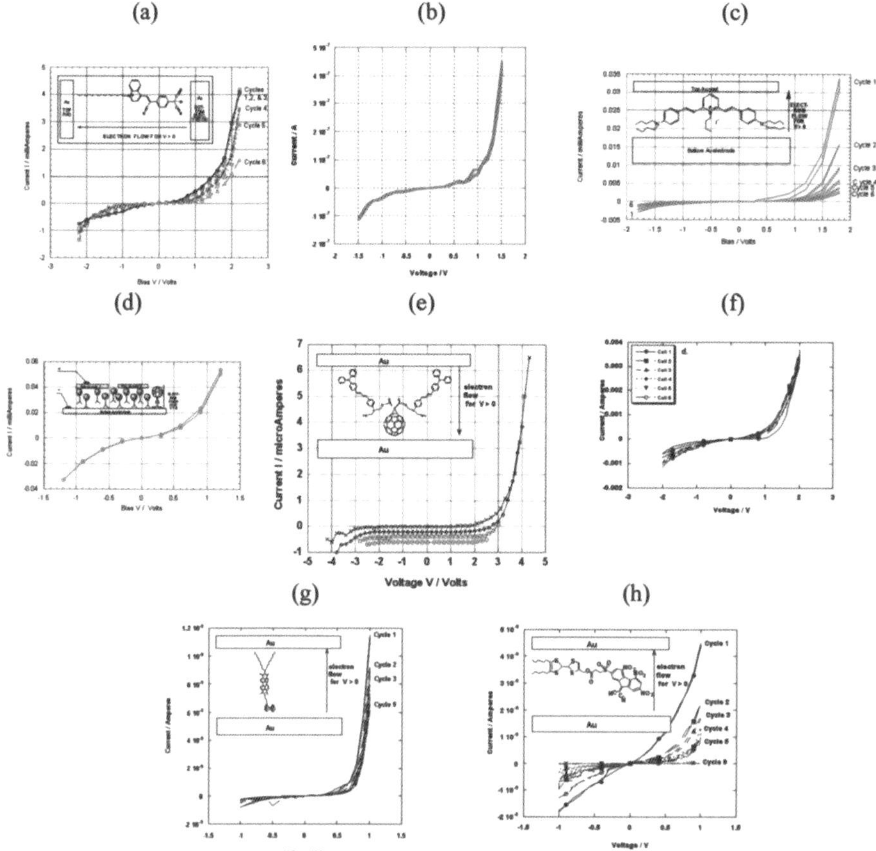

Figure 3. *The IV curves, or rectification of MOM sandwiches consisting of three elements: (i) a macroscopic bottom Au or Al electrode, (ii) an 0.3 mm² top Al or "cold Au" electrode pad, and, between them, (iii) (a) an LB monolayer of 1 [12]; (b) an LB monolayer of 2 [24]; (c) an LB monolayer of 3 [29]; (d) an LB monolayer of 4 [30]; (e) an LS monolayer of 5 [31]; (f) an LB monolayer of 6 [32]; (g) an LS monolayer of 7 [32]; (h) an LS monolayer of 8 [33].*

and how well Ga wets the gold electrodes, cause large variations of the measured currents from pad to pad (from fractions of a μA to several mA at ±1 V). The Schottky barriers between molecule and Au are expected to be very small for **1**, but image dipoles must exist in the Au electrodes. In these two-probe IV measurements, minimizing contact resistances, and avoiding pressure on the monolayer, have been very important, but a better way of making the contacts is always desirable.

  Molecule **2**, a tetrafluoro analog of **1**, is a new rectifier (Figure 3(b))[24].

5 CHEMISORBED MONOLAYER RECTIFIER (A VARIANT OF 1)

A new effort combined the LB and SAM techniques, by measuring two thioacetyl variants of **1**, which could bind covalently to Au[28]. A variant of **1** with an undecyl "tail" followed by a thioacetyl termination disappointed[28]: the PL film collapsed at low pressures, to yield a disordered LB monolayer; physisorption by the dicyanomethide end competed with Au-to-thiolate chemisorption, and the monolayer rectified in either direction, depending on which molecular orientation the STM tip was probing[28].

6 A BACK-CHARGE-TRANSFER RECTIFYING SALT, 3

Molecule **3** transfers to hydrophilic Al as a Z-type multilayer[29]. Its monolayer rectifies (Figure 3(c)[29]. The monolayer thickness is 0.7 nm (SE), 1.15 nm (SPR), and 1.3 nm (X-ray diffraction)[29]. The absorption peak at 490 nm is attributed to iodide-to-pyridinium back-charge-transfer[29]. Some cells have initial RR = 60; RR decreases upon cycling[29].

7 A FULLERENE LB RECTIFIER, 4

Dimethylaminophenylazafullerene **4** has RR $\approx$ 2 (Figure 3(d))[30]. Its IVT peak is at 720 nm[30]. The Langmuir film is very rigid, i.e. the slope of the isotherm is relatively large. However, the molecular areas are only 50 Å^2 at the chosen transfer pressure, half what they should be: the molecules, transferred onto Au on the upstroke, must be staggered[30]. The film thickness is 2.2 nm (XPS)[30].

8 A STURDY FULLERENE LS RECTIFIER, 5

Rectification was seen in an LS monolayer of **5** between Au electrodes (Figure 3(e))[31]. Molecule **5** is based on two triphenylamines (two one-electron Ds) and a single fullerene (weak one-electron A); RR persists upon cycling[31].

9 TWO PERYLENEBISIMIDE-BASED RECTIFIERS, 6 AND 7

Molecules **6** and **7** are D-σ-A molecules, based on the pyrene donor D plus a one-carbon bridge, or a ferrocene donor D plus a two-carbon bridge, and the moderate perylene-bisimide acceptor A[32]. Molecule **7** has an IVT band at 595 nm[32]; its PL isotherm shows that **6** transfers as a monolayer at the fairly high surface pressure of 35 mN m^{-1}, and forms a rectifier; RR $\approx$ 25 -35, and does not change much upon cycling (Figure 3(f))[32].

10 A TETRATHIAFULVALENE-AND-FLUORENONE-BASED D-σ-A RECTIFIER, 8

Molecule **8** is a D-σ-A molecule (D=tetrathiafulvalene, A= dicyano-methylenetrinitro-fluorene): its IVT band is at 1220 nm[33]: the rectification is shown in Figure 3(g)[33].

11 RECENT RECTIFIERS MEASURED BY OTHER GROUPS

The group of Bryce and Petty synthesized a new D-σ-A compound containing the D=TTF and the A=TCNQ[34]: this TTF-σ-TCNQ ester gave strong PL films at the air-water interface. But the TCNQ group lies flat, rather than end-on, on the water surface; the LB multilayers were Y-type, rather than Z-type, and rectification could not be observed[34].

Bryce and Heath and coworkers found a rectifier in an analog of **8**[35].

Ashwell and co-workers studied several zwitterionic systems by scanning tunneling spectroscopy (STS), starting with **1**[36]; for **1**, and for several other zwitterionic systems, the addition of acid could stop, or reverse, the rectification[37-39]. Yu and co-workers reported rectification by STS, which could be reversed by protonation[40].

Weber, Mayor, and co-workers linked bithiols between mechanically controlled Au break junctions: and rectification (RRs scattered between 2 and 10) was seen if the molecule consisted of one tetrafluorophenyl group at one end, and a plain phenyl group at the other end, but no rectfication was seen if the ends were symmetrical [41].

12 CHALLENGES

Table 1 collects some significant data about the eight rectifiers we have studied. Although unimolecular rectification has been established, there are many open questions:

Table 1. Summary data for the eight unimolecular rectifiers **1-8**. All compounds were measured at room temperature in air between Au electrodes inside a Faraday cage (but **1** was also measured between Al electrodes at room temperature[11] also between 105 K and 370 K[22]). The column "# pads" lists how many independent typical MOM pads were discussed in each publication as typically rectifying (out of hundreds measured).

Str.	Type	Transf.Press. (mN/m)	LB or LS?	RR Eq.(1)	# pads	Survives cycling?	U, A, or S?	Refs.
1	D^+-π-A^-	20	LB	2-27	16	no	U,A	[12]
2	D^+-π-A^-	28	LB	3-64	3	no	U,A	[24]
3	D^+ iodide	22	LB	8-60	24	no	A	[29]
4	D-σ-A	22	LB	2	1	no	A	[30]
5	D-σ-A	23	LS	2-16	9	yes	A	[31]
6	D-σ-A	32	LB	2-5	4	yes	U,A	[32]
7	D-σ-A	35	LB	28	1	yes	U,A	[32]
8	D-σ-A	35	LS	3	1	no	U	[33]

(1) Can inelastic electron tunneling spectroscopy (IETS) confirm that the electron tunnels through the molecule (instead of through space)? <u>We are working on it</u>.

(2) When do metal electrodes invade a monolayer during a room-temperature metal vapor deposition[16,17], or during measurements[30]? Can this be limited?

(3) What chemical changes occur during the electrical measurements?

13 ACKNOWLEDGMENTS

This work was generously supported by the United States National Science Foundation, and was made possible by the diligence and insight of many colleagues, students, and post-doctoral fellows.

References

[1]   R. M. Metzger, NATO ASI Series 1991, **B248**, 659.
[2]   A. Aviram and M. A. Ratner, Chem. Phys. Lett. 1974, **29**, 277.
[3]   G. E. Moore, Electronics 10 April 1965, **38**(8), 114-117.
[4]   International Technology Roadmap for Semiconductors, 2001: http://public.itrs.net/Files/2001ITRS/Home.htm
[5]   R. M. Metzger, Acc. Chem. Res. 1999, **32**, 950.
[6]   R. M. Metzger, Chem. Reviews 2003, **103**, 3803.
[7]   R. M. Metzger, Chem. Record 2004, **4**, 291.
[8]   R. M. Metzger, in "Introducing Molecular Electronics" edited by G. Cuniberti, G. Fagas, and K. Richter (Springer, Berlin Heidelberg New York, 2005) pp. 313-349.
[9]   R. M. Metzger, Colloids and Surfaces A, in press.
[10]   M. A. Reed, C. Zhou, C. J. Muller, T. P. Burgin, and J. M. Tour, Science 1997, **278**, 252.
[11]   R. M. Metzger, B. Chen, U. Höpfner, M. V. Lakshmikantham, D. Vuillaume, T. Kawai, X. Wu, H. Tachibana, T. V. Hughes, H. Sakurai, J. W. Baldwin, C. Hosch, M. P. Cava, L. Brehmer, and G. J. Ashwell, J. Am. Chem. Soc. 1997, **119**, 10455.
[12]   R. M. Metzger, T. Xu, and I. R. Peterson, J. Phys. Chem. 2001, **B105**, 7280.
[13]   W. Schottky, Z. Phys. 1942, **118**, 539.
[14]   Y. Liu, Y. Xu, J. Wu, and D. Zhu, Solid St. Commun. 1995, **95**, 695.
[15]   Y. Liu, Y. Xu, and D. Zhu, Synth. Metals 1997, **90**, 143.
[16]   C. Krzeminski, C. Delerue, G. Allan, D. Vuillaume, and R. M. Metzger, Phys. Rev. 2001, **B64**, #085405.
[17]   V. Mujica, M.A. Ratner, and A. Nitzan, Chem. Phys. 2002, **281**, 147.
[18]   M. L. Chabinyc, X. Chen, R. E. Holmlin, H. Jacobs, H. Skulason, C. D. Frisbie, V. Mujica, M. A. Ratner, M. A. Rampi, and G. M. Whitesides, J. Am. Chem. Soc. 2002, **124**, 11730.
[19]   W. B. Davis, M. R. Wasiliewski, M. A. Ratner, V. Mujica, and A. Nitzan, J. Phys. Chem. 1997, **A101**, 6158.
[20]   G. J. Ashwell, J. R. Sambles, A. S. Martin, W. G. Parker, and M. Szablewski, J. Chem. Soc. Chem.Commun. 1990, 1374.
[21]   A. S. Martin, J. R. Sambles, and G.J. Ashwell, Phys. Rev. Lett. 1993, **70**, 218.
[22]   B. Chen and R. M. Metzger, J. Phys. Chem. 1999, **B103**, 4447.
[23]   J. W. Baldwin, B. Chen, S. C. Street, V. V. Konovalov, H. Sakurai, T. V. Hughes, C. S. Simpson, M. V. Lakshmikantham, M. P. Cava, L. D. Kispert, and R. M. Metzger, J. Phys. Chem. 1999, **B103**, 4269.
[24]   A. Honciuc, A. Otsuka, Y.-H. Wang, G. Saito, and R. M. Metzger, submitted for publication.
[25]   G. J. Ashwell and G. A. N. Paxton, Austr. J. Chem. 2002, **55**, 199.
[26]   R.M. Metzger, H. Tachibana, X. Wu, U. Höpfner, B. Chen, M.V. Lakshmikantham, and M. P. Cava, Synth. Metals 1997, **85**, 1359.

[27]   T. Xu, T. A. Morris, G. J. Szulczewski, R. R. Amaresh, Y. Gao, S. C. Street, L. D. Kispert, R. M. Metzger, and F. Terenziani, J. Phys. Chem. 2002, **B106**, 10374.

[28]   A. Jaiswal, R. R. Amaresh, M. V. Lakshikantham, A. Honciuc, M. P. Cava, and R. M. Metzger, Langmuir 2003, **19**, 9043.

[29]   J. W. Baldwin, R. R. Amaresh, I. R. Peterson, W. J. Shumate, M. P. Cava, M. A. Amiri, R. Hamilton, G. J. Ashwell, and R. M. Metzger, J. Phys. Chem. 2002, **B106**, 12158.

[30]   R. M. Metzger, J. W. Baldwin, W. J. Shumate, I. R. Peterson, P. Mani, G. J. Mankey, T. Morris, G. Szulczewski, S. Bosi, M. Prato, A. Comito, and Y. Rubin, J. Phys. Chem. 2003, **B107**, 1021.

[31]   A. Honciuc, A. Jaiswal, A. Gong, K. Ashworth, C. W. Spangler, I. R. Peterson, L. R. Dalton, and R. M. Metzger, J. Phys. Chem. 2005, **B109**, 857.

[32]   W. J. Shumate, D. L. Mattern, A. Jaiswal, D. A. Dixon, T. R. White, J. Burgess, A. Honciuc, and R. M. Metzger, submitted for publication.

[33]   W. J. Shumate, Ph.D. Dissertation, University of Alabama, 2005.

[34]   D. F. Perepichka, M. R. Bryce, C. Pearson, M. C. Petty, E. J. L. McInnes, and J. P. Zhao, Angew. Chem. Int. Ed. 2003, **42**, 4636.

[35]   G. Ho, J. R. Heath, M. Kontratenko, D. F. Perepichka, K. Arseneault, M. Pézolet, and M. R. Bryce, Chem. Eur. J. 2005, **11**, 2914.

[36]   G.J. Ashwell, W. D. Tyrrell, and A. J. Whittam, J. Mater. Chem. 2003, **13**, 2855.

[37]   G.J. Ashwell, A. Chwialkowska, and L. R. Herrmann High, J. Mater. Chem. 2004, **14**, 2848.

[38]   G. J. Ashwell and M. Berry, J. Mater. Chem. 2005, **15**, 108.

[39]   G. J. Ashwell, W. D. Tyrrell, and A. J. Whittam, J. Am. Chem. Soc. 2005, **126**, 7102.

[40]   G. M. Morales, P. Jiang, S. Yuan, Y. Lee, A. Sanchez, W. You, and L. Yu, J. Am. Chem. Soc. 2005, **127**, 10456.

[41]   M. Elbing, R. Ochs, M. Keotopp, M. Fischer, C. von Hänisch, F. Weigend, F. Evers, H. B. Weber, and M. Mayor, Proc. Natl. Acad. Sci. U. S. 2005, **102**, 8815-8820.

MOLECULAR ASSEMBLIES: BRIDGING THE GAP TO FORM MOLECULAR JUNCTIONS

C. R. Kagan, C. Lin[†]

IBM T. J. Watson Research Center, 1101 Kitchawan Road, Yorktown Heights, NY 10598
[†]Current Address: PPG Industries, 440 College Park Drive, Monroeville, PA 15146
Email: cheriek@us.ibm.com, chunlin@ppg.com

1 INTRODUCTION

Probing charge transport through single molecules or molecular assemblies requires the fabrication of electrodes to meet the dimension of the molecule. The 1-2 nm dimensions of most soluble and synthetically tractable molecular systems presents an incredible challenge as reliable junctions are difficult to fabricate. Junctions with dimensions less than ~15 nm are written with only limited statistics by state-of-the-art e-beam lithography. Many creative approaches are being pursued to fabricate 1-2 nm junctions, but knowledge of the structure of the molecule in the junction remains limited.[1] In addition the length of the molecule, which would form the channel of a device, is sufficiently short that tunnelling currents between electrodes may be too large for many applications and the introduction of a strongly coupled third electrode (gate) is prohibited.[2,3] In this proceedings, we describe the physics behind the challenges presented by scaling device dimensions to the molecular scale.

To meet the dimensions of reproducibly fabricated junctions using lithography and to allow the formation of a well-scaled transistor structure, we use chemistry to initiate and construct longer molecular systems from metal-metal bonded coordination compounds to bridge the gap. These metal-metal compounds have previously been studied in solution and as single crystals. They have reversible electrochemical properties and they can be used to engineer crystals with interesting structural motifs.[4,5] We showed that these compounds can be assembled from solution layer-by-layer on the surfaces of substrates, allowing their physical properties to be probed.[6]

2 METHOD AND RESULTS

2.1 Scaling devices to molecular dimensions

Separating electrodes by 1-2 nm, consistent with the dimensions of many molecular systems, presents both fabrication challenges using today's tooling and fundamental limitations for devices. The fundamental limitations arise from both carrier tunnelling, that limits the off current of devices, and electrostatics, if a three-terminal device is desired. Tunnelling currents between electrodes in vacuum depend on the workfunction of the metal electrodes, their separation, and their area.[7] Once a molecule is inserted in the junction, the tunnelling currents are modified and typically increase. The molecule lowers the barrier to carrier tunnelling provided by its electronic levels, but further the interaction between the molecule and the electrodes broadens and shifts the molecular levels, alters the metal workfunction, and creates dipoles at the interface.[8,9]

Introduction of a strongly coupled gate electrode requires fabrication of well-scaled junctions. Figure 1 shows schematics of (left) a well-scaled and (right) a poorly scaled transistor structure (independent of whether the structure is fabricated in a geometry rotated 90° or 180°). The constant potential lines are drawn to indicate the strength of the gate and drain electrodes, as the lines closest to the electrodes are the strongest. In the well-

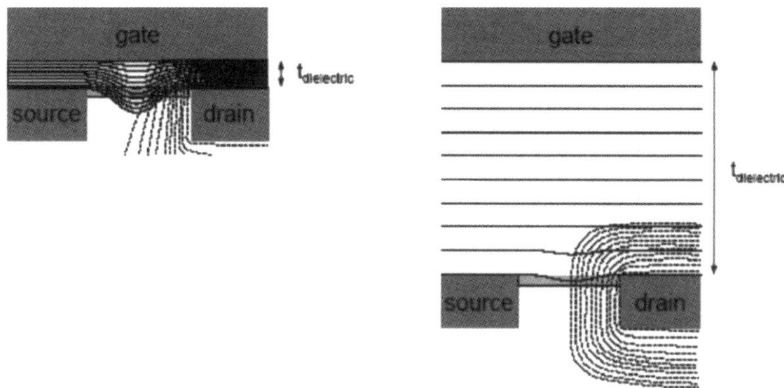

Figure 1 *Schematic of constant potential lines from the gate (solid lines) and drain (dotted lines) electrodes for (left) a well-scaled (right) poorly scaled transistor structure.*

scaled transistor, with a channel length longer than the dielectric thickness ($t_{dielectric}$), the constant potential lines extend well into the channel. In the poorly scaled structure, with a channel length less than the dielectric thickness, the potential lines from the gate barely penetrate into the channel and the lines from the drain are more substantial, overshadowing the influence of the gate.

The electrostatics of transistor scaling is well-known.[10] For a well-scaled device, channel lengths many times the thickness of the dielectric are required to achieve unhampered influence of the gate electrode. Challenges arise when channel lengths approach the molecular scale. The separation between source-gate and drain-gate electrodes must be large enough to limit tunnelling. In addition, fabrication of known dielectrics are >2 nm. Considering both the tunnelling and electrostatic constraints on devices suggests that channel lengths and therefore molecules longer than 3 nm are desired

2.2 Bridging molecular junctions using chemistry

We showed that layer-by-layer assembly may be used to build longer molecular systems to span the gap of nanoscale junctions.[5] The building block for the assembly was based on metal-metal coordination compounds, which are known to form a paddlewheel structure with equatorial and axial ligands (Figure 2).[2,3] In the example described below, we assembled multilayers layer-by-layer in the axial direction of the metal-metal compound as shown in Figure 2. In principle, while not yet demonstrated, the dimetal compounds and organic linkers added in each layer may be chosen to increase the functionality of the assembly, offering modularity in this approach.

To initiate the layer-by-layer assembly off the surfaces of Au electrodes, we assembled 4-pyridylethylmercaptan from 0.1 mM ethanolic solutions. The mercaptan head group selectively binds to the Au electrodes and the pyridine tail group provides the nitrogen donation to coordinate the dimetal compound in the axial direction. The mercaptan functionalized surface is then dipped in 0.1 mM toluene solution of the dirhodium compound **1**. Compound **1** was synthesized according to previous report.[11] The organic linkers were assembled by dipping the substrates in either an 0.1 mM ether solution of *trans*-1,2-bis(4-pyridyl)ethylene, **2**, or an 0.1 mM ethanolic solution of zinc 5,10,15,20-tetra(4-pyridyl)-21*H*,23*H*-porphine, **3**.

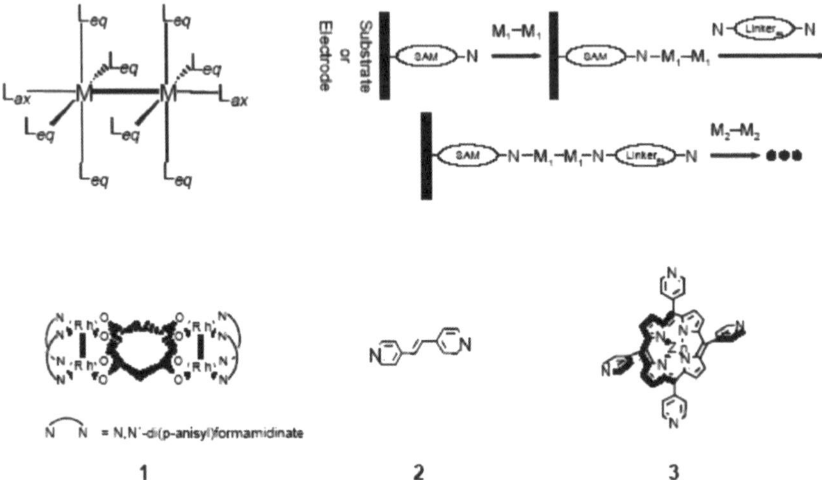

Figure 2   *Left: Paddlewheel structure of metal-metal (M-M) bonded coordination compound showing axial (L_{ax}) and equatorial (L_{eq}) ligands. Right: Scheme for self-assembly of a compound on a substrate or electrode surface with a tail group that initiates the subsequent layer-by-layer assembly of metal-metal coordination compounds and organic linkers. The dimetal or linker groups need not be the same upon deposition of each layer as indicated by M_1-M_1 and M_2-M_2. Compounds 1, 2, and 3 used in the layer-by-layer assembly.*

UV-visible absorption was used to characterize the layer-by-layer assembly of the dirhodium compound **1** and the organic linkers **2** and **3**, shown in Figure 3. The layer-by-layer assembly is followed for the deposition of 10 bilayers. The addition in each step adds to the assembly linearly. Assuming the molar absorption coefficients from solution, for the assembly with linker **2** the surface coverage was estimated at ~90% in comparison with the known molecular density from single crystal solutions of the analogue compound.[11]

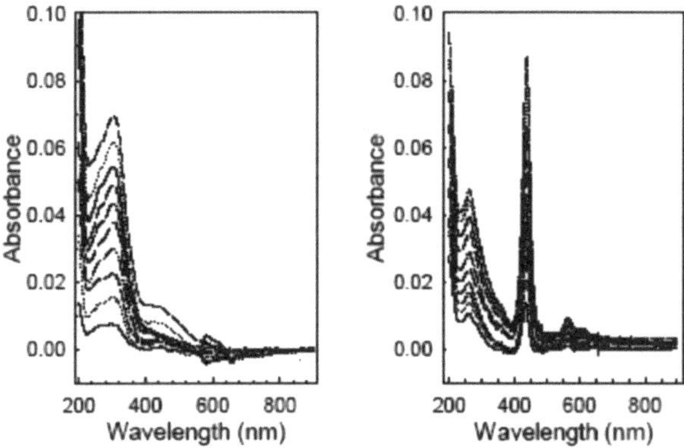

Figure 3 *Absorption spectra following the sequential assembly of 10 bilayers of the dirhodium compound 1 and the linkers (left) 2 and (right) 3.*

In addition the layer-by-layer assembly of the compounds was followed by atomic force microscopy to grow linearly with a spacing commensurate with the single crystal.[6] The electrochemical properties were also recorded and showed reversible oxidation-reduction consistent with Rh_2^{4+} to Rh_2^{5+} as previously measured in solution.[6]

We assembled these compounds layer-by-layer to bridge the gap of 80 nm channel length devices. Once the junctions are spanned, two-terminal I-V characteristics as those shown in Figure 4 are measured at room temperature in vacuum or similarly in nitrogen. At low voltage the characteristics may be cycled without hysteresis. As the voltage increases, the current goes through multiple maxima, returning at a different current level. The I-V

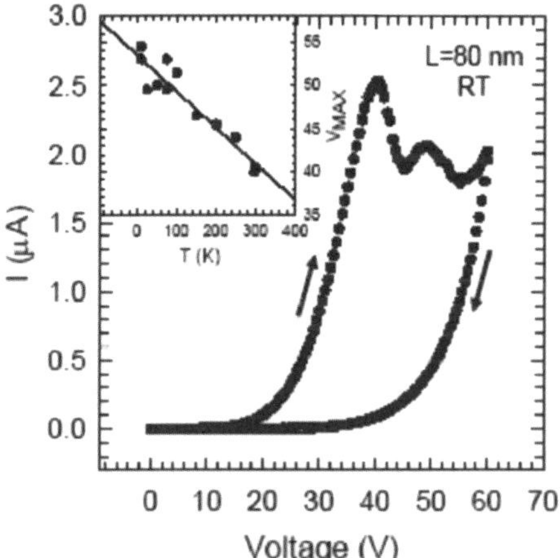

Figure 4 *Two terminal I-V characteristics for the dirhodium compound with linker 2 sweeping the voltage both forward and backward. The compound is assembled layer-by-layer to bridge an 80 nm channel and the trace was measured in vacuum. Inset: The first peak in the I-V characteristics shifts to higher voltages at lower temperatures.*

characteristics are irreversible and cannot be recovered by sweeping the device at opposite polarity. These characteristics were found to be reproducible across the few hundred devices on each sample and from sample-to-sample. Measurements at lower temperatures results in similar I-V characteristics, but shifted to higher voltages, as indicated by the voltage maximum for the first peak as a function of temperature (inset Figure 4). The characteristics are similar to those for samples prepared by spin-coating **1** and **2** together from solution, but the layer-by-layer assembly gives rise to current levels 10^3 higher. We speculate that the compounds oxidize irreversibly in the solid state, but additional experiments are required to understand the current-voltage behaviour.

3 CONCLUSION

Molecular scale devices present both fabrication and fundamental challenges. Devices having low tunnelling currents and that allow the introduction of a strongly coupled third terminal require channel lengths >3 nm, longer than most readily synthesized molecular systems. We showed layer-by-layer assembly of metal-metal coordination compounds may be used to bridge the gap of longer channel length junctions as a route to overcome these limitations. Further manipulation of the chemistry of the monolayers assembled on the electrode surfaces and the metal and organic ligands (both equatorial and axial) are required to understand the observed electrical characteristics.

References

1. C. R. Kagan, M. A. Ratner, *MRS Bulletin*, 2004, **29**, 376.
2. A. W. Ghosh, T. Rakshit, S. Datta, cond-mat/0212166 2002.
3. C. R. Kagan, A. Afzali, R. Martel, L. M. Gignac, P. M. Solomon, A. Schrott, B. Ek, *Nano Letters*, 2003, **3**, 119.
4. F. A. Cotton, R. A. Walton, *Multiple Bonds Between Metal Atoms*, 2nd ed.; Clarendon Press: Oxford, 1993.
5. (a) F. A. Cotton, C. Lin, C. A. Murillo, *Acc. Chem. Res.* 2001, **34**, 759; F. A. Cotton, C. Lin, C. A. Murillo, *Proc. Natl. Acad. Sci. U.S.A.* 2002, **99**, 4810.
6. C. Lin and C. R. Kagan, J. Am. Chem. Soc., 2003, **125**, 336.
7. E. C. Teague, *J. Res. Nat. Bur. Stds*, 1986, **91**, 171.
8. N. D. Lang, Ph. Avouris, *Phys. Rev. B* 2001, **64**, 125323.
9. C. Schönenberger et al, J. Phys. Chem. 1995, **99**, 3259.
10. D. J. Frank, R. H. Dennard, E. Nowak, P. M. Solomon, H-S. P. Wong, *Proc. IEEE* 2001, **89**, 259.
11. F. A. Cotton, C. Lin, C. A. Murillo *Chem. Commun.* 2001, 11.

STUDY OF HOLE INJECTION IN HOLE-ONLY SINGLE-CARRIER DEVICES

Masamichi Fujihira[a] and Chimed Ganzorig

Department of Biomolecular Engineering, Tokyo Institute of Technology, 4259
Nagatsuta, Midori-ku, Yokohama 226-8501, Japan. [a]E-mail: mfujihir@bio.titech.ac.jp;
Fax: +81 45 924 5817

1 INTRODUCTION

After the appearance of the first report of a two-layer organic EL device by Tang and
VanSlyke in 1987,[1] there have been numerous attempts to enhance hole and electron
injection at electrodes. A two-layer device configuration with a tris(8-hydroxyquinoline)
aluminum (Alq_3) electron transport layer (ETL) and an N,N'-diphenyl-N,N'-bis(3-
methylphenyl)-1,1'-biphenyl-4,4' diamine (TPD) hole transport layer (HTL) is illustrated
in Fig. 1a. Injection of charges from most electrodes requires that charges surmount or
tunnel through barrier at the interface[2] as is clear from the energy diagram under forward
bias of a hetero junction device schematically shown in Fig. 1b. Energy barriers for
electron and hole injection (ϕ_e and ϕ_h, respectively) are indicated. The energy barrier
heights for electron and hole injection can be estimated by examining the positions of the
electrode Fermi levels and the positions of the lowest unoccupied molecular orbital
(LUMO) and the highest occupied molecular orbital (HOMO) of Alq_3 and TPD,
respectively. In the absence of dopants as shown in Fig. 1b, HTL and ETL behave almost
like insulator and the applied electric potential changes linearly across these organic
layers,[2] where the built-in potential is neglected. Therefore, low work function cathodes
and high work function anodes are needed to fabricate efficient EL devices by improving
the charge injection.

A crucial process in the operation of organic EL devices is charge injection from
electrodes into organic semiconductors.[2-4] The electrode material for hole injection is
almost invariably indium-tin-oxide (ITO),[1] while a number of low work function metals[1,5-
8] are suitable for electron injection. These metals are highly reactive with oxygen and
moisture, and thus tend to give unstable cathodes. The search for better cathode materials
to replace these air sensitive metals has led to the discovery of bilayer cathodes by
inserting a thin layer of alkali metal halides.[9,10]

ITO is the most widely used as a transparent anode material in organic EL devices[1]
due to its high conductivity, the work function, and transparency in the visible spectral
range. However, the work function of ITO is generally not sufficiently large for the
contact to be ohmic and there is a barrier to hole injection. Thus, various surface
treatments of ITO have been attempted to change the work function of ITO in order to
reduce the hole injection barrier height (reviewed in Ref. 2).

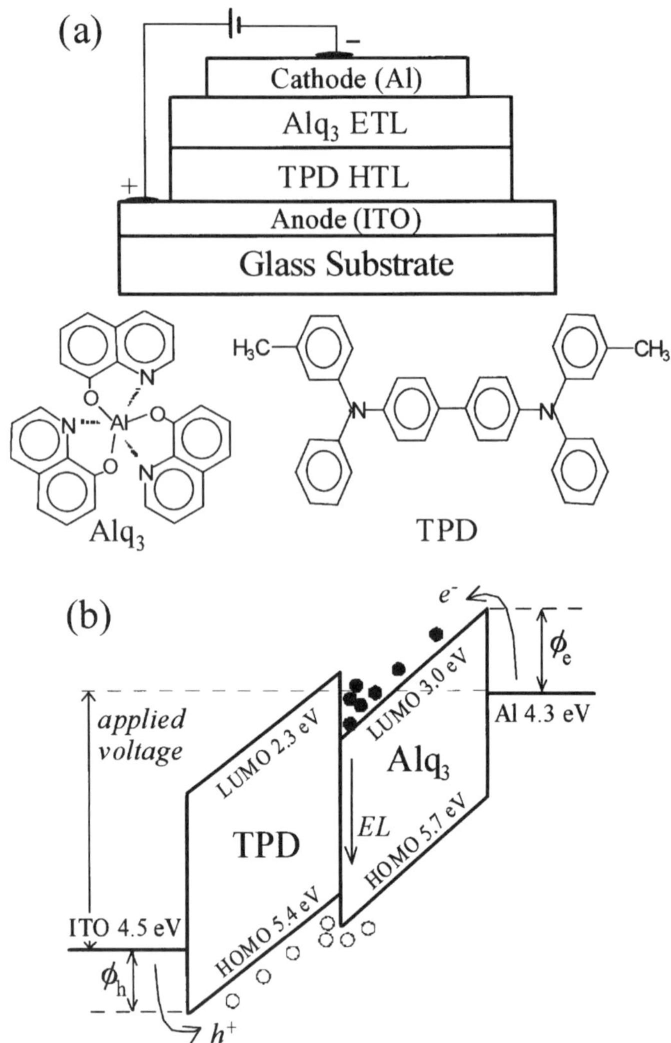

Figure 1   *(a) Typical device configuration and molecular structures of Alq_3 and TPD and (b) schematic energy diagram of a two-layer device without dopants.*

The work function control by chemical modification[11-15] has recently been applied to enhanced charge injection at ITO anodes for organic EL devices.[2,16-25] In our previous paper,[16] we reported that the chemical modification of ITO with different binding groups ($-SO_2Cl$, $-COCl$,[26] and $-PO_2Cl_2$) of *p*-substituted (H−, Cl−, and CF_3−) benzene derivatives leads to enormous increase in ITO work function. Selection of bonding modes of molecules with high permanent dipole moments is important because a dipole layer is created through the bonding[2,16] in the same way as in chemical adsorption of acids and bases.[27,28] We also studied the effect of an additional dipole layer of the terminal group, on the work function of ITO and EL characteristics using *p*-phenylene linkage.[16] In particular, the drive voltages in the EL characterisitics, which were well correlated with the change in the work function, were dramatically improved by the chemical modification.[2,16,17]

In this paper, we report the use of chemically modified ITO with various self-assembling dipole molecules forming effective monolayers for hole injection control at the anode. To study the possibility of achieving an ohmic contact, we examine the behavior of current density-voltage (*J-V*) characteristics caused by varying the ITO work function in hole-only devices with a single TPD HTL with different thickness. In addition, we compare the device performance before (as-cleaned and UV-ozone) and after the surface modifications of ITO surfaces.

2 EXPERIMENTAL

2.1 Materials

Organic semiconductor materials such as TPD are known to be stable during vapor deposition.[8] TPD (purity >99%) was purchased from Aldrich. All other solvents and reagents used were of analytical grade and purchased commercially. Anhydrous grade solvents were used during the chemical modification.

2.2 Substrate Preparation

The substrates used in this work were 160-nm-thick ITO coated glass plates with a sheet resistance of *ca.* 15 Ω/square supplied by Sanyo Vacuum Industries. The substrates were cut into 15×20 mm^2 sample slides, and the ITO layers were etched by an aqua regia solution to form 5-mm-wide stripes for use as anodes. Patterned ITO substrates were cleaned and rinsed by sonication successively in two detergent solutions (Extran MA 03, pH 6.8, MERCK and Kontaminon O, pH 10, WAKO) and deionized water, and then stored under isopropanol until being required. Prior to use, the substrates were further cleaned and ultrasonicated successively in acetone and in isopropanol, and then transferred to boiled isopropanol. The substrates were finally blown dry with pure nitrogen gas before modification. This cleaned ITO will be called hereafter "as-cleaned ITO".

2.3 Electrode Modification

Schematic drawing of the formation of monolayers by self-assembly at ITO electrode surfaces is illustrated in Fig. 2. After cleaning with acetone and isopropanol, the ITO substrates were immersed for 1 - 30 min in dichloromethane solutions containing 1 mM of p-substituted benzoylchloride, or phenylphosphoryl dichloride (Tokyo Chemical Industry).[26] The modified ITO electrodes were rinsed in pure dichloromethane to remove excess unbound molecules and then vacuum dried for 1 h.

2.4 Electrode Characterization

The change in the work function due to the surface modification of ITO was observed by scanning surface potential microscopy (SSPM)[11-13,29] using gold-coated atomic force microscope (AFM) tips derivatized with 1-decanethiol. The modified tips gave reproducible results. For comparison, as-cleaned and UV-ozone (for 20 min)[17] treated ITO electrodes without chemical modification were also used for the measurement of the work function by SSPM. Static water contact angles were measured using a sessile drop method with a contact angle meter CA-X type produced by Kyowa Interface Science Co., Ltd.

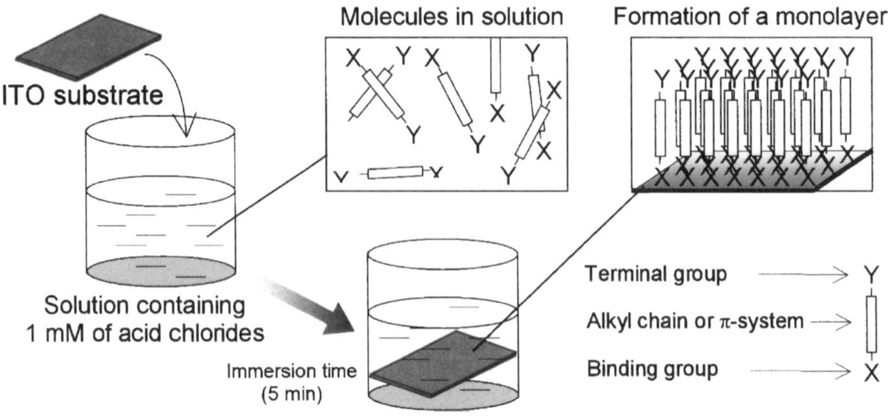

Figure 2 *Schematic drawing of a self-assembled monolayer formed by simply immersing an ITO substrate into a solution containing self-assembling organic molecules.*

2.5 Device Fabrication and Characterization

For the study of hole injection, the typical hole-only single-layer devices of ITO(various treated)/TPD(50, 75, 100, 125, and 150 nm)/Al were fabricated by using a vapor deposition system described previously.[7] The TPD and Al were deposited at a base pressure of $2\text{-}3\times10^{-6}$ Torr with deposition rates of 2.0-2.5 and 9-11 Å s^{-1}, respectively. The film thickness of each layer was monitored by a quartz crystal microbalance with a CRTM 8000 controller (ULVAC Techno. Ltd.) placed in the same height as the substrate. The emitting area was defined by using a shadow mask to be 5×5 mm^2. EL characteristics of the devices were measured using an Advantest 6145 electrometer with a Topcon BM-3 luminance meter at room temperature under ambient atmosphere.

3 RESULTS AND DISCUSSION

3.1 Tuning the Work Function of ITO

Contact potential difference (CPD) for clean metal surfaces is given by the difference in the work functions of the two materials. The work functions can be changed by adsorption of molecules with different dipole moments.[12,29] In a simple model, the CPD between the two materials is $V_{CPD} = - (\phi_1 - \phi_2)/e$, where ϕ_1 and ϕ_2 are the work functions. In the present work, ϕ_1 and ϕ_2 are the work functions of the ITO surface including the change due to the adsorbed layers and the gold coated AFM tip modified with 1-decanethiol, respectively. The CPD decreases with the increase in the work function of ITO.[16,29]

First, it is necessary to determine the optimal immersion time required for monolayer formation at the ITO electrode surface. Figure 3 shows the dependence of CPD values of ITO on the immersion time for *p*-chlorophenylphosphoryl dichloride (at 1 mM). The most effective time of immersion was around 5 min to complete the chemical modification of ITO surfaces. In addition, we also examined the CPD values after 5 min of the surface modification of ITO as a function of the concentration of *p*-chlorophenylphosphoryl dichloride, as shown in Fig. 4. 1 mM was therefore adopted as the best concentration used in this study. Measured contact angles against water on ITO surfaces modified under the optimized conditions, as-cleaned, and UV ozone treated ITO are listed in Table 1.

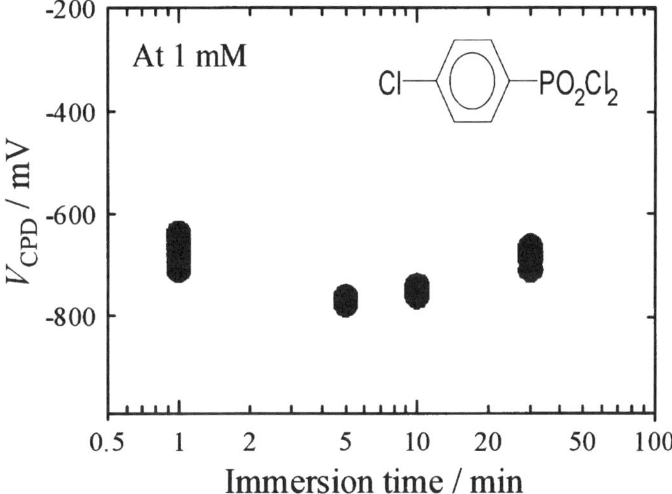

Figure 3 *CPD values as a function of the immersion time for ITO modified with p-chlorophenylphosphoryl dichloride at a concentration of 1 mM.*

Table 1 *Contact angles measured for deionized water on the substrates of as-cleaned ITO and various treated ITO substrates*

Substrate	Contact angle / °
ITO (as-cleaned)	37.3 ± 1.1
ITO (UV ozone treated)	10.9 ± 0.8
ITO (modified with –COCl)	52.8 ± 1.7
ITO (modified with –PO₂Cl₂)	61.8 ± 2.1

In Fig. 5a, CPD values observed are shown for ITO treated at 120 °C after the cleaning, as-cleaned, treated with UV-ozone, and modified with –COCl, and –PO$_2$Cl$_2$ binding groups of *p*-chlorobenzene derivatives. The values of CPDs of these ITO surfaces were measured against the modified gold AFM tip.[16] Then, calculated CPDs of ITO as-cleaned, treated with UV-ozone, and modified with –COCl, and –PO$_2$Cl$_2$ binding sites *with respect to ITO treated at 120 °C after the cleaning* are ~ –150, –280, –410, and –840 mV, respectively. The work function of ITO measured by UV photoelectron spectroscopy has a consistent value of 4.5±0.1 eV, even though originating from different providers.[30-32] The work functions of various ITO shown in Fig. 5b were estimated from the CPD values calculated above together with the assumption that the work function of ITO treated at 120 °C after the cleaning agrees with the consistent value of 4.5±0.1 eV for the work function of as-received bare ITO.[30-32] This type of work function fine tuning of ITO plays a crucial role in balancing the rate at which holes and electrons are injected into two-layer organic thin film with a hetero junction.[2,16,17,33]

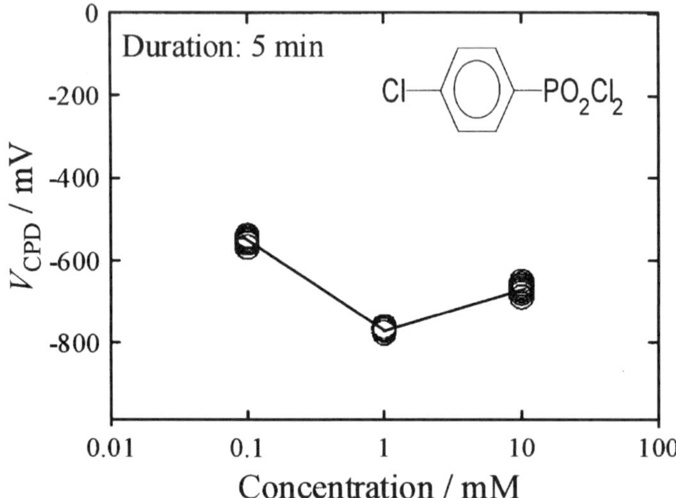

Figure 4   *CPD values as a function of the concentration for ITO modified with p-chlorophenylphosphoryl dichloride for 5 min.*

3.2 Correlation of Device Performance with Changes in the Work Function of ITO

To achieve enhanced hole injection from ITO anode, its Fermi level must be aligned with the HOMO of the adjacent organic film of TPD. The molecular approach allows fine-tuning the work function using organic molecules with dipole moments with different magnitudes and directions.[16] The work function can be changed (the effective work function: ϕ_{eff}) when surface dipole of self-assembled monolayer (SAM) of molecules is oriented as shown in Fig. 6. An interface dipole with its negative end pointing toward the organic layer and its positive end toward the electrode surface increases the ITO work function (i.e. the Fermi level is down) and HOMO energy level in the organic layer is raised by adding an electrostatic energy. As a result, the hole injection barrier ϕ_h is reduced (Fig. 6b). Reversing the direction of the dipole, we can reduce the work function, and thus the electron injection barrier, ϕ_e, as shown in Fig. 6c. Thus, the work function increase (or decrease) is associated with enhanced hole (or electron) injection.[16]

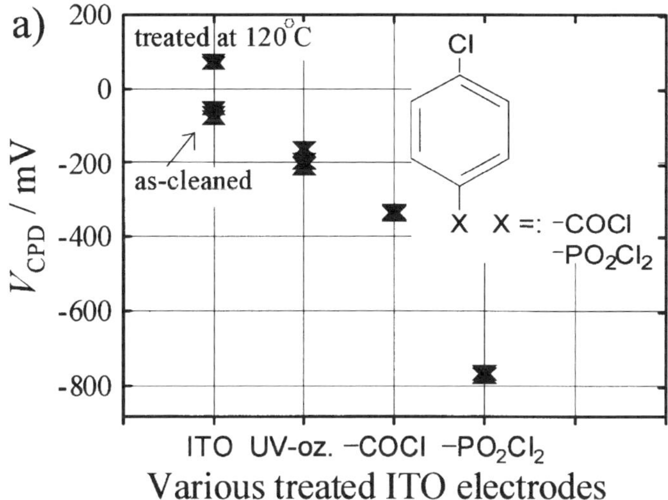

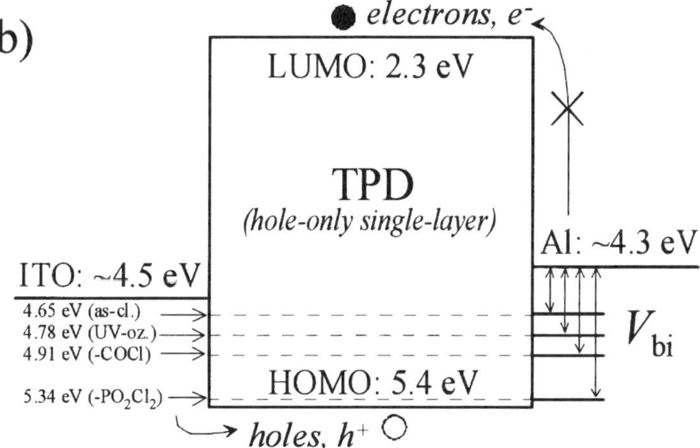

Figure 5   *(a) Change in CPD of variously treated ITO anodes and (b) schematic energy diagram for ITO(variously treated)/TPD(150 nm)/Al hole-only single-layer devices.*

The work function control by the adsorption of SAM is simplified in Fig. 6. Since an interface is newly formed between the SAM and the organic layer, there can be dipolar layer formation at this new organic/organic interface. This effect should be taken into account when the energetics at the interface were examined in more detail. However, the effect is usually small[34] and is neglected in the present work. Various possible origins for interface dipoles have been proposed.[29,35]

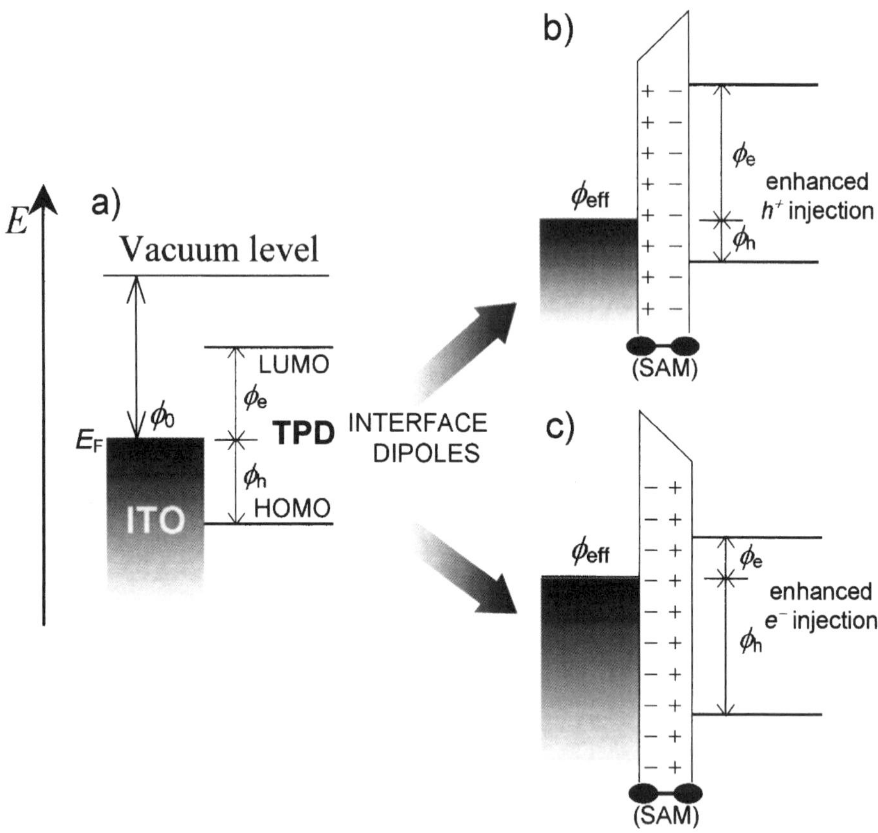

Figure 6   *Schematic energy level diagram showing the impact of the formation of an interface dipole at an ITO/TPD interface: (a) the barrier in the absence of the dipole layer; (b) and (c) reduction of the hole injection and the electron injection barrier, respectively, in the presence of the interface dipole layer on the surfaces of ITO. Here, the interface dipoles with opposite signs are directed toward the ITO surfaces (b) increasing and (c) decreasing the work function of ITO.*

In Fig. 7, J-V_{bias} characteristics of hole-only single-layer devices with ITO modified with p-chlorobenzene derivatives with –COCl and –PO$_2$Cl$_2$ binding groups are compared with those of the devices with as-cleaned and UV-ozone treated ITO. Among them, –PO$_2$Cl$_2$ exhibited the best J-V_{bias} characteristics and gave an extremely low operating voltage. The binding –COCl group gave better J-V_{bias} characteristics than UV-ozone treated ITO. These results suggest that the efficient hole injection into TPD was due to the increase in the work function of ITO covered with the two dipole layers introduced by the chemical modification.[2,16] In particular, the result shows that the selection of the binding sites is also important to tune the total work function change of ITO. Therefore, the device characteristics are strongly correlated with the changes in the observed work functions of various modified ITO (see Fig. 5).

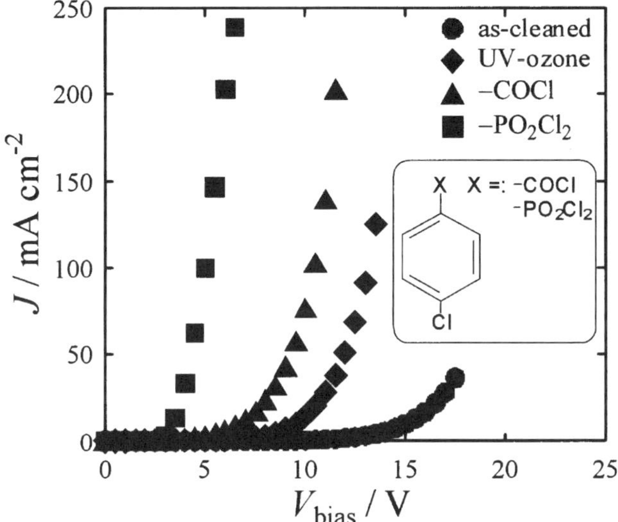

Figure 7   *J-V_{bias} characteristics of the devices with variously treated ITO anodes. The device configuration was ITO(variously treated)/TPD(150 nm)/Al. In the inset, chemical structures of p-chlorobenzoyl chloride and p-chlorophenylphosphoryl dichloride used for the chemical modification are shown.*

The effect of TPD film thickness on the J-V_{bias} characteristics is shown in Fig. 8 for ITO(various treated)/TPD/Al devices. The total thickness of TPD was changed to be 50, 75, 100, 125, and 150 nm in each series. There is negligible luminance observed in all devices studied because of the barriers for hole injection from ITO to TPD are much lower than that of the electron injection from Al to TPD. As the barrier height is increased, the current density is reduced, showing that contacts limit the current flow. Currents are decreased drastically with increase in thickness of TPD, especially for the case with a large barrier height (see Fig. 8a).

The corrected bias voltages (V_c) giving a current density of 10 mA cm^{-2} as a function of the thickness of the TPD layer on variously treated ITO are shown in Fig. 9, where the changes in built-in potential (V_{bi}) are taken into account for the corrected bias voltages ($V_c = V_{bias} - V_{bi}$). In the simplest picture, where the bands of the organic semiconductor remain rigid, V_{bi} is equal to the work function difference between the anode and the cathode.[36] As the studied devices have the same cathode of Al, the changes in V_{bi} obtained for devices with variously treated ITO anodes, are equal to the changes in the ITO work function (see Fig. 5b). All ITO treated surfaces show a linear thickness dependence with different slopes. The results suggest that the same current density was observed on each modified ITO anode when the electric field, V_c/d, was the same. The slopes of $V_c \propto d$ shown in Fig. 9 is highest for as-cleaned ITO and followed by those for ITO with UV-ozone, –COCl, and –PO$_2$Cl$_2$ binding groups of *p*-chlorobenzene derivatives.

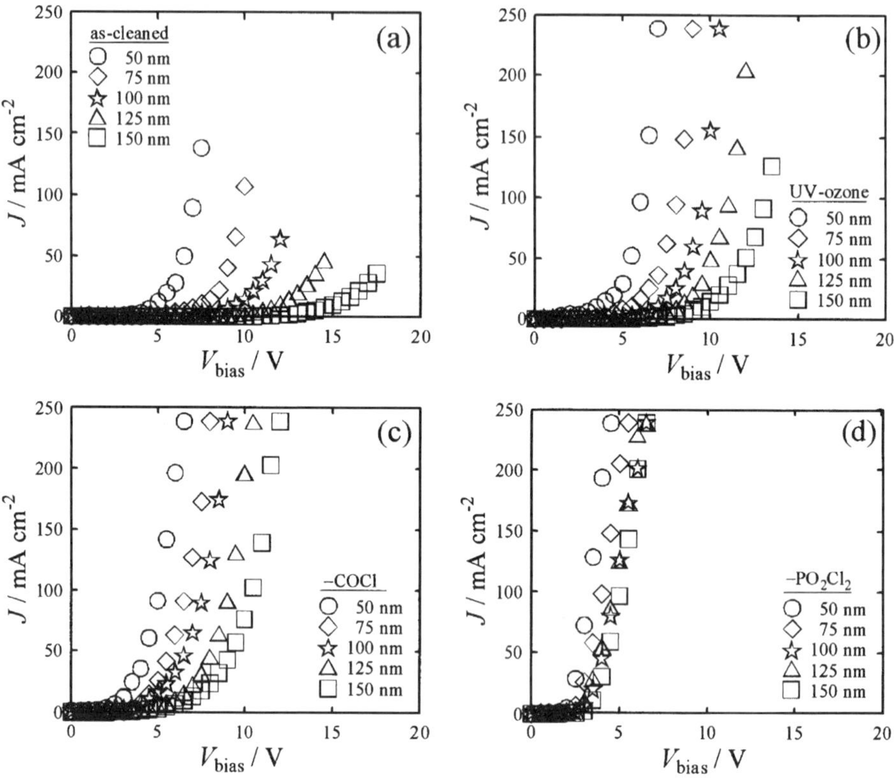

Figure 8   *The effect of TPD film thickness on the J-V_{bias} characteristics of ITO(variously treated)/TPD/Al devices with various surface treatments of ITO: (a) as-cleaned, (b) UV-ozone, (c) p-chlorobenzoyl chloride, and (d) p-chlorophenylphosphoryl dichloride. The thicknesses of TPD were 50, 75, 100, 125, and 150 nm in each series.*

3.3 Analysis of Current Density-Voltage Characteristics

Several recent reviews[37-46] have dealt with fundamental aspects of the basic device physics and related phenomena such as charge injection, transport, and recombination relevant to organic EL devices. Depending on the magnitude of the energy barrier height, the current flowing through the device can either be space-charge limited (transport limited), or injection limited.[47-49] A necessary condition for space-charge limited current (SCLC) is that one of the electrodes can supply more charge carriers per unit time than those can be transported through the organic dielectric.[38,49] A contact that behaves in this way is called an ohmic contact. At an ohmic contact the electric field vanishes owing to screening by the space charge associated with unipolar current flow. This requires an injection barrier small enough to guarantee the efficient injection without the assistance of an external electric field.[38] The problem of SCLC in solids has been extensively treated by Lampert and Mark.[47] Malliaras and coworkers[46] have discussed the forming ohmic contacts and presented examples of contact engineering in organic EL devices.

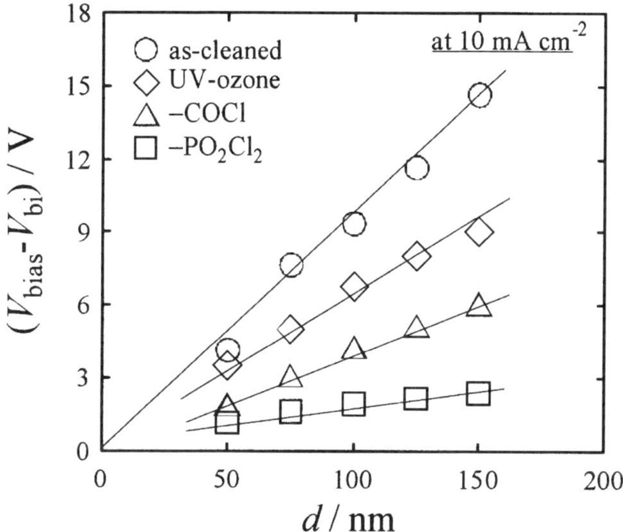

Figure 9 *The corrected bias voltages ($V_c = V_{bias} - V_{bi}$) observed at 10 mA cm^{-2} as a function of TPD film thickness (d) using ITO treated with as-cleaned, UV-ozone, –COCl, and –PO$_2$Cl$_2$ binding groups of p-chlorobenzene derivatives.*

In general, two different mechanisms are considered for the observed injection current. Injection-limited conduction is commonly described either by the Richardson-Schottky model for thermionic injection or by the Fowler-Nordheim model for tunneling injection.[50] It is considered for organic EL devices that the latter model is applicable under high electric fields,[51,52] while the former model is applicable under low electric fields. The injection models described above were developed for band type materials, and it is difficult to rationalize an application of these models to disordered organic materials, where the charge-carrier mobility is low (typically $<<10^{-3}$ cm^2/V s), which means that charge carriers are localized and transport involves discrete hopping within a distribution of energy states.[41,53]

The J-V_{bias} characteristics of holes in TPD are investigated in both the bulk and contact-limited regime using variously treated ITO anodes. Figure 5b schematically shows a band diagram indicating the work functions of the variously treated ITO anodes. Malliaras *et al.*[36] have shown that proper accounting of V_{bi} leads to a consistent analysis of the J-V_{bias} data.

We use the Fowler-Nordheim tunneling theory[54] to calculate the barrier heights at the interfaces of the four hole-only ITO/TPD(150 nm)/Al devices, as shown in Fig. 10, where the changes in V_{bi} are taken into account for the corrected bias voltages ($V_{\text{c}} = V_{\text{bias}} - V_{\text{bi}}$). As predicted, the plots show linear behavior at high fields, i.e., above the threshold voltage. In the linear region, the calculated barrier heights for hole injection into the devices are ~ 0.21 eV for ITO(as-cleaned)/TPD, ~ 0.17 eV for ITO(UV-ozone)/TPD, and ~ 0.14 eV for ITO(modified with –COCl)/TPD interfaces. The plots for these devices at high fields do tend towards a straight line, depending on the work function of ITO. For the device with ITO modified with $-PO_2Cl_2$, the plot is no more linear behavior at high fields. Deviation from linear parts at lower fields on other ITO is likely to be due to a thermionic emission contribution to the current.[51]

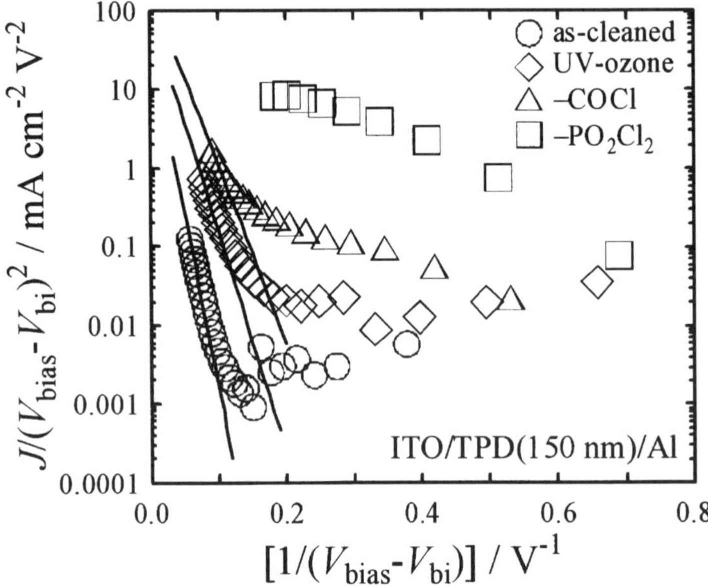

Figure 10 *Fowler-Nordheim plots for tunneling of holes from variously treated ITO into TPD in ITO/TPD(150 nm)/Al devices. Solid lines indicate the linear parts of the curves above the threshold voltages.*

The J-V_{c} characteristics in log-log plot of ITO(modified with $-PO_2Cl_2$)/TPD(50, 75, 100, 125, and 150 nm)/Al devices are shown in Fig. 11, where the changes in V_{bi} are taken into account for the corrected bias voltages ($V_c = V_{bias} - V_{bi}$). The J-V_c characteristics of these devices with different thicknesses exhibit a similar behavior of SCLC. The J-V_c characteristics of the devices show three different regions, and each can be fitted with the power law $J \propto V_c^{l+1}$. These regions are well explained by the standard SCLC theory[47] and a combination model of SCLC and trapped-charge-limited current (TCLC) models.[41,47-50,55-57] In the low bias voltage region ($V_{bias} < 2$ V), the current is proportional to the square of the corrected bias voltage ($J \propto V_c^2$) and then a rapid increase in current density is observed over the threshold voltage of ~ 2 V. In the trap filled region, the J-V curves follow the TCLC model well, with a further increase of the voltage ($J \propto V_c^{4.5-6}$) ($l = 3.5$-5). In the final current region, all traps are filled up and the J-V_c characteristics have become more SCLC behavior. The slope is equal to ~ 2.

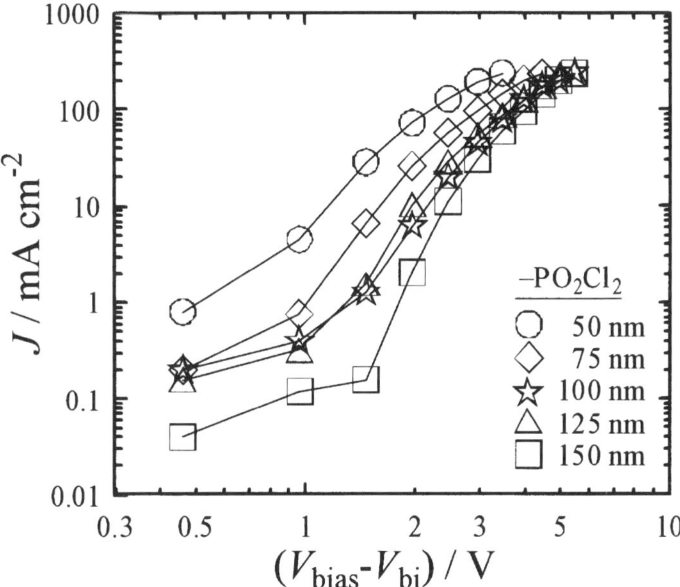

Figure 11 *The log-log plots of the ITO(modified with $-PO_2Cl_2$)/TPD/Al devices with different TPD film thicknesses showing the power-law dependence of current density vs. the corrected bias voltage.*

Thickness dependence of the typical J-V_{bias} characteristics for the ITO/TPD/Al devices is shown in Fig. 8. In general, reducing the thickness of TPD clearly lowers the applied bias voltage in the devices. Figure 12a shows the thickness dependence of TPD film on the J-V_c characteristics in the log-log plot for ITO/TPD/Al devices with as-cleaned ITO anodes. The current density vs. electric field dependence for these devices is plotted in Fig. 12b. Here, all plots at high electric fields merge into one straight line. The J-V_c characteristics of these devices with as-cleaned ITO are dependent on the electric field and not on the applied bias voltage. Such field-dependent behavior is usually associated with tunneling processes.[51,52] Field-dependent behavior is likely to be due to the injection limitation to the current.

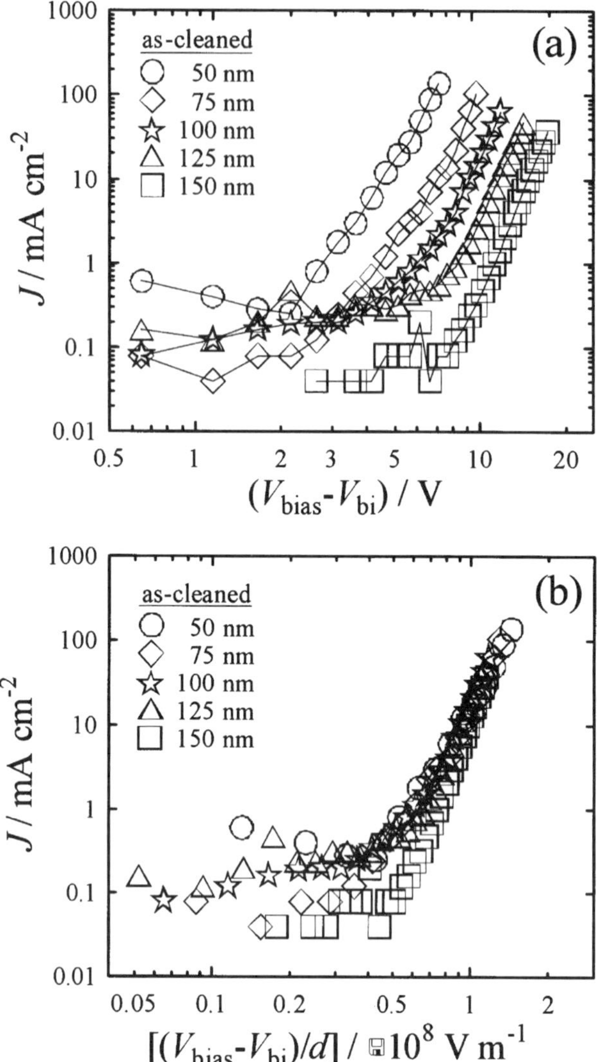

Figure 12 *(a) The effect of TPD film thickness on the J-V_c characteristics for the ITO/TPD/Al devices with as-cleaned ITO and (b) the current density vs. the corrected electric field [$(V_{bias} - V_{bi})/d$] dependence.*

A range of different bulk and injection dominated mechanisms have been proposed to interpret the J-V_{bias} characteristics in the EL devices. According to the theory formulated by Lampert and Mark,[47] in the case of trap-free SCLC currents under single carrier injection (and electric field independent drift mobility), the current density follows a V^2 dependence. For the single-layer devices, the TPD was assumed to be trap-free[41,58,59] against our present observation. Campbell *et al.*[60] have shown that TPD based single-layer devices display purely injection-limited behavior with cleaned ITO. In the present work, however, both injection- and transport-limited behaviors were observed on the chemically modified ITO.

4 CONCLUSION

We have reported the use of chemically modified ITO with various self-assembling monolayers as effective dipole layers for hole injection at the ITO anode. To investigate the possibility of achieving an ohmic contact, we have examined the J-V_{bias} characteristics caused by varying the work function of ITO in the ITO/TPD/Al devices with different TPD film thicknesses. In addition, we have also examined how the behavior of the J-V_{bias} curves in these devices is interpreted by the transport- and injection-limited conduction mechanisms. Upon grafting with *p*-chlorophenylphosphoryl dichloride, the J-V_{bias} characteristics become a SCLC behavior. Tuning work function of ITO by the chemical modification is a key parameter, which is useful for interpreting the experimental data, and leads to improvements in the J-V_{bias} characteristics of the devices.

Acknowledgements

We would like to gratefully acknowledge and thank financial support of this work provided by the Grant-in-Aid for Creative Scientific Research on "Devices on molecular and DNA levels" (No. 13GS0017) from the Ministry of Education, Culture, Sports, Science and Technology of Japan.

References

1   W. Tang and S. A. VanSlyke, *Appl. Phys. Lett.*, 1987, **51**, 913.

2   M. Fujihira and C. Ganzorig, 'Molecular Control of Electron and Hole Injection at Electrodes and at Organic Layer Interfaces in Organic Electroluminescent Devices' in *Conjugated Polymer and Molecular Interfaces*, eds., A. Kahn, J. J. Pireaux, W. R. Salaneck and K. Seki, Marcel Dekker, New York, 2002, Chapter 26, pp. 817-858.

3   T. M. Brown, R. H. Friend, I. S. Millard, D. J. Lacey, T. Butler, J. H. Burroughes and F. Cacialli, *J. Appl. Phys.*, 2003, **93**, 6159.

4   T. Yokoyama, D. Yoshimura, E. Ito, H. Ishii, Y. Ouchi and K. Seki, *Jpn. J. Appl. Phys.*, 2003, **42**, 3666.

5   D. Braun and A. J. Heeger, *Appl. Phys. Lett.*, 1991, **58**, 1982.

6   J. Kido, K. Nagai and Y. Okamoto, *IEEE Trans. Electron Dev.*, 1993, **40**, 1342.

7   C. Ganzorig and M. Fujihira, *Jpn. J. Appl. Phys.*, 1999, **38**, L1348.

8   C. Ganzorig and M. Fujihira, *Appl. Phys. Lett.*, 2004, **85**, 4774.

9   T. Wakimoto, Y. Fukuda, K. Nagayama, A. Yokoi, H. Nakada and M. Tsuchida, *IEEE Trans. Electron Dev.*, 1997, **44**, 1245.

10   L. S. Hung, C. W. Tang and M. G. Mason, *Appl. Phys. Lett.*, 1997, **70**, 152.

11   M. Fujihira, H. Kawate and M. Yasutake, *Chem. Lett.*, 1992, **21**, 2223.

12   M. Fujihira and H. Kawate, *Thin Solid Films*, 1994, **242**, 163.
13   M. Fujihira and H. Kawate, *J. Vac. Sci. Technol. B*, 1994, **12**, 1604.
14   I. H. Campbell, S. Rubin, T. A. Zawodzinski, J. D. Kress, R. L. Martin, D. L. Smith, N. N. Barashkov and J. P. Ferraris, *Phys. Rev. B*, 1996, **54**, 14321.
15   I. H. Campbell, J. D. Kress, R. L. Martin, D. L. Smith, N. N. Barashkov and J. P. Ferraris, *Appl. Phys. Lett.*, 1997, **71**, 3528.
16   C. Ganzorig, K. J. Kwak, K. Yagi and M. Fujihira, *Appl. Phys. Lett.*, 2001, **79**, 272.
17   C. Ganzorig and M. Fujihira, *Appl. Phys. Lett.*, 2002, **81**, 3137.
18   Y. Enatsu, C. Ganzorig and M. Fujihira, *Mat. Res. Soc. Sym. Proc.*, 2003, **771**, 111.
19   F. Nüesch, F. Rotzinger, L. Si-Ahmed and L. Zuppiroli, *Chem. Phys. Lett.*, 1998, **288**, 861.
20   L. Zuppiroli, L. Si-Ahmed, K. Kamaras, F. Nüesch, M. N. Bussac, D. Ades, A. Siove, E. Moons and M. Grätzel, *Eur. Phys. J. B*, 1999, **11**, 505.
21   S. E. J. Appleyard, S. R. Day, R. D. Pickford and M. R. Willis, *J. Mater. Chem.*, 2000, **10**, 169.
22   P. Sigaud, J. N. Chazalviel and F. Ozanam, *J. Appl. Phys.*, 2002, **92**, 992.
23   J. Morgado, A. Charas and N. Barbagallo, *Appl. Phys. Lett.*, 2002, **81**, 933.
24   Q. Huang, J. Cui, H. Yan, G. C. Veinot and T. J. Marks, *Appl. Phys. Lett.*, 2002, **81**, 3528.
25   S. Khodabakhsh, D. Poplavskyy, S. Heutz, J. Nelson, D. D. C. Bradley, H. Murata and T. S. Jones, *Adv. Funct. Mater.*, 2004, **14**, 1205.
26   M. Fujihira, T. Kubota and T. Osa, *J. Electroanal. Chem.*, 1981, **119**, 379.
27   F. Nüesch, L. J. Rothberg, E. W. Forsythe, Q. T. Le and Y. Gao, *Appl. Phys. Lett.*, 1999, **74**, 880.
28   L. Swint and P. W. Bohn, *Appl. Phys. Lett.*, 2004, **84**, 61.
29   M. Fujihira, *Annu. Rev. Mater. Sci.*, 1999, **29**, 353.
30   Y. Park, V. Choong, Y. Gao, B. R. Hsieh and C. W. Tang, *Appl. Phys. Lett.*, 1996, **68**, 2699.
31   J. S. Kim, M. Granström, R. H. Friend, N. Johansson, W. R. Salaneck, R. Daik, W. J. Feast and F. Cacialli, *J. Appl. Phys.*, 1998, **84**, 6859.
32   K. Sugiyama, H. Ishii, Y. Ouchi and K. Seki, *J. Appl. Phys.*, 2000, **87**, 295.
33   C. Ganzorig and M. Fujihira, *Mat. Res. Soc. Sym. Proc.*, 2003, **771**, 105.
34   C. Shen, A. Kahn and I. Hill, 'Organic Molecular Interfaces: Investigations of Electronic Structure, Chemistry, and Carrier Injection Properties' in *Conjugated Polymer and Molecular Interfaces*, eds., A. Kahn, J. J. Pireaux, W. R. Salaneck and K. Seki, Marcel Dekker, New York, 2002, Chapter 11, pp. 351-400.
35   H. Ishii, K. Sugiyama, F. Ito and K. Seki, *Adv. Mater.*, 1999, **11**, 605.
36   G. G. Malliaras, J. R. Salem, P. J. Brock and J. S. Scott, *Phys Rev. B*, 1998, **58**, R13411.
37   N. C. Greenham and R. H. Friend, in *Solid State Physics*, Vol. 49, Academic Press, New York, 1995, pp. 1-149.
38   H. Bässler, *Polym. Adv. Technol.*, 1998, **9**, 402.
39   J. Kalinowski, *J. Phys. D*, 1999, **32**, R179.
40   P. W. M. Blom and M. C. J. M. Vissenberg, *Mater. Sci. Eng. R*, 2000, **27**, 53.
41   W. Brütting, S. Berleb and A. G. Mückl, *Org. Electron.*, 2001, **2**, 1.
42   I. H. Campbell and D. L. Smith, in *Solid State Physics*, Vol. 55, Academic Press, New York, 2001, pp. 1-117.
43   M. A. Baldo and S. R. Forrest, *Phys. Rev. B*, 2001, **64**, 085201.
44   J. C. Scott, *J. Vac. Sci. Technol. A*, 2003, **21**, 521.

45 D. Braun, *J. Polym. Sci. B*, 2003, **41**, 2622.

46 Y. Shen, A. R. Hosseini, M. H. Wong and G. G. Malliaras, *ChemPhysChem.*, 2004, **5**, 16.

47 M. A. Lampert and P. Mark, *Current Injection in Solids*, Academic Press, New York 1970.

48 K. C. Kao and W. Hwang, *Electrical Transport in Solids*, Pergamon Press, Oxford, 1981.

49 M. Pope and C. E. Swenberg, *Electronic Processes in Organic Crystals and Polymers*, 2nd edn., Oxford University Press, New York, 1999.

50 S. M. Sze, *Physics of Semiconductor devices*, Wiley & Sons, New York, 1969.

51 I. D. Parker, *J. Appl. Phys.*, 1994, **75**, 1656.

52 A. J. Heeger, I. D. Parker and Y. Yang, *Synth. Met.*, 1994, **67**, 23.

53 S. Barth, U. Wolf, H. Bässler, P. Müller, H. Riel, H. Vestweber, P. F. Seidler and W. Rieß, *Phys. Rev. B*, 1999, **60**, 8791.

54 R. H. Fowler and L. Nordheim, *Proc. R. Soc. London, Ser. A*, 1928, **119**, 173.

55 M. A. Abkowitz, H. A. Mizes and J. S. Facci, *Appl. Phys. Lett.*, 1995, **66**, 1288.

56 P. E. Burrows, Z. Shen, V. Bulovic, D. M. McCarty, S. R. Forrest, J. A. Cronin and M. E. Thompson, *J. Appl. Phys.*, 1996, **79**, 7991.

57 H. Yamamoto, H. Kasajima, W. Yokoyama, H. Sasabe and C. Adachi, *Appl. Phys. Lett.*, 2005, **86**, 083502.

58 H. Antoniadis, J. N. Miller, D. B. Roitman and I. H. Cambell, *IEEE Trans. Electron. Dev.*, 1997, **44**, 1289.

59 C. Giebeler, H. Antoniadis, D. D. C. Bradley and Y. Shirota, *J. Appl. Phys.*, 1999, **85**, 608.

60 A. J. Campbell, D. D. C. Bradley, J. Laubender and M. Sokolowski, *J. Appl. Phys.*, 1999, **86**, 5004.

SCHOTTKY-GATED POLYMER FET FABRICATED BY MULTIPLE LASER PRINTING METHOD

H. Okuzaki and S. Ashizawa

Interdisciplinary Graduate School of Medicine and Engineering, University of Yamanashi, 4-4-37 Takeda, Kofu 400-8511, Japan

1 INTRODUCTION

Poly(3,4-ethylenedioxythiophene) doped with poly(4-styrenesulfonate) (PEDOT/PSS), commercially available in the form of a water suspension, has attracted considerable attention because of its superior electrochemical stability and transparency in the doped state,[1] which provides potential applications in electrical and optical devices, such as antistatic coatings,[2] electro-chromic windows,[3,4] and hole transport layers in organic light-emitting diodes.[5,6] MacDiarmid and co-workers have developed the "line patterning", namely, "laser printing method", one of the most simple, rapid, and inexpensive methods capable of making a pattern of thin coating of PEDOT/PSS on a plastic substrate using a standard non-modified laser printer or copying machine,[7] which can be applied to a variety of plastic electronics devices, such as polymer dispersed liquid crystal display,[8] push button array for keypad,[8] and field-effect transistor (FET).[9]

Since the first observation of a curious phenomenon whereby an electric field affects the conductivity of PEDOT/PSS in an all-polymer FET,[10,11] some devices with doped conducting polymers as an active channel material have been reported.[12-14] Unlike in the conventional organic FETs using organic semiconductors, such as pentacene, thiophene oligomer, or undoped conjugated polymers,[15-17] our device has the following features: (1) the source, drain, gate electrodes and the channel are made of the same doped conducting polymer; (2) the source, channel, and drain are continuous without boundaries; (3) it is easy to fabricate due to its simple configuration. The FET is a "normally-on" type device and operates in the depletion mode by application of a positive gate voltage. A similar response was observed for the devices using polyaniline doped with hydrochloric acid or polypyrrole doped with chloride (PPy/Cl), representing a general phenomenon characteristic of doped conducting polymers. In the earlier studies, the device consisted of two substrates: one was source/channel/drain electrode and another was gate electrode both prepared on poly(ethyleneterephthalate) (PET) substrates by the laser printing method. An optical adhesive was used as an insulator inserted between two substrates. Because of the difficulty in preparing very thin and uniform insulating layer (< 1 μm), the device required higher gate voltages (> 10 V) and response was quite slow (> 1 min). Previously, we have reported flexible and transparent all-polymer FETs fabricated on one substrate,[9] where a thin insulating layer was prepared by spin coating. It was found that the device operated at lower voltages both in the depletion and enhancement modes in response to positive and negative gate voltages, respectively. The response was relatively faster (0.4 s-1 min) and on/off ratio reached $> 10^3$. However, the detailed mechanism is unclear since the application of conventional field effect theory yields surprisingly large carrier mobility (μ). Indeed, (1) the response is slower than that expected from the value of μ, (2) the gate leakage is considerably large (1-100 nA), (3) the I-V characteristics show hysteresis, (4) the device fabricated on SiO_2 as an insulator gives a small value of μ (~ 0.1 cm^2/Vs), and (5) the device is unstable where threshold voltage moves and response becomes slower with time, which distinguish the mechanism from conventional field effect. Epstein et al. proposed a model of electric field induced ion-leveraged insulator-metal transition (IMT),[12] in which a positive gate voltage led to motion of the mobile counter ions, such as H^+ or Na^+ in PEDOT/PSS or Cl^- in PPy/Cl, which decreased the localization length and increased the activation energy for hopping of the charge carrier.

In this study, we have fabricated Schottky-gated FET devices by multiple laser printig method, using an interface between PEDOT/PSS and Al instead of a polymer insulator such as polyvinylphenol.

2 METHOD AND RESULTS

2.1 Multiple Laser Printing Method

A widely applicable method, laser printing method, can be used to obtain conductive polymer patterns on various substrates, e.g. plastic or paper. Since the laser printing method does not incorporate printing of conductive ink, but plainly coating with a dilute polymer solution or dispersion, most hurdles of the processing of conducting polymers are bypassed. The process of laser printing method can be summarized in three major steps: (a) Design and printing a pattern. The pattern is to be designed reverse ('negative') to the resulting conducting polymer pattern. The design is printed on a substrate, e.g. on paper and PET overhead transparency film using an office-type laser printer or Xerox machine; (b) Coating of the printed substrate with a conducting polymer solution. The printed substrate is coated with the solution or dispersion of a conducting polymer such as PEDOT/PSS using spray, dip or roller coating. Exploit the fact that the printing (toner or ink) exhibits a strong difference in hydrophobicity as compared to the substrate, the conducting polymer solution is only attracted to one, either printed line or substrate, but repelled by the other. In case of the water dispersion, a conducting polymer pattern spontaneously forms as reverse to the printed design. Subsequently, the coating and substrate are dried; (c) Removal of the reverse printed toner pattern. If necessary, the printing can be removed by, e.g. sonication in an appropriate solvent. As many conducting polymers are inert to most solvents, but toner is sensitive to toluene or acetone, only the conducting polymer pattern remains. In applications where the strongly coloured print may remain, this step is optional. Here, we extend this technique not only for the patterning of conducting polymers, but also for the patterning of insulating polymers and metals, where different materials can be stacked layer by layer by repeating the same process.[18] The multiple laser printing method significantly shortens the time for fabrication of the device because additional annealing or curing after each step of the coating is not necessary. This means that the removal of the residual solvent in the coating and/or improvement of flatness and homogeneity of the coating can be performed simultaneously with fixing of toner while the substrate passes through the thermal rollers in a fuser unit of the laser printer.

Figure 1 shows optical and AFM images of a 100 µm-wide line pattern with gaps of 100 µm. Dark lines represent toner lines, light areas represent the PEDOT/PSS coating on the transparency substrate. The printed toner lines appear inhomogeneous, show holes, coarse edges and randomly distributed toner. After removal of the toner (right) the blue-grey PEDOT/PSS coating becomes clearly visible, as reverse to the exposed PET substrate. Again, coarse edges appear, and most removed toner particles are represented by holes in the film. Here, a standard laser printer, having a resolution of 1200 dpi, is considered to print about 20 µm features. The real resolution is, however, lower than the theoretical value: the minimum line width of the conductive pattern is about 50 µm, which depends on both hardware (laser printer) and software (printer driver and drawing software). The line pattern with 50 µm features is still conductive but the resistance varies more than twice, which is ascribed to the short between the adjacent lines and/or inhomogeneity of conductive lines with holes and coarse edges due to the distributed toner particles. Since mostly the accuracy of the printing process determines the quality of the conductive pattern, it is expected that use of high-resolution, high-quality printing methods will yield in high-resolution conductive patterns.

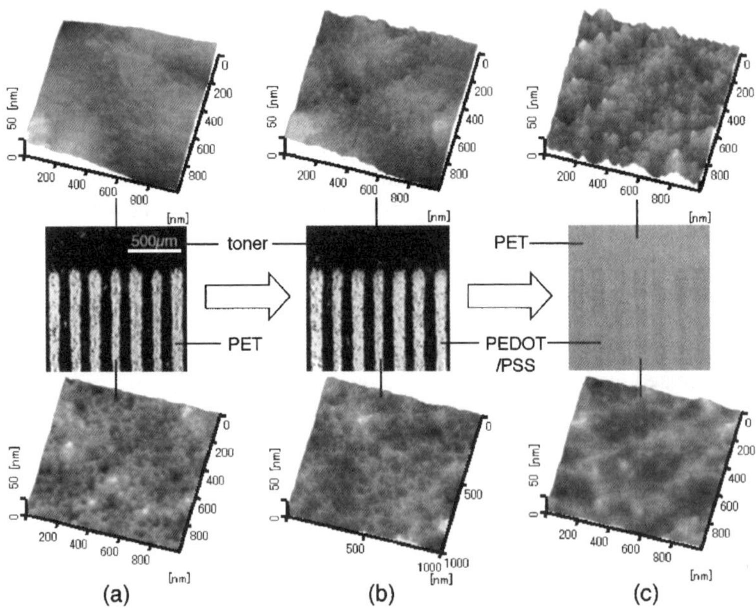

Figure 1 *Optical and AFM images of a line pattern of PEDOT/PSS at each step of the process. (a) a negative pattern of 100 µm-wide lines was printed on PET substrate with a laser printer (1200 dpi), (b) the printed pattern was coated with PEDOT/PSS solution by bar coating and dried with a hot air gun, and (c) the printed toner was removed by sonication in toluene and dried at 160°C in a vacuum for 1 h. The AFM images of the substrate, printed toner, and PEDOT/PSS were taken with SPI3800 (SII NanoTechnology Inc.) using a commercial silicon tip in tapping mode.*

2.2 Solvent Effect

The solution for coating was prepared by mixing the aqueous solution of PEDOT/PSS with various solvents, such as pure water, ethanol, isopropanol, acetonitrile, DMSO, ethylene glycol, and poly(ethylene glycol) (PEG), containing 0.1% of dodecylbenesulfonic acid to improve the adhesion of coating to the PET film (100 μm thick) used as a substrate. The coating of the mixed solution was carried out by a bar coating: a drop of the PEDOT/PSS solution was deposited on one end of the PET substrate and swiped over the substrate using a glass rod. Immediately after coating, the film was dried with a hot air gun, and then heat treated at 100-200°C in a vacuum for 1 h to remove the residual solvent from the coating.

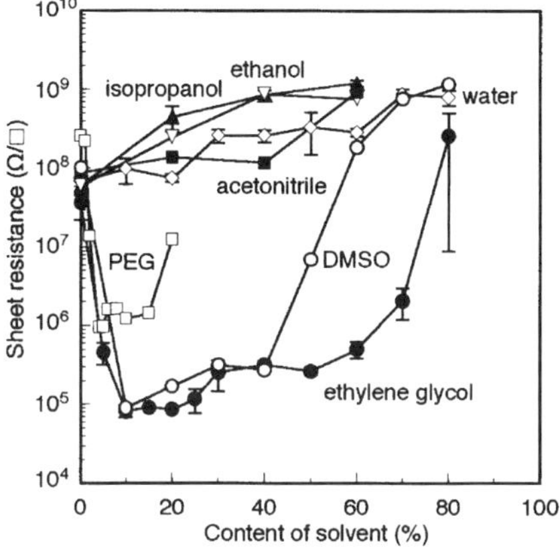

Figure 2 *Effect of solvent on sheet resistance of PEDOT/PSS coating measured at 25°C by a four-probe method.*

Figure 2 shows the effect of solvent on sheet resistance of the PEDOT/PSS coating. The sheet resistance for the coating of PEDOT/PSS pristine solution is 10^7-10^8 Ω/□ and dilution with pure water, ethanol, isopropanol, or acetonitrile results in a monotonous rise of the value, which can be explained by thinning the film and/or inhomogeneity of coating. On the other hand, a significant drop of the sheet resistance by two or three orders of magnitude is seen for ethylene glycol.[19] A similar tendency is observed for other organic solvents, such as DMSO,[20] glycerol, PEG,[21] and sorbitol[22] regardless of the substrate and the method of coating. The thickness of the PEDOT/PSS coating is nearly constant about 20-50 nm regardless of the ethylene glycol content up to 80%, while the electrical conductivity rises by two orders of magnitude from 0.015 S/cm for the pristine solution to 3.5 S/cm for the solution with 20% of ethylene glycol. We consider the mechanism as follows: During the drying process of the wet coating, the concentration of ethylene glycol gradually increases until all solvent is evaporated. Since ethylene glycol is considered to be a 'good-solvent' for the dispersed PEDOT/PSS, and causes the polymer chains to extend in the film. Thus, individual PEDOT chains remain in close contact, which enhances the hopping of charge carrier between the grains. Greczynski et al. have found that phase segregation occurs in the PEDOT/PSS system and PSS is predominant in the surface region, where the layer of excess PSS surrounding the PEDOT/PSS grains is estimated to be about 37 Å using XPS and UPS.[23] The increase of conductivity is associated with the reduction of the insulating PSS shell surrounding PEDOT/PSS grains, which improves the connectivity between such grains in the coating that enhances the transportation of charge carriers between the grains. Indeed, the mixed solution of PEDOT/PSS and ethylene glycol can form a gel, which is a clear evidence of formation of a 3D network of the PEDOT/PSS. On the other hand, a further increase of ethylene glycol content results in a decrease of the conductivity. This may be ascribed to the decrease of PEDOT/PSS concentration, which lowers the density of connections between the grains and/or increases the inhomogeneity of coating.

2.3 Schottky-Gated FET

Figure 3 shows I-V characteristics of Schottky-gated FETs fabricated by the multiple laser printing method, in which the current flow between the source and drain contacts at a distance of 7 mm is modulated by the gate voltages (V_G). It is seen from Figure 3a, an "on" state of the FET is at $V_G = 0$ V because the doped conducting channel (C, 135 long and 1 mm wide), made of PEDOT/PSS, is already built in between the source and drain electrodes but not be induced by the gate voltage, i.e., the device is "normally-on". The drain current is depleted and becomes nearly zero at the gate voltages higher than 2 V, which is called "off" state of the FET. The on/off ratio is about 100. This phenomenon is reproducible in ambient air for a few days and can be explained by the formation of a Schottky barrier at the interface between PEDOT/PSS and Al. On the other hand, normally-off type FET can be produced by controlling the channel size (511 μm long and 275 μm wide). As shown in Figure 3b, the channel resistance is high enough to cut off the drain current without gate voltage where the thickness of the depletion layer is considered to be comparable to the channel thickness (34 nm) and the drain current is enhanced by the negative gate voltages. Thus, the FET operates both in the depletion and enhancement modes in response to the positive and negative gate voltages, respectively.

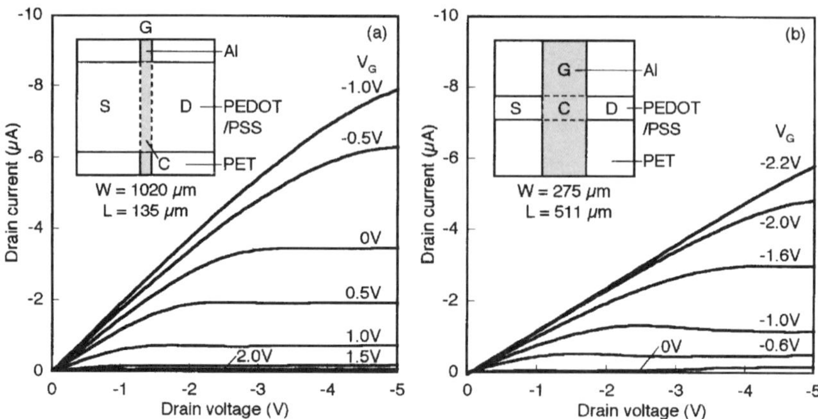

Figure 3 *Drain current-drain voltage characteristics of normally-on (a) and normally-off (b) type Schottky-gated polymer FETs under various gate voltages.*

Using the data of built-in voltage (0.88 V) and relative dielectric constant (3),[24] the carrier density (N_A) at $V_G = 0$ V is calculated to be about 10^{18} cm^{-3}, which is consistent with that obtained from C-V characteristics of the Schottky diode[24] but smaller than that of other heavily doped conducting polymers ($> 10^{20}$ cm^{-3}). At $V_G = 0$ V, the value of μ is calculated as 10^{-1}-10^{0} cm^2/Vs that is still higher than that measured for metal-oxide-semiconductor FET using SiO_2 as a gate insulator, which may be ascribed to the small value of N_A and/or electrochemical effect since the typical leakage current is about 50 nA.

3 CONCLUSION

We have shown that conductive polymer patterns for Schottky-gated FET applications can be prepared from the commercial conducting polymer dispersion by the process of multiple laser printing method. Since the process utilizes mostly standard office equipment, FETs can be prepared using minimal efforts and expenses. We expect that the multiple laser printing method will find primary application in low-cost, low-resolution plastic electronics.

References

1   G. Heywang and F. Jonas, *Adv. Mater.*, 1992, **4**, 116.
2   F. Jonas and J. T. Morrison, *Synth. Met.*, 1997, **85**, 1397.
3   Q. Pei, G. Zuccarello, M. Ahlskog and O. Inganäs, *Polymer*, 1994, **35**, 1347.
4   H.W. Heuer, R. Wehrmann and S. Kirchmeyer, *Adv. Funct. Mater.*, 2002, **12**, 89.
5   M. Granström, M. Berggren and O. Inganäs, *Science*, 1995, **267**, 1479.
6   Y. Cao, G. Yu, C. Zhang, R. Menon and A.J. Heeger, *Synth. Met.*, 1997, **87**, 171.
7   D. Hohnholz and A.G. MacDiarmid, *Synth. Met.*, 2001, **121**, 1327.
8   D. Hohnholz, H. Okuzaki and A.G. MacDiarmid, *Adv. Funct. Mater.*, 2005, **15**, 51.
9   H. Okuzaki, M. Ishihara and S. Ashizawa, *Synth. Met.*, 2003, **137**, 947.
10   A.G. MacDiarmid, *Angew. Chem. Int. Ed.*, 2001, **40**, 2581.
11   This work was done at the University of Pennsylvania on sabbatical leave from the University of Yamanashi.
12   A.J. Epstein, F.-C. Hsu, N.-R. Chiou and V.N. Prigodin, *Curr. Appl. Phys.*, 2002, **2**, 339.
13   A.J. Epstein, F.-C. Hsu, N.-R. Chiou and V.N. Prigodin, *Synth. Met.*, 2003, **137**, 859.
14   J. Lu, N.J. Pinto and A.G. MacDiarmid, *J. Appl. Phys.*, 2002, **92**, 6033.
15   A. Tsumura, H. Koezuka and T. Ando, *Appl. Phys. Lett.*, 1986, **49**, 1210.
16   H.E. Katz and Z. Bao, *J. Phys. Chem.*, 2000, **104**, 671.
17   G. Horowitz, *Adv. Mater.*, 1998, **10**, 365.
18   S. Ashizawa, Y. Shinohara, H. Shindo, Y. Watanabe and H. Okuzaki, *Synth. Met.*, 2005, **153**, 41.
19   S. Ashizawa, R. Horikawa and H. Okuzaki, *Synth. Met.*, 2005, **153**, 5.
20   J.Y. Kim, J.H. Jung, D.E. Lee and J. Joo, *Synth. Met.*, 2002, **126**, 311.
21   S.K.M. Jönsson, J. Birgerson, X. Crispin, G. Greczynski W. Osikowicz, A.W. Denier van der Gon, W.R. Salaneck and M. Fahlman, *Synth. Met.*, 2003, **139**, 1.
22   S. Ghosh and O. Inganäs, *Synth. Met.*, 2002, **121**, 1321.
23   G. Greczynski, Th. Kugler, M. Keil, W. Osikowicz, M. Fahlman and W.R. Salaneck, *J. Elect. Spectr. Rel. Phenom.*, 2001, **121**, 1.
24   G. Liang, T. Cui and K. Varahramyan, *Solid-State Electr.*, 2003, **47**, 691.

MECHANISM OF AMBIPOLAR FIELD-EFFECT TRANSISTORS ON ONE-DIMENSIONAL ORGANIC MOTT INSULATORS

K. Yonemitsu[1,2]

[1]Institute for Molecular Science, Okazaki 444-8585, Japan. E-mail: kxy@ims.ac.jp
[2]Graduate University for Advanced Studies, Okazaki 444-8585, Japan

1 INTRODUCTION

Field-effect transistors are fabricated on many molecular materials. Their characteristics are not always explained in the conventional manner. Indeed, ambipolar characteristics are reported in metal-insulator-semiconductor field-effect transistor device structures based on organic single crystals of the quasi-one-dimensional Mott insulator (BEDT-TTF)(F$_2$TCNQ) [BEDT-TTF=bis(ethylenedithio)tetrathiafulvalene, F$_2$TCNQ=2,5-difluorotetracyanoquinodimethane].[1] In quasi-one-dimensional systems with coherent band transport, the insulator-electrode interface barrier potentials are crucial to the current-voltage characteristics. For example, carbon nanotube field-effect transistors operate as Schottky barrier transistors, in which transistor action occurs primarily by varying the contact resistance rather than the channel conductance.[2,3] The Schottky barriers are sensitive to the work-function difference between the channel and the source/drain electrodes. As long as the work functions of the electrodes are different from that of the channel, the characteristics are unipolar in general. By matching their work functions, the ambipolar field-effect characteristics are achieved.[2,3]

Thus, the ambipolar characteristics of the organic Mott insulator[1] imply that electron correlations are crucial. In organic insulators and conductors, the importance of electron correlations is acknowledged in the equilibrium phase diagram and the bulk transport properties.[4] For example, the motion of carriers is collective and sometimes confined into chains, which is qualitatively different from the conventional behavior of individual quasiparticles. Because of the collective nature, the motion of carriers inside the Mott insulator is correlated with that near the insulator-electrode interface. In order to clarify the relationship between the electron correlations and the Schottky barriers, we calculate the current-voltage characteristics using the one-dimensional Hubbard model for Mott insulators and adding to it potentials that originate from the long-range Coulomb interaction and are modified by the work-function difference, the gate bias and the drain voltage. The time-dependent Schrödinger equation is combined with the Poisson equation and numerically solved self-consistently at each site and time within the unrestricted Hartree-Fock approximation. The ambipolar field-effect characteristics are shown to be caused by balancing the correlation effect with the barrier effect. For the gate-bias polarity with higher Schottky barriers, the correlation effect is weakened accordingly.

2 METHOD AND RESULTS

2.1 One-Dimensional Model for Field-Effect Transistors

Many works employ the one-dimensional Hubbard model for a Mott insulator, to which the tight-binding model is attached for metallic electrodes.[5,6] We add a scalar potential v_l,[7]

$$H = \sum_l (\varepsilon_l + v_l) n_l - \sum_{l,\sigma} \left(t_{l,l+1} c_{l,\sigma}^+ c_{l+1,\sigma} + \text{h.c.} \right) + \sum_l U_l \left(n_{l\uparrow} - 1/2 \right) \left(n_{l\downarrow} - 1/2 \right), \qquad (1)$$

where $c_{l,\sigma}^+$ ($c_{l,\sigma}$) creates (annihilates) an electron with spin σ at site l, $n_{l,\sigma} = c_{l,\sigma}^+ c_{l,\sigma}$, and $n_l = \sum_\sigma n_{l,\sigma}$. The site energy ε_l is set at 0 in the crystal and at ϕ in the electrodes. The absolute value of the transfer integral $|t_{l,l+1}|$ is set at t_c if either l or $l+1$ is in the crystal and at t_e otherwise. The on-site repulsion U_l is set at U in the crystal and at 0 in the electrodes. When we consider band insulators instead, we use $U = 0$ and replace t_c by $t_c - (-1)^l \delta t$. The total number of electrons is the same as the number of sites. The left and right electrodes are regarded as the source and drain, respectively. The periodic boundary condition is imposed by introducing the Peierls phase, which is proportional to the vector potential, to the transfer integral for finite drain voltage V_D. The scalar potential v_l obeys the Poisson equation on the discrete lattice (l=1, 2, ... L),

$$v_{l+1} - 2v_l + v_{l-1} = -V_{Pl} \left(\langle n_l \rangle - n_{Bl} \right), \qquad (2)$$

where the parameter V_{Pl} comes from the long-range Coulomb interaction and n_{Bl} from the background charge. The parameter V_{Pl} is set at V_{Pc} in the crystal and at V_{Pe} in the electrodes, with $V_{Pe} \ll V_{Pc}$. The gate bias U_G, which is relative to the middle of the source-drain potential,[1] is assumed to appear with the gate efficiency factor α in the boundary value of the Poisson equation,

$$v_0 - v_{L/2} = v_L - v_{L/2} = \alpha U_G - \phi, \qquad (3)$$

where l=$L/2$ is set at the middle of the crystal and l=0 (L) in the source (drain) electrode is the furthest point from the interfaces. Hereafter we rewrite αU_G as U_G for simplicity.

The time-dependent Schrödinger equation is numerically solved in the unrestricted Hartree-Fock approximation simultaneously with the Poisson equation. The self-consistency is imposed at each time and site in these equations. The drain current I_D is obtained by averaging the current density,[8]

$$J(t) = -(1/L) \sum_{l,\sigma} \left[i t_{l+1,l}(t) c_{l+1,\sigma}^+ c_{l,\sigma} - i t_{l,l+1}(t) c_{l,\sigma}^+ c_{l+1,\sigma} \right], \qquad (4)$$

over the period $0 < t < \Delta t$ with $\Delta t = 2\pi L/(4V_D)$. This definition leads to positive (negative) I_D for positive (negative) V_D in general. For calculations, we set t_e=t_c=1, V_{Pc}=0.05, V_{Pe}=10^{-3}, and L=100 with 51 crystal sites. Because of finite V_{Pe}, a charge is induced at the metallic surface. Its detail is beyond the scope of the Poisson equation, but it does not affect the

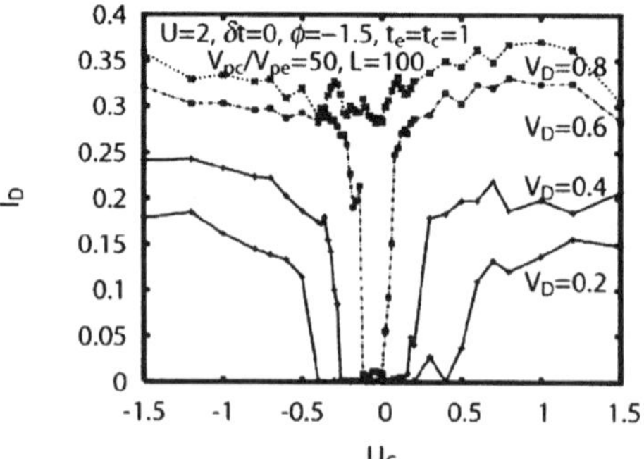

Figure 1   *I_D-U_G characteristics at various V_D for the Mott insulator with U=2, δt=0 and the work-function difference φ=−1.5. The other parameters are t_e=t_c=1, V_{Pc}/V_{Pe}=50, and L=100.*

conclusion. Both U=2 used for the Mott insulator and δt=0.17 used for the band insulator give a charge gap of about Δ_{CG}=0.68. The work-function difference ϕ is set at −1.5, so that the electrodes have a higher work function than the crystal. Then, the Schottky barriers appearing at the insulator-electrode interfaces become higher for the electron injections U_G>0 and lower for the hole injections U_G<0.

The I_D-U_G characteristics of the Mott insulator are nearly symmetric (i.e., even functions) with respect to the polarity of U_G (Figure 1). This is the case even if the work-function difference makes the Schottky barriers higher for U_G>0. For small V_D and | U_G |, I_D is suppressed by the charge gap, and I_D increases with U_G>0 and with U_G<0 in a very similar manner. These ambipolar characteristics are achieved, for the gate-bias polarity with the higher Schottky barriers (U_G>0), by weakening the correlation: the deviation from half filling inside the insulator is larger, reducing the umklapp scattering strength.

The I_D-U_G characteristics of the band insulator are very asymmetric with respect to the polarity of U_G (Figure 2). This is reasonable, from the viewpoint of individual particles,

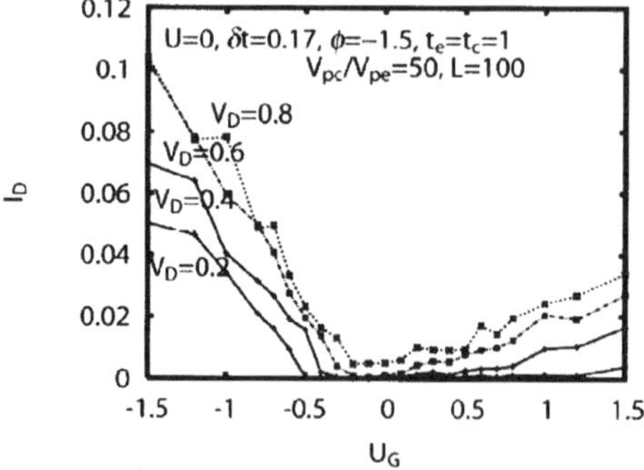

Figure 2   *I_D-U_G characteristics at various V_D for the band insulator with U=0, δt=0.17 and the work-function difference φ=−1.5. The other parameters are the same as in Figure 1.*

because the high Schottky barriers for $U_G>0$ simply enhance the backward scatterings at the insulator-electrode interfaces and consequently reduce the current density for $U_G>0$. For further electron injections, I_D remains suppressed, so that the characteristics are unipolar. Thus, the field-effect characteristics depend largely on the origin of the charge gap in the insulator. The potential distribution and the local density of states in the band insulator are similar to those in the Mott insulator as long as the gate bias is so small that the current does not flow. Therefore, the differences are manifested only in the dynamical and non-equilibrium condition.

2.2 Insulators Attached to Electrodes with Different Work Functions

In the model above, the gate electrode modulates only the potential at the midpoint of the crystal. Therefore, it is quite reasonable for one to regard only one insulator-electrode interface as essential to the peculiar characteristics. Namely, the above field-effect characteristics would be approximately given by the difference between two current-voltage relations for systems with one interface each. However, two interfaces are needed for the current to be measured. Here, we consider insulators attached to two electrodes with different work functions, where only one insulator-electrode interface gives a significantly high Schottky barrier.

We employ the model (1), where the site energy ε_l is set at 0 in the crystal, at ϕ_L in the left electrode (for small l's), and at ϕ_R in the right electrode (for large l's). Otherwise common parameters are assigned to the two electrodes. The periodic boundary condition is imposed again by introducing the Peierls phase to the transfer integral for finite voltage V, which corresponds to V_D in the field-effect transistors. Now, the boundary values of the Poisson equation are set by

$$v_0 - v_{L/2} = -\phi_L , \tag{5}$$

which compensates for the work-function difference between the insulator and the left electrode, and by

$$v_L - v_{L/2} = -\phi_R , \tag{6}$$

which compensates for the work-function difference between the insulator and the right electrode. The definition of the current I is the same as that of I_D in the field-effect transistors.

We numerically calculate the time-evolution of the system with the same parameters as before unless otherwise noted. Both $U=2$ used for the Mott insulator and $\delta t=0.17$ used for the band insulator give a charge gap of about $\Delta_{CG}=0.68$; $U=1.9$ and $\delta t=0.145$ give $\Delta_{CG}=0.58$; and $U=1.8$ and $\delta t=0.1225$ give $\Delta_{CG}=0.49$. The parameter ϕ_R is so set that the Fermi level of the right electrode, if isolated, is set at the top of the lower Hubbard band or the valence band. The parameter ϕ_L is so set that the Fermi level of the left electrode, if isolated, is set at about 3/4 times the charge gap higher than the bottom of the upper Hubbard band or the conduction band. Then, only the Schottky barrier at the left interface is significantly high. It becomes higher for the left-going current $V<0$ and lower for the right-going current $V>0$.

The *I-V* characteristics of the Mott insulator are rather anti-symmetric (i.e., odd functions) with respect to V (Figure 3). This is the case even if the work-function

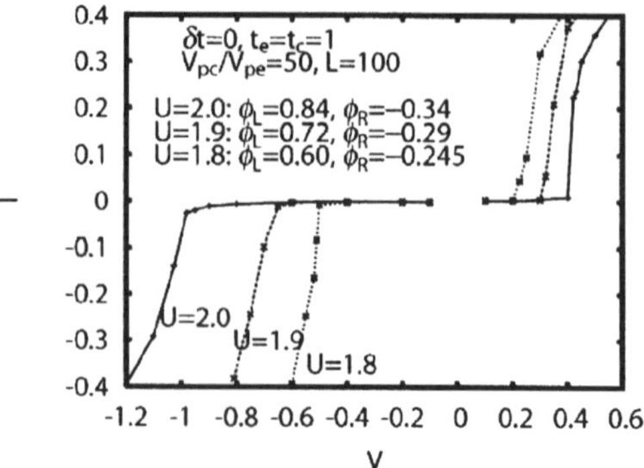

Figure 3   *I-V characteristics for Mott insulators with various U and δt=0. The work-function differences ϕ_R and ϕ_L are such that the Fermi level of the right electrode, if isolated, is set at the lower Hubbard band and that of the left electrode at about 3/4 times the charge gap higher than the upper Hubbard band. The other parameters are the same as in Figure 1.*

difference at the left interface makes the Schottky barrier higher for $V<0$. For small V, $|I|$ is suppressed by the charge gap, and $|I|$ increases with $V<0$ and with $V>0$ in a similar manner above the threshold. For $V<0$, the correlation is accordingly weakened.

The *I-V* characteristics of the band insulator are far from anti-symmetric with respect to V (Figure 4). Simply because the Schottky barrier at the left interface is higher for $V<0$, the enhanced backward scattering suppresses the current density for $V<0$ only. In short, the field-effect characteristics can be interpreted as the difference between the current-voltage relation governed by the insulator-source-electrode interface and that by the insulator-drain-electrode interface (V_D times the derivative in the limit of small V_D).

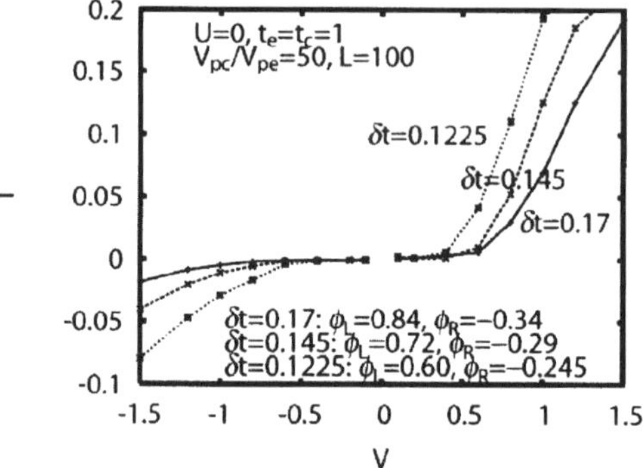

Figure 4   *I-V characteristics for band insulators with various δt and U=0. The work-function differences ϕ_R and ϕ_L are such that the Fermi level of the right electrode, if isolated, is set at the valence band and that of the left electrode at about 3/4 times the charge gap higher than the conduction band. The other parameters are the same as in Figure 1.*

3 CONCLUSION

In the one-dimensional Hubbard model for a Mott insulator attached to the tight-binding model for electrodes, both with scalar and vector potentials for Schottky barriers at interfaces, the field-effect characteristics are calculated by solving the time-dependent Schrödinger and Poisson equations simultaneously and self-consistently. They are always ambipolar for Mott insulators even if the work function of the crystal is quite different from that of the electrodes. This is in contrast to the generally unipolar characteristics for band insulators. In Mott insulators, for gate-bias polarity with the higher Schottky barriers, compared with the opposite polarity, we find that the deviation from half filling inside the insulator is larger, which reduces the umklapp scattering strength. What does this mean? Electrons are scattered backwards at the insulator-electrode interface. If they behaved individually as in band insulators, the charge-density distribution would be largely altered at high cost owing to the on-site repulsion. To avoid this situation, all the electrons are so correlated that they flow collectively. This is made possible by the fact that the strong backward scattering at the interface is counterbalanced by weakening the umklapp scattering inside the insulator. Therefore, the robust ambipolar field-effect characteristics in Mott insulators are due to balancing the correlation effect with the barrier effect, which allows the collective charge transport. The ambipolar (i.e., nearly even functions with respect to U_G) field-effect characteristics can be explained by the difference between the anti-symmetric (i.e., nearly odd functions with respect to V) I-V characteristics at the two insulator-electrode interfaces. The latter is a sufficient condition for the former, and indeed theoretically realized for Mott insulators attached to electrodes with quite different work functions, where only one insulator-electrode interface gives a significantly high Schottky barrier. The qualitative differences between the non-equilibrium properties of Mott insulators and those of band insulators are also found in photoinduced phase transitions between these insulators: the dynamics are cooperative in Mott insulators[9] and uncooperative in band insulators.[10]

Acknowledgements

The author is grateful to T. Hasegawa for showing his data prior to publication and for enlightening discussions including systems attached to electrodes with different work functions. This work was supported by Grants-in-Aid and NAREGI Nanoscience Project from the Ministry of Education, Culture, Sports, Science and Technology, Japan.

References

1   T. Hasegawa, K. Mattenberger, J. Takeya and B. Batlogg, *Phys. Rev. B*, 2004, **69**, 245115.
2   S. Heinze, J. Tersoff, R. Martel, V. Derycke, J. Appenzeller and Ph. Avouris, *Phys. Rev. Lett.*, 2002, **89**, 106801.
3   A. Javey, J. Guo, Q. Wang, M. Lundstrom and H. Dai, *Nature*, 2003, **424**, 654.
4   J. Kishine and K. Yonemitsu, *Int. J. Mod. Phys. B*, 2002, **16**, 711.
5   A. Oguri, *Phys. Rev. B*, 2001, **63**, 115305.
6   Y. Tanaka and A. Oguri, *J. Phys. Soc. Jpn.*, 2004, **73**, 163.
7   K. Yonemitsu, *J. Phys. Soc. Jpn.*, 2005, **74**, 2544.
8   T. Oka, R. Arita and H. Aoki, *Phys. Rev. Lett.*, 2003, **91**, 066406.
9   K. Yonemitsu, *J. Phys. Soc. Jpn.*, 2004, **73**, 2868.
10 K. Yonemitsu, *J. Phys. Soc. Jpn.*, 2004, **73**, 2879.

SELF-ASSEMBLED MONOLAYERS MODIFICATION ON ELECTRODES OF ORGANIC THIN-FILM TRANSISTORS

N. Hiroshiba[1], R. Kumashiro[1], H. Ishii [2], M. Yamashita[3], S. Takaishi[3], K. Tukagoshi[4], M. Okinaka[4] and K. Tanigaki[1]

[1]Department of Physics, Graduate School of Science, Tohoku University and CREST-JST 6-3 Aoba Aramaki Aoba-ku, Sendai, Miyagi 980-8578, Japan
[2]Research Institute of Electrical Communication, Tohoku University, 2-1-1 Katahira, Aoba-ku, Sendai 980-8577, Japan
[3]Department of Chemistry, Graduate School of Science, Tohoku University 6-3 Aoba Aramaki Aoba-ku, Sendai, Miyagi 980-8578, Japan
[4]RIKEN, Hirosawa 2-1, Wako, Saitama 351-0198, Japan

1 INTRODUCTION

In organic field effect transistors (OFETs), one of the most important key issues will be the interfacial problems between organic thin-films and gate-insulators and/or electrodes, and this is now drawing much intense attention. The surface control using self-assembled-monolayers[1] (SAMs) is a well-known good technique for such surface modifications and provides microscopically good interface regulations. Better surface modifications have been reported to be achieved on SiO_2 gate insulators using silane molecules[2] and on gold electrodes using thiol ones[3]. Although recent advanced works have clarified important effects obtained by SAMs modifications on the SiO_2 gate insulator[4], little has been still known about the issues of SAMs modification on the Au electrodes.

In the present study, interfacial modifications of source and drain gold electrodes using thiol molecules have been studied in detail for pentacene-FETs, using three benzene thiol (BzT) molecules as SAMs with different functional groups, 4-methyl (4MeBzT), 4-nitro (4-NBzT) and 4-amino (4-ABzT).

2   EXPERIMENTAL

The Au electrodes were modified by 4-NBzT and 4-AbzT by insertion of the FET substrate into the solution of these thiol compounds for a few minutes to 24 hours after the cleaning treatment with a piranha solution.The FET actions of pentacene thin films deposited on this modified substrate, with a deposition rate of $0.1 nm\ sec^{-1}$ to a thickness of 50nm, have been studied in a high vacuum chamber at room temperature. Work functions of the modified FET electrodes were measured by a home made Kelvin Probe apparatus in air and in vacuum. The charge transfer degree between SAMs and the Au electrodes was calculated using Gaussian 98.

3   RESULT & DISCUSSION

FET properties of pentacene fabricated on SiO_2 substrates have been measured. As seen in Fig.1. (left figure), the source-drain current (I_{SD}) observed was indeed weak. Because this was a little bit surprise to us, we have compared the experiments to those using the FET substrates without any thiol-SAMs modifications. We immediately observed a standard FET action and it is shown together in Fig.1. This experimental observations has led us to the thought that the thiol-SAMs would probably block the carrier injection from the source electrode by performing as an insulator. Accordingly, we have shorten the immersion time of the FET substrates in a thiol-SAMs solution and have carried out the same experiments. As can be seen in Fig.1 (right figure), we were able to observe large I_{SD} current for the pentacene FETs prepared on the substrates treated both by 4-NBzT and 4-AbzT. It is very interesting to note here that the I_{SD} current observed in these experiments wes in turn greatly larger than those without any SAMs treatment.

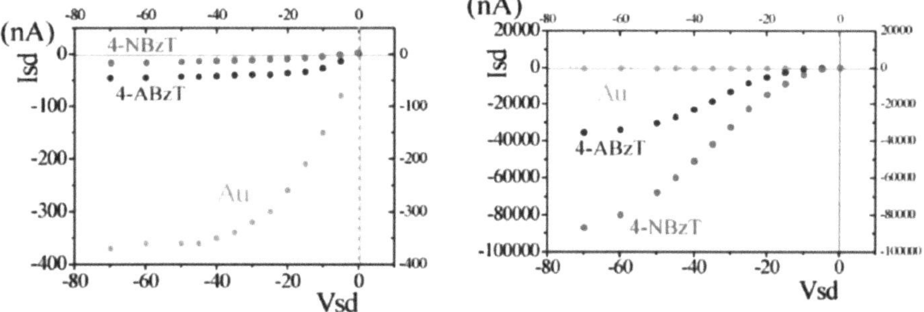

Figure 1: FET characteristics of pentacene FETs at V_G=70V with gold electrodes and the modified electrodes by thiols with 1hour(left) and 5minutes (right) preparation time.

The situation of the surface of the Au electrodes modified by thiol-SAMs has been studied by a contact angle method. The immersion time was varied from a few minuets to one hour and a water droplet was softly put down on the thiol-SAMs treated Au surfaces. The situation of the Au surfaces has been studied from the viewpoint of the contact angles.

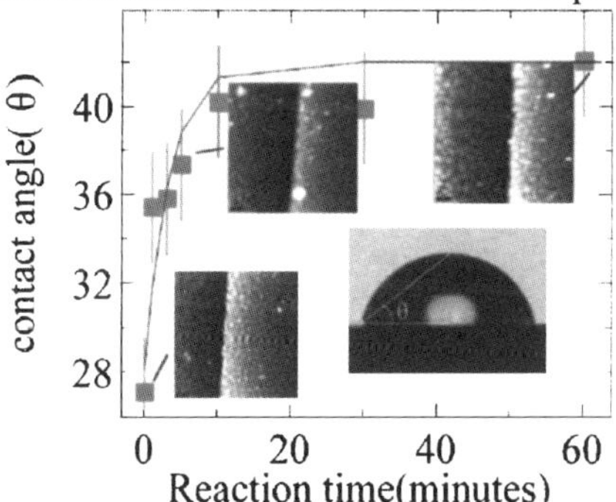

Figure 2: The evolution of the contact angle of a water droplet on the modified gold surface by SAMs as a function of immersion time in SAMs solutions.

The contact angles with different immersion time of the substrates in SAMs solutions were plotted as a function of time in Fig.2. As is seen in this figure, the contact angle became large after a very short immersion time of around a few minutes, then shows saturation at 42 degree after 10 minutes. There was almost no change in the contact angle after 10 minutes. This indicates that the Au surface is completely modified after 10 minutes. It can be understood that we observe little I_{SD} current when the Au surface is perfectly covered by thiol-SAMs. Differently from the situation of the SAMs treatment in the case of SiO_2 gate insulators, the best modification of the Au electrode surfaces by thiol-SAMs should not be perfect in order to have a good FET performance.

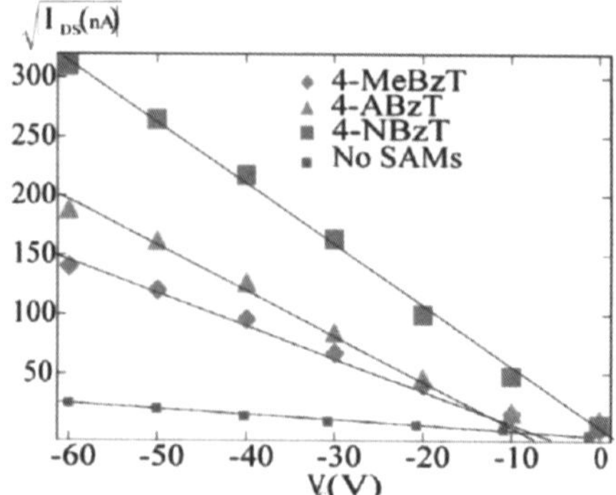

Figure 3: The $\sqrt{I_{SD}}$ -V_G plots for FETs with SAMs electrodemodifications.

Another important parameter, the threshold voltage V_{th} has been studied before and after the SAMs treatment on the Au electrodes using pentacene thin films. In the $\sqrt{I_{SD}}$ -V_G plots in the region of $V_{DS} >> (V_G-V_{th})$, the shape of the plots can be described by $\sqrt{I_{SD}} = \sqrt{\frac{WC}{2L}}\mu$ (V_G-V_{th}), where L is the channel length and W the channel width, and C the capacity of the gate insulator. It is apparent that the electric field mobility has been improved greatly by the SAMs surface modification on the Au electrodes. The μ values observed are 1.9×10^{-3} cm^2/Vs for 4-ABzT, 1.7×10^{-3} cm^2/Vs for 4-MeBzT, 5.7×10^{-3} cm^2/Vs for 4-NBzT SAMs and ~10^{-4} cm^2/Vs for FET with Au electrodes at the same condition. The on/off ratios observed are 3.1×10^{2} for 4-ABzT, 1.9×10^{4} for 4-MeBzT and 2.1×10^{4} for 4-NBzT SAMs. The subthreshold slopes observed are 11.7 V/dec. for 4-ABzT, 6.8 V/dec. for 4-MeBzT and 6.8 for 4-NBzT SAMs. One may imagine that the grain size may be influenced by the SAMs modifications when the pentacene thin films were grown. However, as can be seen in the atomic force microscopy (AFM) pictures, the grain size does not change very much by the SAMs modifications. Therefore, the improvement in field effect mobilities μ should be ascribed to the other factors.

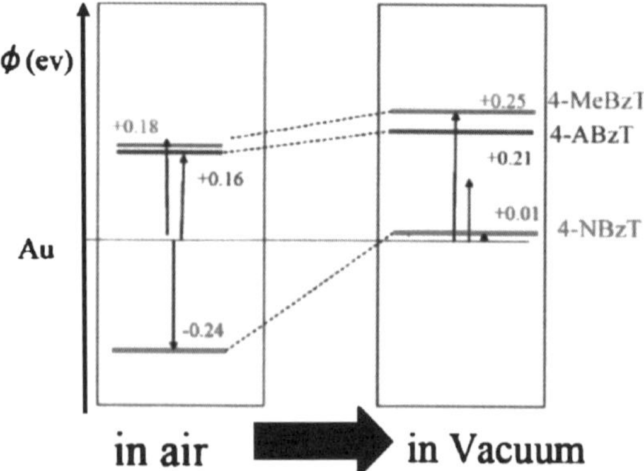

Figure 4: The work functions of the modified gold surface by SAMs measured by a Kelvin probe method.

We have hence measured the work function of Au modified by the thiol SAMs by a Kelvin probe method[5] and the results are shown in Fig.4. Actually as can be seen in Fig.4, the values changed from those of the prepared Au electrodes. It can also be observed that the workfunction changes between in air and in vacuum. This may be due to the influences of adsorbed gas species on the SAMs surface in air. In vacuum, these values of the Au electrodes modified by the three types of SAMs have +0.24 eV for 4-MeBzT, +0.10 eV for 4-ABzT and +0.01 eV for 4-NBzT SAMs and these are less than that of the Au electrodes. Since the FET experiments were made in vacuum, the values should be taken into consideration for understanding the V_{th} results. The V_{th} values observed are -8.5 V for 4-ABzT, -5.0 V for 4-MeBzT and 0.4 V for 4-NBzT SAMs, and these values seem to reflect the values measured in vacuum at first glance. However, if one looks into more detail, the V_{th} observed from 4-MeBzT and 4-ABzT are very close to each other and thoroughly the order of the V_{th} differs from what expected from the experimental values.

Considering the situations described above, one may consider the charge transfer between the interfacial contacting layer as to be another possible explanation accounting for the V_{th}s. We have made preliminary molecular orbital calculations using pentacene and the three thiol-SAMs supposing the ideal geometrical positions using gaussian 98 with the 6-31G base function set. The calculations indicated that a small amount of charges by less than 0.1 electrons per pentacene transfers in the case of NH_2- and CH_3-SAMs, but 0.74 electron transfer in the case of NO_2-SAMs. This charge transfer seems to be important to explain the fact that was observed in the V_{th} experiments. In the equilibrium situation in the course that holes and electrons carriers transfer from SAMs to pentacene, impurity carriers can be accumulated in the pentacene thin films and the Fermi level of the organic semiconducting thin films can be tuned to be matched with the Fermi level of the electrodes when the both levels are not far away. This in turn causes a change in work functions. As a results, as far as the Fermi levels between the semiconducting thin film and the electrodes are close to each other and the carrier transfer is possible in their interfacial contact layers, the shift in V_{th} values can be observed.

4 CONCLUSION

The influences by the SAMs modifications on the gold electrodes were studied in detail using benzene thiols with three types of substituent groups. It can be clarified that imperfect modifications with a short preparation time is better when the thiol compounds work as a charge blocking layer. Due to the carrier transfer in the interfaces leading to a equilibrium state of the Fermi levels of both semiconducting thin films and the electrodes, V_{th} can be controlled by the types of substituents when the Fermi levels are not apart from each other. If one can keep these experimental results in mind, search for conducting thiol compounds for surface modifications on the gold electrodes will be intriguing in the future to have further improvements.

REFERENCE

1. Frank Schreiber, *Progress in Surface Science* **65**, 151(2000)
2. J. Wang, D. J. Gundlach, C. C. Kuo and T. N. Jackson, *41st Electronic Materials Conf. Dig.*, 16 (1999)
3. Ashkenasy G.*et al,Acc. Chem. Res.* **35** (2002), 121–128
4. S. Kobayashi *et al., Nature Materials* **3**, 317 (2004)
5. H. Ishii *et al., Adv. Mater.* **11**, 605 (1999)

ORGANIC FIELD-EFFECT TRANSISTORS BASED ON PHENYL SUBSTITUTED TETRATHIAFULVALENE DERIVATIVES

B. Noda, H. Wada, M. Katsuhara, I. Aoyagi, T. Mori, T. Taguchi, T. Kambayashi, K. Ishikawa and H. Takezoe
Department of Organic Materials, Tokyo Institute of Technology, 2-12-1 O-okayama, Meguro-ku, Tokyo, Japan 152-8552

1 INTRODUCTION

Organic field-effect transistors (OFETs) are of interest for flexible, low-cost, large-area electronic applications such as flat-panel displays, electronic papers, and chemical sensors.[1] To date, OFETs with the best performance using thin films have been based on pentacene, with the field-effect mobility exceeding 1 $cm^2V^{-1}s^{-1}$.[2] For improving the performance of devices, the amount of π-overlap is expected to play an important role. In this respect tetrathiafulvalenes (TTFs), which form many good conductors and superconductors,[3] have much larger bandwidth (typically 1 eV)[4] than pentacene (0.2 eV).[5] Although high-mobility OFET (1.4 $cm^2 V^{-1}s^{-1}$) has been reported using a single-crystal TTF derivative,[6] the film formability of TTF derivatives is usually poor.[7] Moreover, we have reported that an alkyl- and cyano-substituted TTF (C_6EDT-DC) functions as an active layer of OFETs (μ_{FET} = 0.02 $cm^2 V^{-1}s^{-1}$).[8] Biphenyl substituted bithiophene[9] as well as phenyl fused TTF[10] makes good OFETs showing high mobility of 0.66 and 1 $cm^2 V^{-1}s^{-1}$, respectively. These results show that the substitution of phenyl groups tends to improve the OFETs. In an attempt to improve the device characteristics by introducing phenyl groups to TTF based semiconductors, we have prepared ten TTF derivatives (Scheme 1), and used them as active layers of OFETs. In this paper, we report the chemical and thin-film properties, and device characteristics of these compounds.

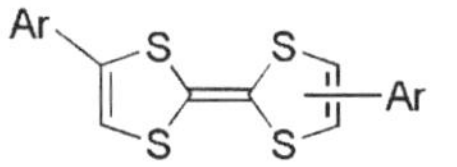

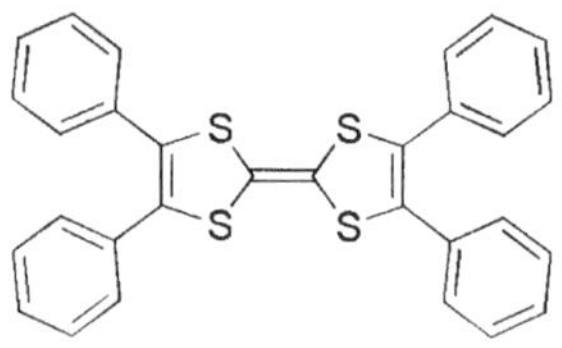

Ar— =	
Ph-	DP-TTF (**1**)
p-methylphenyl-	DC_1P-TTF (**2**)
p-ethylphenyl-	DC_2P-TTF (**3**)
p-propylphenyl-	DC_3P-TTF (**4**)
p-butylphenyl-	DC_4P-TTF (**5**)
biphenyl-	DBP-TTF (**6**)
2-naphthyl-	DN-TTF (**7**)
p-trifluoromethylphenyl-	DFC_1P-TTF (**8**)
perfluorophenyl-	DF_5P-TTF (**9**)

TP-TTF (**10**)

Scheme 1

2 EXPERIMENTAL PROCEDURES

The synthetic methods of TTF derivatives **1**, **2**,[11] **5**,[12] **6**,[11] **7**,[13] and **10**[11] have been described in the previous literatures. The new derivatives **3**, **4**, **8**, and **9** are prepared according to Scheme 2. The oxidation potentials are measured by cyclic voltammetry. The first oxidation potentials are as follows: TTF(0.38 V), **1** (0.42 V), **2-5** (0.40 V), **6** (0.44 V), **7** (0.43 V), **8** (0.52 V), **9** (0.58 V), **10** (0.46 V). Introduction of phenyl rings increases the oxidation potentials, and electron-withdrawing fluoro groups further increases them.

The crystal structures of **1-3**, **8**, and **10** are determined by single crystal X-ray structure analysis. The one-dimensional uniform-stacking structure of **5** has been reported.[12] Single crystals were obtained by slow sublimation for **1-3** and slow evaporation of the toluene solutions for **8** and **10**.

The FET devices were fabricated with bottom contact geometry. Thermally grown SiO_2 substrates were used as gate insulators with vacuum deposited gold or TTF-TCNQ thin-film electrodes.[14] The derivatives were deposited on the substrates at high vacuum (< 10^{-7} torr), and at the same time on another SiO_2 substrate without electrodes, used for XRD measurements.

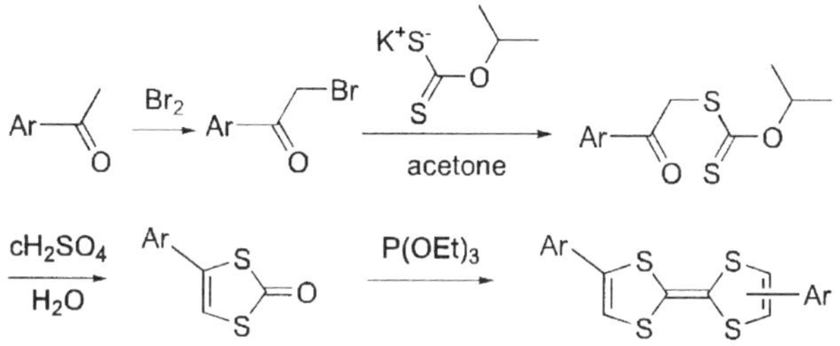

Scheme 2

3 RESULTS AND DISCUSSION

The crystal structure analysis revealed bisphenyl and tetraphenyl derivatives **1** and **10** belong to a triclinic crystal system, space group $P1$-. Compound **1** has two crystallographically independent half molecules in a unit cell, which are oriented in different directions. Compound **10** has one independent molecule, and the four phenyl groups are tilted largely from the TTF plane. The TTF parts are dimerized strongly. On the other hand **1-3** and **8** belong to a monoclinic crystal system with space group $P2_1$ for **1** and **2** and $P2_1/c$ for **3** and **8**. The molecules stack uniformly and the overlap integral calculation shows almost one-dimensional stacks like **5**. All compounds whose crystal structures are analyzed are composed of only anti forms with respect to the phenyl groups.

XRD profiles have afforded the *d*-spacings of these thin films, and the tilt angles are evaluated by using the molecular lengths calculated by MOPAC. The results are summarized in Table 1. Compounds **2** and **6** have relatively small tilt angles. Compounds **5** and **9** have two *d*-spacings, indicating the existence of two crystalline orientations. Compound **7** gives weak peaks in the XRD profile, indicating a disordered morphology and a small *d*-spacing. Compound **8** has a small *d*-spacing corresponding to the *b* axis along the molecular stacking direction of the single crystal. This shows that the conductive direction is perpendicular to the substrate.

The FET characteristics are listed in Table 1. Compounds **1-4** and **6** show p-type FET behavior. The other compounds do not show FET characteristics either with gold or with TTF-TCNQ electrodes. Compound **1** shows the lowest mobility among these compounds, reflecting the two different stacks observed in the single crystal. Mobilities of **2-4** are two orders improved in comparison with **1**, but the values are still not high, indicating that the simple uniform stacks give rise to the FET characteristics but the one-dimensional conductive pass restricts the transport of the carriers. Although the crystal structure of **4** is not known, the crystal of **4** is expected to have a monoclinic crystal system in analogy with **2** and **3**. However, compound **5**, which has a similar crystal structure to **2-4**, does not show FET characteristics. Two *d*-spacings revealed by XRD measurements may make phase boundaries, which prevent the carriers transport. Biphenyl substituted compound **6** shows the highest mobility among these compounds due to the extension of the π skelton in analogy with the moderately rigid substitution like biphenyl substituted bithiophene and the phenyl fused TTF derivatives. Compound **7** does not show FET characteristics, which may be affected by disordered morphology. Compound **8** shows no FET characteristics. This is explained by the XRD results that the most conductive direction of the crystal is perpendicular to the substrate. For compound **9**, the existence of two crystalline orientations prevents the carrier transport similarly to **5**. Compound **10** does not show FET characteristics due to the dimerization.

Table 1 FET characteristics, *d*-spacings of the thin films, molecular lengths (calculated by MOPAC), and tilt angles.

Compound	Mobility $(cm^2V^{-1}s^{-1})$	On / off ratio	*d*-spacing (Å)	Molecular length (Å)	Tilt angle (°)
DP-TTF (**1**)	3.0×10^{-6}	10	No clear peak	17.0	-
DC$_1$P-TTF (**2**)[a]	3.0×10^{-4}	40	18.3	18.4	5
DC$_2$P-TTF (**3**)[a]	5.0×10^{-4}	30	19.5	21.0	22
DC$_3$P-TTF (**4**)	1.0×10^{-4}	1.02	21.8	22.9	18
DC$_4$P-TTF (**5**)	-	-	24.0 / 19.7	25.9	22
DBP-TTF (**6**)	0.11	100	24.9	25.2	10
DN-TTF (**7**)	-	-	9.2	21.2	-
DFC$_1$P-TTF (**8**)	-	-	9.3	19.3	-
DF$_5$P-TTF (**9**)	-	-	20.4 / 13.4	17.4	-
TP-TTF (**10**)	-	-	13.4	14.5	22

[a]FET characteristics are measured with TTF-TCNQ electrodes, the others are with gold electrodes

4 CONCLUSION

We have prepared several TTF derivatives, and OFETs of five of them have shown p-type FET characteristics. A comparative study of the bulk crystal structures and the FET characteristics shows that the compounds with dimerized structure and with two crystalline orientations in the films do not show FET characteristics, whereas those with one-dimensional columns afford moderate FET performance and two-dimensional compounds leads to the best results.

This work demonstrates that the substitution by phenyl groups is a versatile tool to modify molecules for the use of thin-film OFET and proves that modified TTFs are, at least, as promising materials as oligothiophenes for active layers of OFET.

References

1 (a) H. E. Katz, Z. Bao and S. L. Gilat, *Acc. Chem. Res.*, 2001, **34**, 359. (b) A. Facchetti, M. -H. Yoon, C. L. Stern, H. E. Katz and T. J. Marks, *Angew. Chem. Int. Ed.*, 2003, **42**, 3900. (c) C. D. Dimitrakopoulos and P. R. L. Malenfant, *Adv. Mater.*, 2002, **14**, 99. (d) S. R. Forrest, *Nature*, 2004, **428**, 911.

2 (a) Y.-Y. Lin, D. J. Grundlach, S. F. Nelson and T. N. Jackson, *IEEE Electron Device Lett.*, 1997, **18**, 606. (b) H. Klauk, M. Halik, U. Zschang, G. Schmid and W. Radik, *J. Appl. Phys.*, 2002, **92**, 5259.

3 T. Ishiguro, K. Yamaji and G. Saito, *Organic Superconductors* (Springer, Berlin, 1998).

4 T. Mori, *Bull. Chem. Soc. Jpn.* 1999, **72**, 2011.

5 R. C. Haddon, X. Chi, M. F. Itkis, J. E. Anthony, D. L. Eaton, T. Siegrist, C. C. Mattheus and T. T. M. Palstra, *J. Phys. Chem. B*, 2002, **106**, 8288.

6 (a) M. Mas-Torrent, M. Durkut, P. Hadley, X. Ribas and C. Rovira, *J. Am. Chem. Soc.*, 2004, **126**, 984. (b) M. Mas-Torrent, P. Hadley, S. T. Bromley, X. Ribas, J. Tarrés, M. Mas, E. Molins, J. Veciana and C. Rovira, *J. Am. Chem. Soc.*, 2004, **126**, 8546.

7 I. Aoyagi, M. Katsuhara and T. Mori, *Sci. Tech. Adv. Mater.*, 2004, **5**, 443.

8 M. Katsuhara, I. Aoyagi, H. Nakajima, T. Mori, T. Kanbayashi, M. Ofuji, Y. Takanishi, K. Ishikawa, H. Takezoe and H. Hosono, *Synth. Met.*, 2005, **149**, 219.

9 M. Ichikawa, H. Yanagi, Y. Shimizu, S. Hotta, N. Suganuma, T. Koyama and Y. Taniguchi, *Adv. Mater.*, 2002, **14**, 1272.

10 M. Mas-Torrent, P. Hadley, S. T. Bromley, N. Crivillers, J. Veciana and C. Rovira, *Appl. Phys. Lett.*, 2005, **86**, 012110.

11 G. Schukat and E. Fanghänel, *J. Prakt. Chem.*, 1985, **327**, 767.

12 C. Polycarpe, E. Torreilles, L. Giral, A. Babeau, N. Tinh and H. Gasparoux, *J. Heterocyclic Chem.*, 1984, **21**, 1741.

13 J. Hellberg and M. Moge, *Synth. Met.*, 1993, **55-57**, 2124.

14 Y. Takahashi, T. Hasegawa, Y. Abe, Y. Tokura, N. Nishimura and G. Saito, *Appl. Phys. Lett.*, 2005, **86**, 063504.

ORGANIC SEMICONDUCTORS BASED ON NICKEL COMPLEXES FOR THIN-FILM TRANSISTORS

H. Wada, T. Taguchi, B. Noda, T. Kambayashi, T. Mori, K. Ishikawa, and H. Takezoe

Department of Organic Materials, Tokyo Institute of Technology, 2-12-1 O-okayama, Meguro-ku, Tokyo152-8552, Japan

1 INTRODUCTION

Organic field-effect transistors (OFETs) are receiving considerable attention due to their potential applications as flexible displays, RF-ID tags and smart cards.[1] Recently various new materials have been synthesized aiming at high performance OFETs. However, most of these materials are p-channel semiconductors, and very few n-channel semiconductors that are stable in air are reported.[2] In order to develop stable n-channel semiconductors, we have explored planar nickel complexes, which have low reduction potentials (easily reducable), and have stable anionic forms. As a metal complex, it has been reported that benzendiaminado nickel complex shows p-type FET characteristics,[3] and n-type FET characteristics also reported for a Ni(dmit)$_2$ complex.[4] Recently we have found that bis(dithiobenzil)nickel **3a**, where four sulfur atoms coordinate to the nickel atom, shows n-channel behaviour. To improve the FET performance, fluoro-substituted nickel complex **3b** has been prepared, and the thin film is fabricated onto SiO$_2$ substrates by vacuum deposition. This compound shows slight negative shift of redox potentials in comparison to **3a**, and the device shows n-type behaviour.

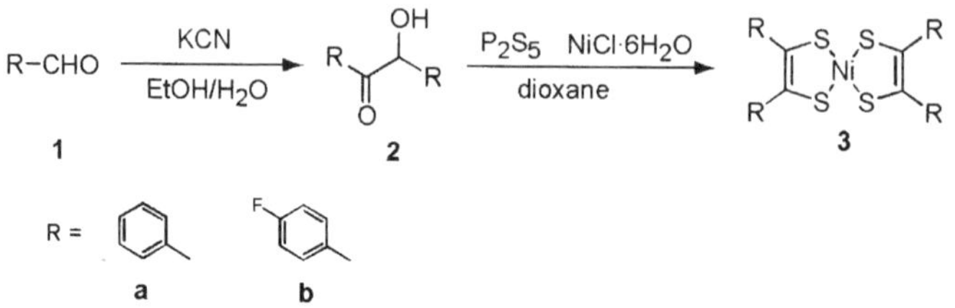

Scheme 1 Preparation of nickel complexes

2 EXPERIMENTAL

2.1 Synthesis

The nickel complex substituted with fluoro phenyl groups **3b** was prepared as shown in Scheme 1. Compound **2** was prepared according to the benzoin condensation.[5] The obtained benzoin was treated with P_2S_5.[6] The product was purified by sublimation in a good yield.

Synthesis of compound 2b. 2.47 g *p*-fluorobenzaldehyde, 5 ml EtOH, and 0.23 g potassium cyanide dissolved in 4 ml water were refluxed for 3 h. After the solution color changed red to pale orange, the reaction mixture was cooled, extracted with ether, washed with water, and dried with $MgSO_4$. After the solvent was removed by evaporation, obtained oil was distilled under reduced pressure, to give 1.90 g of a pale yellow solid.

Synthesis of compound 3b. To 0.50 g **2b** and 0.66 g P_2S_5 was added 6 ml dioxane under nitrogen, and heated at 100 °C for 2 h. After cooling, the mixture was filtered, and $NiCl_2.6H_2O$ (0.24 g) dissolved in 1.5 ml water was added to the filtrate. The solution color changed to black. The mixture was refluxed at 100 °C for 2 h. The reaction mixture was filtered, and the residue was washed with acetone, to give 0.14 g of a black solid.

Syntheses of compounds 2a and 3a. The procedures were the same as for **2b** and **3b**.

2.2 Electrochemistry

Redox potentials of these nickel complexes are measured by cyclic voltammetry as shown in Figure 1. The oxidation of **3a** to the monoanion (1-) is observed at -0.06 V (vs. Ag/AgCl), and the oxidation of **3b** to the monoanion is observed at 0.05 V. Introduction of the fluoro groups slightly stabilizes the 1- state.

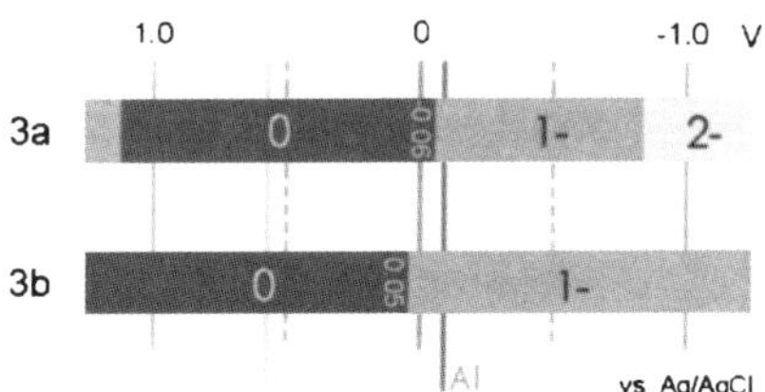

Figure 1 Redox potentials of **3a** and **3b**

2.3 FET Device Fabrication

Bottom-contact devices were fabricated. Gold, aluminium and tetrathiafulvalene-tetracyanoquinodimethane (TTF-TCNQ) were vacuum-deposited for source and drain electrodes with $W/L = 10 \sim 20$ onto highly n-doped silicon substrates used as gate electrodes with 300-nm silicon dioxide dielectric layer with a capacitance per unit area of 13.3×10^{-9} F/cm^2. Thin films of nickel complexes (50 nm) were formed by vacuum evaporation under a pressure of $\sim 5 \times 10^{-6}$ Torr. FET characteristics were measured at room temperature in air and under vacuum.

3 RESULTS AND DISCUSSION

The FET properties of **3a** and **3b** measured are shown in Figures 2 and 3. A thin-film FET of **3a** shows n-type characteristics with Al electrodes, where the mobility and the on/off ratio are 2.0 x 10^{-5} cm^2/Vs and 10^2 respectively. Good saturation property is observed under vacuum. N-type FET is not stable owing to the oxidation by atmospheric oxygen.[7] In air, the current decreases by one order of magnitude, and the threshold voltage increases from 10 V to 68 V (Figure 2 (b)).

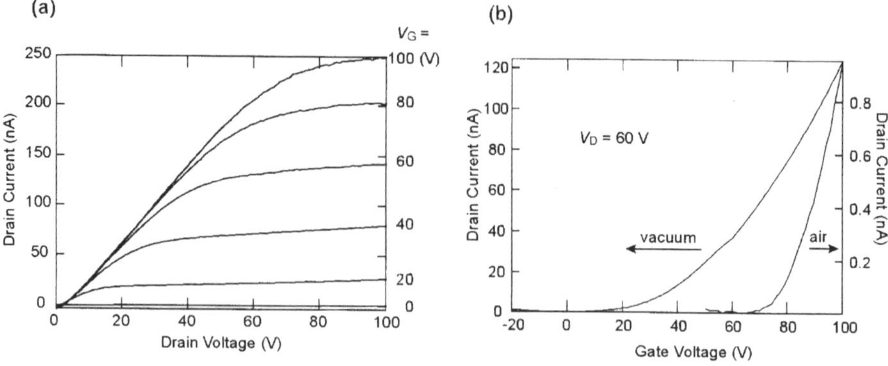

Figure 2   (a) Output characteristics of **3a** under vacuum, and (b) transfer characteristics under vacuum and in air.

Compound **3b** also shows n-type FET characteristics with TTF-TCNQ electrodes, where the mobility and the on/off ratio are 5.0 x 10^{-5} cm^2/Vs and 10^2 respectively (Figure 3). For this compound, a device with Al electrodes does not show FET characteristics. The current decreases one order in air, but the threshold voltages, -12 V and -5 V, are about the same under vacuum and in air.

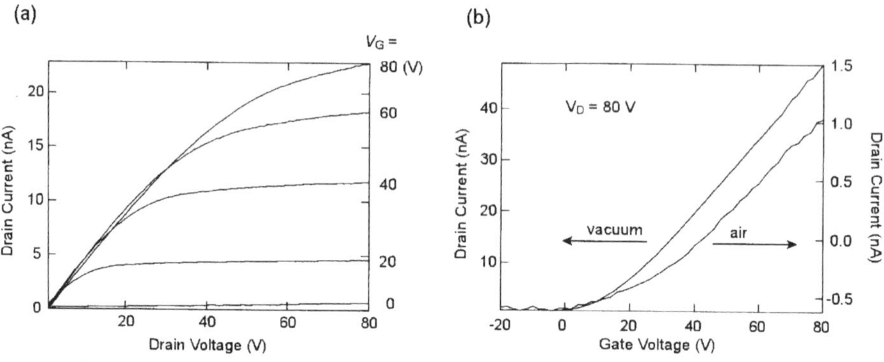

Figure 3   (a) Output characteristics of **3b** under vacuum, and (b) transfer characteristics under vacuum and in air.

AFM images are shown in Figure 4 for thin films of **3a** and **3b**. The grains are small, but the surface is covered by the semiconductors. However, XRD does not show clear peaks. These results indicate that the molecular orientation does not seem to be controlled very well in these thin films.

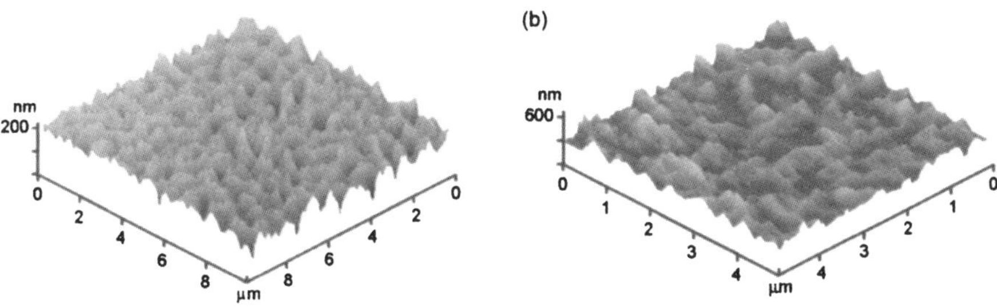

Figure 4   AFM images of (a) **3a** and (b) **3b**.

4 CONCLUSION

Compounds **3a** and **3b** are moderately strong electron acceptors. The present study shows that these compounds exhibit n-channel FET characteristics. The improved stability of **3b** FET in air may be associated with the slightly higher oxidation potential of **3b** than that of **3a**. This suggests the possibility of more air-stable OFET by using a compound with a higher oxidation potential. In addition, molecular design for better molecular arrangement and morphology is necessary to improve the device performance.

References

1. (a) C. D. Dimitrakopoulos, P. R. L. Malenfant, *Adv. Mater.* 2002, **14**, 99. (b) H. E. Katz, Z.Bao, and S. L. Gilat, *Acc. Chem. Res.* 2001, **34**, 359.
2. (a) Z. Bao, A. J. Lovinger and J. J. Brown, *J. Am. Chem. Soc.*,1998, **123**, 207. (b) H. E. Katz, J. Johnson, A. J. Lovinger and W. Li, *J. Am. Chem. Soc.*, 2000, **122**, 7787. (c) Y. Takahashi, T. Hasegawa, Y. Abe, Y. Tokura, K. Nishimura, and G. Saito, *Appl. Phys. Lett.*, 2005, **86**, 063504.
3. S. Noro, H. C. Chang, T. Takenobu, Y. Murayama, T. Kanbara, T. Aoyama, T. Sassa, T. Wada, D. Tanaka, S. Kitagawa, Y. Iwasa, T. Akutagawa, and T. Nakamura, *J. Am. Chem. Soc.*, 2005, **127**, 10012.
4. C. Pearson, A. J. Moore, J. E. Gibson, M. R. Bryce, and M. C. Petty, *Thin Solid Films*, 1994, **244**, 932.
5. B.S. Furniss, A. J. Hannaford, P. W. G. Smith, A. R. Tatchell, *Vogel's Textbook of Practical Organic Chemistry*, 5th Edn., Longma, New York, 1989, ch.6-12, p.1045.
6. G. N. Schrauzer and V. P. Mayweg, *J. Am. Chem. Soc.*, 1965, **87**, 1483.
7. D. M. de Leeuw, M. M. J. Simenon, A. R. Brown, and R. E. F. Einerhand, *Synth. Met.*, 1997, **87**, 53.

POLY(*p*-PHENYLENEVINYLENE) NANOFIBER YARNS FABRICATED BY ELECTROSPINNING

T. Takahashi and H. Okuzaki

Interdisciplinary Graduate School of Medicine and Engineering, University of Yamanashi, 4-4-37 Takeda, Kofu 400-8511, Japan. E-mail: okuzaki@yamanashi.ac.jp

1 INTRODUCTION

Poly(*p*-phenylenevinylene) (PPV) has been paid considerable attention due to its properties of electrical conductivity,[1,2] electro- or photoluminescence,[3] and non-linear optical response,[4] which have potential applications in electrical and optical devices, such as light-emitting diodes or solar cells.[5] On the other hand, PPV is expected to be a high-performance polymer with high modulus and strength and superior thermal stability,[6] because of its rigid structure consisting of alternating phenylene and vinylene groups. Previously, we have developed a zone-drawing/zone-reaction method capable of orienting the molecular chains simultaneously with thermal reactions, and applied it to the PPV precursor films.[7,8] It was found that the zone-reacted PPV film had extremely high orientation of molecular chains in the drawing direction with Young's modulus and tensile strength of as high as 70 GPa and 1.3 GPa, respectively. Most of the work with regard to PPV exploits cast films or thin coatings, while few reports have been investigated on PPV fibers. It is of importance to produce fibers of conducting polymers not only from the fundamental viewpoints for understanding the electrical and mechanical properties in the bulk state but also from the practical use for electromagnetic interference shielding or conducting textiles.

We have succeeded in fabricating conducting microfibers of poly(3,4-ethylenedioxythiophene) doped with poly(4-styrenesulfonate), having a diameter ranging between 4.6-16 μm, by a wet-spinning technique, where the microfibers have no notable orientation of polymer chains with poor crystallinity and electrical conductivity is in the order of 10^{-1} S/cm regardless of the diameter.[9] Furthermore, MacDiarmid and coworkers have fabricated nanofibers of conducting polymers and their blends with conventional polymers using a non-mechanical electrospinning technique.[10,11]

This paper is the first study on fabrication of PPV nanofibers by electrospinning of precursor that is subsequently converted into PPV by heat treatment. The changes in the diameter and morphology of the resulting nanofibers electrospun under various conditions have been investigated.

2 METHOD AND RESULTS

2.1 Precursor Yarns Fabricated by Electrospinning

Poly(p-xylenetetrahydrothiophenium chloride), the precursor of PPV, was commercially available in the form of 0.25% aqueous solution from Aldrich Inc.[12,13] A schematic diagram of apparatus for the electrospinning was shown in Figure 1. About 1 ml of an aqueous solution of the precursor containing different methanol contents was poured into a glass syringe (12 mm in diameter) and an electric field was applied to a single-hole spinneret (350 μm in diameter) from a variable high-voltage power supply (Towa Keisoku Inc.) capable of applying positive voltages (E) up to 30 kV. The electrospun fibers were collected on the surface of an aluminum foil placed on a stainless target electrode (10 x 10 cm), where the distance between the tip of the needle and the target electrode was 20 cm. The electrospinning of as-purchased aqueous solution of precursor brings about no fiber formation under the voltage up to 30 kV, as the viscosity and surface tension of the solution are too high to maintain a continuous jet from the spinneret. Dilution with pure water can not improve it significantly but numerous small droplets of the solution are collected on the surface of the aluminum foil, which may be ascribed to the strong retention of water by polyelectrolyte in the droplet and/or to the low viscosity and surface tension not enough to maintain a stable drop at the spinneret.

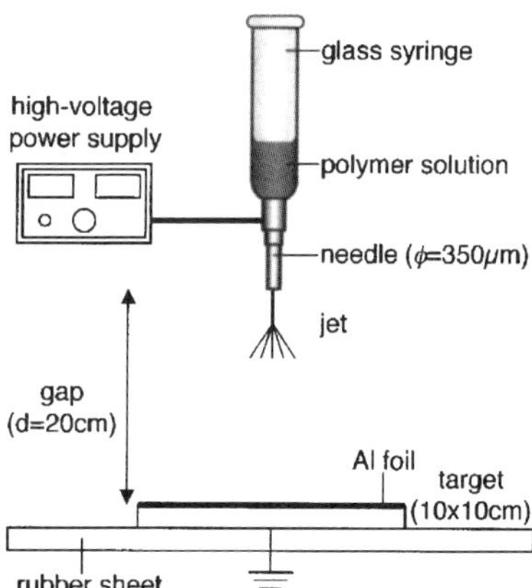

Figure 1 *Schematic diagram of apparatus for electrospinning.*

An addition of organic solvent, such as methanol, enables to form fibers where the increase of applied voltage elongates the drop of the precursor solution at the spinneret from a hemispherical shape into a Taylor cone[14] and results in a stable jet underneath the droplet. As shown in Figure 2, a number of precursor fibers grow up from the surface of the aluminum target that spontaneously spun into yarn of about 50-200 μm in diameter. At E = 20 kV, PPV nanofiber yarns are obtained in a range of methanol content (C) between 40% and 80%, while at C > 80% the viscosity and surface tension of the solution are not high enough to maintain a stable drop at the spinneret and/or the solvent is not evaporated completely. On the other hand, at C < 40% the viscosity is too high to maintain a continuous jet from the spinneret as described above. The diameter of fibers is less dependent of the applied voltage between 10-30 kV.

Figure 2   *Optical image of typical precursor fibers electrospun into yarns at 20 kV under 50%RH (methanol content 70%).*

2.2 Thermal Conversion to PPV Nanofiber Yarns

Thermogravimetric (TG) analysis was performed with a TG-DTA 2000S (Bruker axs Inc.) measured in the temperature range from room temperature to 800°C at a heating rate of 10°C/min under a nitrogen atmosphere. The as-spun precursor fibers has three major steps of the weight loss at TG thermograms: The first small weight loss of 13% up to 100°C is associated with the elimination of sorbed moisture, and the second one at 100-300°C with the elimination of tetrahydrothiophene and hydrochloric acid. On the other hand, the third weight loss above 500°C is related to the decomposition of polymer chains. The conversion of precursor into PPV was carried out at 250°C for 12 h in a vacuum by elimination of tetrahydrothiophene and hydrochloric acid. It is found that the PPV fibers exhibit no notable weight loss in the temperature range of the first and second steps of the weight loss, demonstrating that the precursor unit is effectively changed into PPV by the heat treatment at 250°C for 12 h in a vacuum.

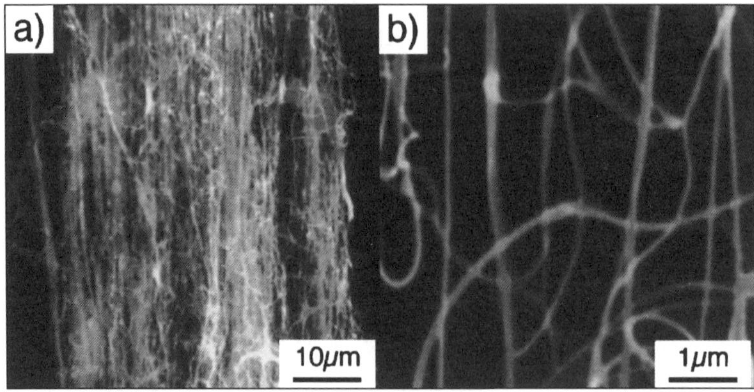

Figure 3   *SEM micrographs of PPV yarn (a) and nanofibers (b).*

It is seen from Figure 3 that the fibers having a diameter ranging between 50 nm and 200 nm preserve their shape even after the elimination reaction. Interestingly, the PPV nanofibers are aligned to each other along the axis of the yarn.

3 CONCLUSION

We have succeeded in fabricating PPV nanofiber yarns by electrospinning of the precursor solution and subsequent heat treatment. The electrospinning is strongly affected by methanol content, where the precursor fibers are spontaneously electrospun into yarn. The fibers preserve their shape even after the thermal conversion to PPV and the resulting nanofibers, having a diameter of 50-200 nm, are aligned along the axis of the yarn.

Acknowledgements

This work was supposed in part by Grant for Promotion of Research from Okaya City (Nanokako). H.O. acknowledges Dr. A.G. MacDiarmid for giving him an opportunity to work on electrospinning.

References

1   I. Murase, T. Ohnishi, T. Noguchi and M. Hirook, *Polym. Commun.*, 1984, **25**, 327.
2   G.E. Wnek, C.W. Chien, E. Frank. E. Karasz and C.P. Lillya, *Polym. Commun.*, 1979, **20**, 1441.
3   J.H. Burrourghes, D.D.C. Bradley, A.R. Brown, R.N. Marks, K. Mackay, R.H. Friend, P.L. Burns and B. Holmes, *Nature*, 1991, **347**, 539.
4   T. Kaino, K. Kubodera, H. Kobayashi, T. Kurihara, S. Saito, T. Tsutsui and S. Tokito, *Electron. Lett.*, 1987, **23**, 1095.
5   N.S. Sariciftci, D. Braun, C. Zhang, V.I. Srdanov, A.J. Heeger, G. Stucky and F. Wudl, *Appl. Phys. Lett.*, 1993, **62**, 585.
6   T. Ohnishi, I. Murase, T. Noguchi and M. Hirooka, *Synth. Met.*, 1986, **14**, 207.
7   H. Okuzaki, Y. Hirata and T. Kunugi, *Polymer*, 1999, **40**, 2625.
8   H. Okuzaki, N. Ikeda, I. Kubota and T. Kunugi, *Macromolecules*, 1999, **32**, 5606.
9   H. Okuzaki and M. Ishihara, *Macromol. Rapid Commun.*, 2003, **24**, 261.
10 I.D. Norris, M.M. Shaker, F.K. Ko and A.G. MacDiarmid, *Synth. Met.*, 2000, **114(2)**, 109.
11 A G. MacDiarmid, W.E. Jones, Jr., I.D. Norris, J. Gao, A.T. Johnson, Jr., N.J. Pinto, J. Hone, B. Han, F.K. Ko, H. Okuzaki and M. Llaguno, *Synth. Met.*, 2001, **119**, 27.
12 R.W. Lenz, C.-C. Han, J.S. Smith and F.E. Karasz, *J. Polym. Sci., Polym. Chem.*, 1988, **26**, 3241.
13 P.M. Lahti, D.A. Modarelli, F.R. Denton, III, R.W. Lenz and F.E. Karasz, *J. Am. Chem. Soc.*, 1988, **110**, 7258.
14 D.H. Reneker and I. Chun, *Nanotechnology*, 1996, **7**, 216.

PAPER ACTUATORS WITH CONDUCTING POLYMERS

H. Okuzaki and T. Saido

Interdisciplinary Graduate School of Medicine and Engineering, University of Yamanashi, 4-4-37 Takeda, Kofu 400-8511, Japan

1 INTRODUCTION

Conducting polymers have attracted considerable attention because dimensional changes resulting from electrochemical or chemical doping, characterized by the transportation of solvated ions between the inside of the polymer matrix and the surrounding electrolyte solution, electrostatic repulsion, and/or structural distortion through oxidation of polymers, can be applied to produce microactuators[1] or artificial muscles.[2] Most of them exhibit dimensional changes in an electrolyte solution or in a swollen state, while few reports have been investigated on solid state polymers in a redox gas atmosphere[3] or that employ a polyelectrolyte solid, gel, or ionic liquid.[4]

In the preceding papers,[5] we have reported that electrochemically synthesized polypyrrole films undergo quick and intensive bending in ambient air as a result of anisotropic dimensional change due to the sorption of water vapor from one side of the film and have constructed polymeric motors that work in air.[6] Furthermore, we have found that polypyrrole film exhibits contraction in air under application of an electric field,[7] where the contraction was strongly affected by applied voltage and ambient humidity.

Although the applied voltage plays a predominant role in causing a dimensional change of the film in air, little is known about electro-active behavior of the polypyrrole film. This system is of interest not only in proposing a new class of electro-responsive conducting polymers but also for understanding of the electromechanical properties of humido-sensitive conducting polymers. In this paper, electrically induced dimensional change of polypyrrole films in air and in a vacuum have been investigated and applied to a paper actuator exhibiting a large dimensional change.

2 METHOD AND RESULTS

2.1 Preparation of PPy Film

Pyrrole and propylene carbonate were purified by distillation before use. Tetraethylammonium tetrafluoroborate for use as an electrolyte was commercially available and used without further purification. Polypyrrole (PPy) films doped with tetrafluoroborate were electrochemically synthesized by anodic oxidation of pyrrole as described elsewhere.[5] A constant current was applied through a platinum plate as anode and Al foil as cathode with a potentiostat at -20°C. After polymerization, the PPy film was peeled from the platinum electrode, soaked in a large amount of propylene carbonate, and dried overnight in a vacuum.

2.2 Measurements

The degree of contraction was measured using a laser displacementometer, where the strain of film at relative humidity (RH) of 50% was adjusted to zero. The electric field was applied to the film with a potentiostat through copper wires attached to both ends of the film with silver paste. The temperature at the film surface was measured with an infrared thermometer. The RH in the vicinity of the film was measured with a hygrometer, the probe of which was located 1 mm from the film surface. The absolute humidity (AH) near the film surface was calculated using RH and the saturated water vapor pressure at each temperature.

2.3 Electromechanical Properties of PPy Film

It is seen from Figure 1 that when dc 2 V is applied to the film (35 mm long, 5 mm wide, and 30 μm thick) for 10 min, the degree of contraction reaches to 1.2%, which is comparable to that caused by electrochemical or chemical doping of conducting polymers in an electrolyte solution. The PPy film in this system, however, exhibits contraction in ambient air without the use of an electrolyte solution or a redox gas. This clearly demonstrates that the mechanism of motion is different from the electrochemical doping.

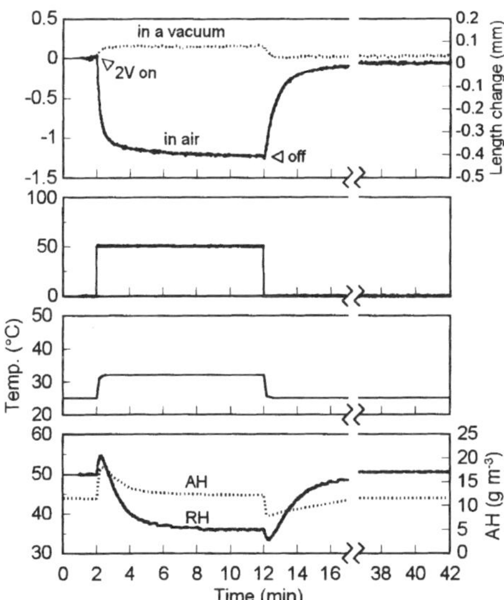

Figure 1 *Time profiles of strain, electric current, and surface temperature of PPy film and of AH and RH in the vicinity of the film surface measured at 25°C under 50% RH.*

The electric current passing through the film is about 50 mA and the temperature at the film surface increases from 25 to 32°C due to the Joule heating. It is noted that AH in the vicinity of the film surface abruptly increases by the electric field, and then recovers to the original level with time. This fact demonstrates that water vapor sorbed in the film desorbs and scatters into ambient air. Indeed, elemental analysis indicates that the film contains 4.6% of moisture in air. On the other hand, a drop of the AH when the electric field is turned off, can be explained as the resorption of water vapor from the air surrounding the film. Because the AH change coincides with the dimensional change of the film, the contraction of the film will be associated with desorption of water vapor induced by the electric field.

It is found that the degree of contraction tends to level off at 3-4 V, while a further increase in the applied voltage results in a decrease of the value. In contrast, the PPy film in a vacuum undergoes elongation by the electric field, which increases as the applied voltage becomes higher. Thus, the apparent strain of the film under the electric field can be expressed by two processes both caused by the Joule heating: one is the contraction due to the desorption of water vapor and another is the thermal expansion of polymer chains.

2.4 Paper Actuator ('Origami' Actuator)

The electrically induced dimensional change of the PPy film is, however, only 1-2%, which is one order of magnitude smaller than that of skeletal muscle in animals.[8] To actuate a large strain under an electric field, we fabricated a paper actuator as shown in Figure 2. In the manner of 'origami', two strips of PPy films are repeatedly folded so as to cross each other into the figure of a paper spring. When 2 V is applied through copper wires attached to top and bottom ends of the spring, the paper actuator undergoes significant elongation of about 30% (Figure 3). It is worth to note that the application of an electric field causes contraction of the PPy film but elongation of the paper actuator. This can be explained by the modulation of elasticity: Upon application of an electric field, the desorption of water vapor increases the elastic modulus of PPy[9] since water vapor molecules may behave as a plasticizer and enhance the motion of polymer chains. Thus, the PPy becomes less deformative that tends to unfold the creases in the spring, leading to elongate the paper actuator.

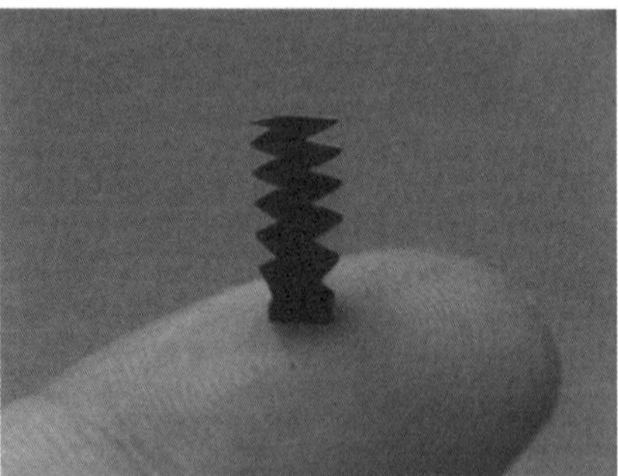

Figure 2 *Photograph of paper actuator fabricated in the manner of folding PPy films into the figure of a paper spring.*

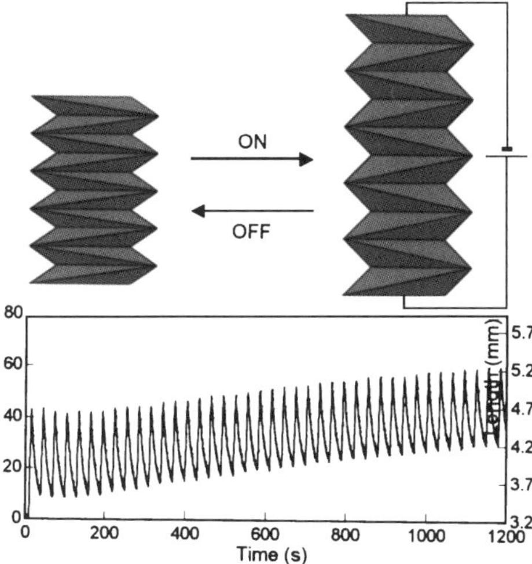

Figure 3 *Electrically induced elongation of PPy paper actuator under an application of 2 V measured in air at room temperature.*

3 CONCLUSION

Unlike in the conducting polymers based on the electrochemical doping, this system can serve as a new type of electrically driven muscle that works in air without using an electrolyte solution or counter and reference electrodes, where the applied voltage is capable of controlling the sorption equilibrium to desorb water vapor from the film. Furthermore, this system can work at voltages one or two orders of magnitude lower than piezoelectric or electrostatic actuators because the dimensional change is owing to the electric current but the electric field. Moreover, various conducting polymers or their derivatives that undergo dimensional changes due to the sorption of water vapor can be employed on the same principle.

References

1 E. Smela, O. Inganäs and I. Lundström, *Science*, 1995, **268**, 1735.
2 R.H. Baughman, L.W. Shacklette, R.L. Elsenbaumer, E.J. Plichta and C. Becht, *Molecular Electronics*, Kluwer Academic Pub., Netherlands, 1991.
3 Q. Pei and O. Inganäs, *Synthetic Metals*, 1993, **55-57**, 3730.
4 W. Lu, A.G. Fadeev, B. Qi, E. Smela, B.R. Mattes, J. Ding, G.M. Spinks, J. Mazurikiewicz, D. Zhou, G.G. Wallace, D.R. MacFarlance, S.A. Forsyth and M. Forsyth, *Science*, 2002, **297**, 983.
5 H. Okuzaki and T. Kunugi, *J. Polym. Sci. Part B: Polym. Phys.*, 1996, **34**, 1747.
6 H. Okuzaki, T. Kuwabara and T. Kunugi, *Polymer*, 1997, **38**, 5491.
7 H. Okuzaki, T. Kunugi, *J. Polym. Sci. Part B: Polym. Phys.*, 1998, **36**, 1591.
8 R.M. Alexander, *Exploring Biomimetics: Animals in Motion*, W.H. Freeman and Company Pub., New York, 1992.
9 H. Okuzaki and K. Funasaka, *Macromolecules*, 2000, **33**, 8307.

Subject Index